PRODUCTIVITY AND REUSE IN LANGUAGE

PRODUCTIVITY AND REUSE IN LANGUAGE

A Theory of Linguistic Computation and Storage

Timothy J. O'Donnell

The MIT Press
Cambridge, Massachusetts
London, England

Library of Congress Cataloging-in-Publication Data

O'Donnell, Timothy J., 1977-

Productivity and reuse in language : a theory of linguistic computation and storage / Timothy J. O'Donnell.
pages cm

Includes bibliographical references and index.
ISBN 978-0-262-02884-4 (hardcover : alk. paper) 1. Psycholinguistics–Mathematical models. 2. Memory. 3. Language. 4. Cognitive grammar. 5. Recognition. 6. Psycholinguistics. I. Title.

P37.5.M46O36 2015
410.1'51–dc23

2014029651

10 9 8 7 6 5 4 3

Contents

Preface

The work reported in this book was originally presented in my 2011 Harvard University dissertation in Psychology. My thesis was written as a single study of the problem of productivity and reuse, rather than as a collection of stand-alone papers as is more typical in the field of psychology. In the summer of 2013, Ted Gibson suggested to me that this research was best presented together, maintaining the structure of my dissertation. This book is a result of following his advice.

In the time since I completed my dissertation, I have continued to think about the problem of productivity and reuse, and have pursued a number of research projects that have extended this work both theoretically and empirically. This process has made me aware of earlier work which I missed in completing my dissertation, deepened my understanding of the theory and results presented here, and provided practical experience with communicating them to a scientific audience. In preparing this manuscript, I took the opportunity to revise my dissertation to incorporate some of these lessons.

Chapter 1 covers mostly the same content as the introduction to my thesis. However, based on suggestions by Steven Pinker and Jesse Snedeker, I have reorganized and expanded several parts to clarify the crucial psychological assumptions underlying the model. Based on conversations with Mark Johnson, Bob Berwick, and others, I have also included discussion of some earlier literature which I was not aware of when I wrote my dissertation. Chapters 2 and 3, co-authored with Noah Goodman, present the mathematical and computational theory behind the model, some of which was also published in O'Donnell et al. (2009). These chapters have remained largely intact, aside from the removal of some discussion that we deemed superfluous to the results that followed.

Chapters 4 through 7 discuss empirical phenomena, theoretical approaches, and modeling results for domains of the English past tense (Chapters 4–5) and English derivational morphology (Chapters 6–7). These chapters have been substantially revised. The main changes have been to reorganize the chapters to emphasize common empirical problems (e.g., productivity, frequency, processing, etc.) between the two domains, and to clarify the relationship to earlier theories—for example, the blocking-and-retrieval-failure hypothesis of Marcus et al. (1992), the relative frequency hypothesis of Hay (2001), and the hapax-based productivity statistics of Baayen (1993).

In preparing this manuscript, I did a complete reanalysis of the empirical results presented in Chapters 5 and 7. This has resulted in several differences between the results reported here and those in my dissertation (and in earlier publications such as O'Donnell et al., 2011). First, based on a suggestion of Mark Johnson, in order to make the comparison between fragment grammars and the other models more interpretable, I used a single point-estimated approximation to the maximum a posteriori fragment grammar when computing the probabilities used in empirical analyses (rather than the selective model-averaging approach used in my dissertation). This change led to several quantitative differences in results, but did not change any qualitative patterns. Second, I discovered a bug in the scripts that compute the maximum a posteriori adaptor grammar from an input dataset. Specifically, I discovered that these scripts were not recursively storing all subtrees of each input form. This problem was corrected by Thang Minh Luong at Stanford, and I made use of his updated scripts in revising this manuscript. Again, while this affected some of the specific numbers reported in Chapter 7, it did not affect the qualitative pattern of results. Finally, I also discovered a low-level accounting error in the way in which fragment grammar probabilities were accumulated. Correction of this error led to a qualitative change in the analysis reported in Section 5.3.2.1. Whereas in my dissertation I had reported that fragment grammars counter-intuitively outperformed adaptor grammars even on attested past-tense items, this turned out to be the result of this bug.

The conclusion in Chapter 8 replaces that of my dissertation with a short summary and discussion of the main results of the book.

Acknowledgments

All work such as this is the product of the efforts of numerous people besides the one whose name appears on the title page. This book grew out of my dissertation, and I feel that I have been unusually lucky to have so many remarkable people help me both in the process of completing the dissertation and in the process of revising that work into the present form.

First, I would like to thank the people who advised and guided this project from the beginning. Noah Goodman, with whom I developed the mathematical framework and the fragment grammars model—and whose name appears alongside mine as author on Chapters 2 and 3—taught me to do probabilistic modeling and has been a source of constant encouragement, research ideas, intellectual clarity, and friendship. Mark Johnson, who developed the adaptor grammars model, upon which the model in this book is based, has generously provided detailed advice about countless technical and non-technical aspects of this work (and computational linguistics, in general) both during the writing of my dissertation and since. Josh Tenenbaum shares a large part of the responsibility for any clarity in the presentation of these ideas and their relationship to bigger questions of cognitive science. I continue to be inspired by the research program he has developed, grateful for his intellectual feedback, and thankful for the environment he has provided at MIT. Steven Pinker suggested that I look at the past tense, asked crucial questions about the psychological interpretation of the theory, and gave detailed comments on all aspects of this work, which—it is no exaggeration to say—have shaped the direction of this entire project and improved its theoretical grounding and its presentation. Finally, my dissertation advisor, Jesse Snedeker's intellectual (and editorial) influence can be found on every page of this book. Her deep insight, rigor, and thoughtfulness made the arguments tighter, the ideas clearer, and the presentation more coherent both in the version of this work presented in my thesis and in its current manifestation. Her generosity, sense of humor, and humanity have made her a great advisor and her lab a wonderful place to work.

I would like to thank Ted Gibson for suggesting that I revise my dissertation into this book and encouraging me along the way. I would also like to thank my editors at the MIT Press for their help during this lengthy process.

A number of colleagues and friends have gone above and beyond the call of duty in helping me with various aspects of this project, for which I am deeply grateful. Manizeh Khan, Josh Hartshorne, and Amy Geojo all read large parts of the thesis upon which this book was based. Rebecca Nappa read large parts of the revisions that turned my thesis into this book, and contributed many helpful comments at all levels. Adam Albright, Bob Berwick,

Peter Graff, Vikash Mansinghka, Andrew Nevins, Dan Roy, Josh Tenenbaum, and Jelle Zuidema, provided technical expertise on various topics, including Bayesian non-parametrics, probabilistic programming, minimum-description length modeling, natural language processing models (especially data-oriented parsing), and the details of phonological and morphological theory. I thank them for patiently answering my more naive or incoherent questions. I thank Bob Berwick for pointing me to crucial earlier related work. Miriam Meyerhoff deserves special mention for supporting me in my intellectual pursuits for the years before I was able to get into graduate school and the years after, and for being a friend. I would also like to thank Thang Minh Luong especially for helping correct errors in several of the scripts used to compute the results presented here.

I would like to thank my friends and colleagues in Snedlab, Tedlab, Cocosci lab, the MIT linguistics department, and elsewhere for making the easy parts fun and the hard parts bearable, and my parents and brothers for providing support and encouragement along the way. Other collaborators, friends, and family to whom I am grateful for help, advice, and encouragement include: Greg Aloe, Asaf Bachrach, Regina Barzilay, Leon Bergen, Erin Conwell, David Craft, Peter Dirix, Frederik Durant, Sven Duytschaever, Ev Fedorenko, Edward Fleming, Cameron Freer, Michelle Fullwood, Richard Futrell, Mark Gregor, John Hale, Ray Jackendoff, Frank Jaekel, Roni Katzir, Jackie Lee, Kyle Mahowald, Vikash Mansingkha, Clemens Mayr, Charles O'Donnell, David Pesetsky, Steve Piantadosi, Martin Rohrmeier, Sam Scarano, Kenny Smith, Scott Smith, Mahesh Srinivasan, Ellen Stepansky, Andreas Stuhlmüller, Donca Steriade, Melissa Troyer, and Nathan Wilson.

Finally, I would like to thank Marjorie Freedman who has participated in every stage of this work, tolerating the worst, sharing in the best, and reading everything many times—all with insight, patience, and good humor. Both the thesis on which this book was based, and the book itself, would not have been written without her. Thanks, Marj.

I MODEL BACKGROUND AND DEVELOPMENT

1 Introduction

1.1 The Problem of Productivity

Consider the following words: *coolness*, *circuitousness*, *grandness*, *orderliness*, *pretentiousness*, *cheapness*, and *warmness*. Most native speakers of English will immediately recognize that all of these words end in the suffix *-ness*, and—upon a moment's reflection—will be able to explain that this suffix carries a meaning like "the abstract quality of being X," where X is some property such as *cool* or *circuitous*. They might also observe that this suffix generally attaches to adjectives to form nouns, and that, when it does, it usually does not change the way that the adjective base is pronounced (although it introduces some standard spelling changes). There are a number of exceptions to these generalizations. For example, the word *forgiveness* ends in *-ness*, but *forgive* is not an adjective. Similarly, the word *business* is historically derived from *busy* and *-ness*, but over time its pronunciation and the meaning have changed. Nevertheless, such exceptions are rare. The suffix *-ness* is overwhelmingly **regular** in its phonology and **transparent** in its meaning.

The suffix *-ness* has another very important property. It can be easily used to construct new words. Take any adjective—common, rare, or even novel—and a corresponding abstract noun can be formed from it: *pine-scented*/*pine-scentedness*, *sketchy*/*sketchiness*, *Lady Gagaesque*/*Lady Gagaesqueness*. Many new words formed with *-ness* seem so natural that an English speaker may even have trouble knowing whether he or she has heard the word before. This property of *-ness* is what linguists and psychologists call **productivity**—the ease with which a linguistic process gives rise to new forms—and it is this property that is the subject of this book.

Now, consider another set of words: *verticality*, *tractability*, *severity*, *seniority*, *inanity*, and *electricity*. These words share the suffix *-ity*. Like *-ness*, *-ity* attaches to adjectives to form nouns, and its meaning is approximately the same: the abstract quality of possessing the property described by the adjective. However, closer examination reveals that *-ity* also differs from *-ness* in a number of important ways. While words such as *curiosity*, *ability*, and *ethnicity* can be synonymous with *curiousness*, *ableness*, and *ethnicness*, they can also mean something that the *-ness* words cannot. The words ending in *-ity* can be used to refer to particular specialized instances of the abstract quality they describe: an albino rhinoceros is a *curiosity* because it may inspire *curiousness*; juggling is an *ability* that reflects *ableness*; Italian is an *ethnicity* and an example of *ethnicness*. Unlike *-ness*, *-ity* can attach to bases that are not words (e.g., *hilarity*, *unity*, *charity*). The suffix *-ity* also differs from *-ness* in that it often causes systematic changes in the pronunciation of the base to which it attaches. For example, when it attaches to words ending in *-ic*, the pronunciation of the final /k/ changes to /s/ (e.g., *electric/electricity*, *historic/historicity*). In some words, it causes a vowel in the base to be pronounced differently (e.g., *inane/inanity*, *sane/sanity*).

Most importantly, *-ity* is far less productive than *-ness*. It cannot be used to turn arbitrary adjectives into abstract nouns: *pine-scented/*pine-scentedity*, *sketchy/*sketchity*, *Lady Gagaesque/*Lady Gagaesqueity*. There are occasional exceptions to this generalization. For example, the Urban Dictionary defines *coolity* to mean "of or pertaining to being cool or the act of being cool."[1] However, this is intentionally humorous, and its humor follows precisely because we know that *-ity* usually cannot be generalized productively. To complicate matters, however, there are several morphological contexts where *-ity* can be generalized productively: for example, after the suffixes *-able*, *-ic*, and *-al*. If one uses the term *googleable*, the associated abstract noun is *googleability*.

Finally, consider the following words: *warmth*, *width*, *truth*, and *depth*. These words all end in the suffix *-th*, which also turns adjectives into abstract nouns, and means approximately the same thing as *-ity* and *-ness*. However, this suffix is very different from the first two. Most native speakers are not even aware of its existence until it is explicitly pointed out to them. Furthermore, many words ending in *-th* involve substantial changes to the pronunciation of their base adjectives. In many cases, the changes result in a relationship between the base adjective and the abstract noun that is very obscure: *vile/filth*, *slow/sloth*. There are even a few words that appear to end in *-th* but don't seem to have any corresponding adjective in modern English: *weal?/wealth*, *wroth?/wrath*, *merry?/mirth*.

The suffix *-th* is even less productive than *-ity*. While it may be possible to interpret **pine-scentedity*, **sketchity*, and **Lady Gagaesqueity*; **pine-scent(ə)th*, **sketchi(ə)th*, and **Lady Gagaesqueth* are hardly more than gibberish. Nevertheless, *-th* can be used occasionally to form new words for humorous or poetic effect. George Elliot wrote of the *gleams and greenth of summer* in *Daniel Deronda* (Eliot, 1876); Steve Allen wrote a book called *Dumbth: The Lost Art of Thinking with 101 Ways to Reason Better & Improve*

1. http://www.urbandictionary.com/define.php?term=coolity

Your Mind (Allen, 1998);[2] and on the internet we can find statements such as *Many enjoy the warmth, Vikings prefer the coolth.*[3]

Patterns like *-ness/-ity/-th*—three ways of forming adjectives into abstract nouns that mean approximately the same thing, but differ greatly in their productivity, semantic transparency, and phonological regularity—raise several serious problems for linguistic and psycholinguistic theory, as do many similar cases throughout English and other languages.

First, what kind of representations can account for this mixture of productive generalization in some cases (e.g., *-ness*) and reuse of specific stored forms (e.g., *warmth*) in others? What is it that speakers know that allows them to recognize that *coolth* is intended to be humorous while *pine-scentedness* is a perfectly well-formed word?

Second, how does the language learner arrive at this knowledge? The affix *-th* was productive at some point in the history of English. There is nothing cross-linguistically unusual about its phonology or semantics and it appears in dozens of English words. How and why does the child learn that it cannot be used to form new words despite these facts? By contrast, how and why do they learn that *-ness* can? What sort of evidence is available and how is it used?

Finally, why do several ways of expressing (roughly) the same meaning exist in English in the first place? How do speakers resolve competition between these alternate means of expression? Why does English continue to maintain multiple possibilities?

In this book, we propose a formal framework that provides a partial solution to these problems. The model treats productivity and reuse as the target of inference in a probabilistic framework. Situating this question in a probabilistic framework allows us to the ***rational analysis*** methodology of Anderson (1990), which asks how an agent would best make use of patterns in the linguistic input to infer the true distribution of productive word-formation processes and reusable forms in their language.

The specific formal model presented in this book, ***fragment grammars***, is a generalization of a model known as ***adaptor grammars*** (Johnson et al., 2007a). Adaptor grammars provide a probabilistic formalization of the idea that anything that can be computed can be stored. Fragment grammars generalize adaptor grammars by allowing productive computation to be interleaved with the reuse of stored items in arbitrary ways.

2. Thanks to Steven Pinker for the example.

3. http://www.coolth.ca/

In this book, we adopt a modeling approach situated in the framework known as *probabilistic programming* (Erwig and Kollmansberger, 2006; Goodman et al., 2008; Kiselyov and Shan, 2009; Mansinghka, 2009; Pfeffer, 2001; Roy, 2011). In probabilistic programming, models are formalized as computer programs that generate observable data. In order to capture uncertainty in the data—which can arise from both noise in measurement and ambiguity over the multiple causal paths that lead to the same observations—the generative processes captured by probabilistic programs are inherently stochastic. Learning and reasoning in the framework is implemented using probabilistic conditioning, a form of hypothetical reasoning that allows us to make inferences about the different ways in which observed data might have been generated.

In Chapters 2 and 3, we formalize a number of models, including fragment grammars and adaptor grammars, from both probabilistic programming and more traditional perspectives. The result will be a mathematical framework that integrates linguistic computation with memory in a coherent and unified way. As we will explain below, the key to understanding learning and inference in this framework is to see how it gives rise to a trade-off between productivity on one hand and reuse on the other.

Consider two word-formation processes, such as the affixation of *-ness* and *-th* to an adjective. How can the learner use the evidence available in the primary linguistic input to determine that *ness*-affixation is productive while *th*-affixation is not? We can recast this question: How can the learner predict whether the next morphologically complex word will be a new form, computed productively, or an existing form, reused as a whole from memory?

In this book, we focus on evidence arising from the distribution of words and word parts, setting aside phonological and semantic structure. In this setting, there are only two kinds of relevant information. For productivity, the relevant evidence is how often word parts (like *-ness* and *-ity*) have been combined in the past to form new words. For reuse, the relevant evidence is how often particular combinations like *warm⁀th* have appeared as a single unit.[4]

The first kind of evidence—the use of word parts like *-ness* to form novel words—is the same kind of information used by language scientists to probe the productivity of word-formation processes. Confronted with the question

4. Throughout the book, we will use the symbol ⁀, as in *warm⁀th*, to denote the combination of two logically possible morphemes which may be stored together in the minds of speakers (or may not even be segmented at all). When we wish to express the active composition of two morphemes, we will use +, as in *pine-scented+ness*

of whether *-ness* or *-th* is productive, the scientist will see whether they can construct a new word using that suffix, for example, do speakers judge *pine-scentedness* to be acceptable? Thus, the proposal in this book is just that the child learner is like the language scientist. The child learns that certain grammatical operations are productive by observing their productive use.

The second kind of evidence—repeated reuse of combinations of word parts like *warm⁀th*—is the same kind of frequency information widely accepted to modulate memory formation and consolidation in psychology and neuroscience. However, when combined with evidence of novelty, there is a twist. The two kinds of information are mutually exclusive. The next utterance that a child hears cannot be both a new word and a repeated use of an earlier word. This leads to a logically necessary trade-off between productivity and reuse. As we will show in subsequent chapters, this has surprisingly deep consequences for language structure, learning, and processing. In the next sections, we discuss the historical setting in which the trade-off between productivity and reuse has been studied and then provide an informal overview of the modeling framework.

1.2 The Study of Productivity and Reuse

The creative potential of natural language has been recognized and celebrated for centuries. The term ***productivity***, in the sense of the ability to compute novel expressions, is due to the German grammarian Diez (Bauer, 2001; Diez, 1838; Schultink, 1992). The concept, however, has been around much longer, making appearances in the work of both ancient Sanskrit grammarians (see Bauer, 2001; Schultink, 1992; for detailed discussions of the history of the concept) and enlightenment thinkers, such as Wilhelm von Humboldt who famously described language as making infinite use of finite means (Chomsky, 1966; von Humboldt, 1836). Nevertheless, linguistic productivity resisted formal explanation until the second half of the 20th century. During the 1920s, '30s, and '40s, research on the foundations of mathematics and logic led to the development of a number of mathematical systems that captured the intuitive notion of computation. This work showed how an unbounded number of computations could be described using a finite system (e.g., Church, 1932, 1933; Post, 1943, 1944; Turing, 1937). Starting in the 1950s, researchers began to adapt these tools to provide formal accounts of linguistic computation (Bar-Hillel, 1953a,b; Chomsky, 1955; Hockett, 1955; Lambek, 1958). Natural language, however, differs in one critical respect from the mathematical problems that had inspired research into the theory of computation. It is riddled with exceptions and irregularities, generalizations which apply only in restricted contexts, and structure-building processes that differ in their degree of productivity. Thus a central challenge for linguistic and psychological theories is to explain why some logically possible linguistic processes generalize productively while others do not.

At first glance, the answer to this question seems simple: words are stored while phrases and sentences are computed. However, as many researchers have noted, it is difficult to find a clean dividing line between units that must be stored and those which must be computed (see, e.g., Di Sciullo and Williams, 1987; Jackendoff, 2002b; Nooteboom et al., 2002). Many linguistic units larger than words must be stored. For example, the meaning of verbal idioms like *leave no stone unturned* cannot be completely predicted from their individual component words. Nevertheless, the phrase can be used as a part of more complex expressions such as *the investigators left no legal stone unturned.* The syntactic flexibility of such idioms shows that they have complex internal structure, but their idiosyncratic meaning shows that in some respect they must be stored as coherent units (see Nunberg et al., 1994; for detailed discussion). There is accumulating evidence that even fully compositional phrasal structures can be stored (e.g., Arnon and Snider, 2010; Bannard and Matthews, 2008; Berant et al., 2008; Bod, 2001; Conklin and Schmitt, 2008; Gurevich and Goldberg, 2010; Tremblay and Baayen, 2010).

On the other end of the spectrum, it is also clear that many words and word parts are computed productively. For example, examination of any newspaper will reveal novel words formed using the suffix *-ness*. We know that a *pine-scented* candle exhibits the property of *pine-scentedness* without ever having heard either word before.

In this book, we provide a new perspective on these long-standing theoretical issues by suggesting that the problem of productivity can be treated as a structure-by-structure inference in a probabilistic framework, balancing a trade-off between productivity on one hand, and reuse on the other. Historically, there are two alternate approaches to the problem of explaining which structure-building processes are productive and which are not. The first, which we call the ***modularity-based*** approach, identifies computation and storage with different modules of grammar. The second, which we call the *idiosyncrasy-based* approach, identifies productivity with regularity and storage with idiosyncrasy (Bloomfield, 1933; Chomsky, 1965; Sweet, 1913). In the following sections, we discuss each approach, and identify why both ultimately fail. Along the way, we emphasize the close relationship between non-productivity and storage on one hand and productivity and computation on the other.

The problems with the modularity-based and idiosyncrasy-based views have led many contemporary theorists to converge on theories that posit a *heterogeneous lexicon*, in the terminology of Jackendoff (2002a,b). This theoretical move introduces a new set of problems for representation, processing, and learning, which we discuss below.

1.2.1 Modularity-Based Approaches to Productivity and Reuse

Language can be organized into a number of different kinds of units. Words are made up of morphemes; phrases are made of words, and so on. Although there is considerable debate about the precise makeup of this hierarchy, it is an obvious truism that smaller units, such as words or morphemes, are fewer, more idiosyncratic, and more likely to be reused, while larger units, such as phrases, are more numerous, more regular and transparent, and more likely to be novel. Modularity-based theories of productivity elevate this observation to an architectural principle of grammar. Productivity is a property of some modules of grammar (e.g., phrase structure) while reuse is a property of others (e.g., the lexicon).

Because linguistic units from different modules of grammar can be characterized by properties independent of productivity, modularity-based approaches appear to provide an appealing solution to the problem of learning the productivity of different linguistic units. For example, words are sometimes defined as the maximal domains exhibiting certain phonological phenomena (such as vowel harmony) or the minimal units visible to syntactic operations (such as those implementing movement or binding). If productivity were a property only of the phrase structure module, and the minimal units visible to this module of grammar were words, then the problem of learning which units combine productively in a language reduces to identifying the phonologically- and/or syntactically-defined words. Depending on our assumptions about the accessibility of properties like phonological structure or syntactic visibility, this may simplify the learning problem.

The existence of novel words (e.g., *pine-scentedness*) and stored phrases (e.g., *leave no stone unturned*) obviously defeats the simplest versions of this kind of account. Nevertheless, differences in productivity have often been invoked as evidence that different linguistic units are handled by different modules of grammar. Such arguments tacitly assume that productivity and reuse are properties of modules. For example, in linguistics, arguments of this kind have been prominent in the debate over **lexicalism**, the idea that the linguistic system consists of two distinct computational components: a component for phrase structure and a component for word structure.[5]

Much of the initial motivation for lexicalist theories grew out of the famous discussion of English nominalization in Chomsky (1970). Chomsky observed that there are important differences between gerundive nominals (e.g., *destroy/destroying*) and derived nominals (e.g., *destroy/destruction*). In particular, the former class are fully productive and semantically regular, while the latter vary in productivity and, in general, are more idiosyncratically related to the verbs they nominalize, both semantically and morphologically. He concluded that nominalization should not be handled by the transformational component of syntax (as was standard in earlier theories). Chomsky (1970) was widely interpreted as an argument that word-formation processes and syntactic processes should take place in separate components of the grammar and be handled by different kinds of rules. Theorists immediately began developing accounts of this less-productive, lexical component of the grammar (influential early works include, Aronoff, 1976; Halle, 1973; Jackendoff, 1975).[6]

The research in this tradition developed a crucial insight that was incorporated into all subsequent theories: that there is a tight inverse relationship between productivity and storage. Beginning with works such as Halle (1973), morphology came to be understood as the theory of possible words. A theory of possible words should be able to explain both structure in existing words, and

5. It should be emphasized that the term *lexicon* when used in reference to lexicalist theories refers to a computational component of the grammar—the one which computes the structure of words. It is not used in the more familiar sense of the inventory of primitive items stored in memory and manipulated by rules. Some authors (e.g., Di Sciullo and Williams, 1987) are careful to distinguish these two ideas, using terms such as *listed* and *listedness* rather than *lexical* to refer to storage of primitive units.

6. As Marantz (1997) pointed out, only a small part of Chomsky (1970) concerned arguments against the transformational analysis of nominalizations, and Chomsky presented no theory of derived nominal forms. Instead, Chomsky's main purpose was to provide a theory that explains the commonalities between clauses and noun phrases in the absence of a transformational account. Nevertheless, whether or not it was the intention of the original paper, Chomsky (1970) has been widely understood as promoting the need for a separate lexical module for word structure.

the possibility of productive generalization. However, as we discussed above, morphological systems contain many more logically possible generalizations than can actually occur in practice. Any morphological theory, therefore, must contain mechanisms to constrain the productivity of some of the rules which describe word structure.

A potential solution to this problem is provided if rules in the lexical component of the grammar always store their outputs in memory, or only operate on stored forms. For example, Jackendoff (1975) argued that shared structure between word forms should be accounted for via ***lexical redundancy rules***. These are rules expressing the generalizations inherent in stored lexical items but differing from the productive rules of syntax. Unlike the latter, they operate on stored items, and serve mainly during learning (by reducing the amount of new information needed to learn morphologically complex words) and in cases of analogy (e.g., *coolth* by analogy with *warmth*), but do not necessarily function during normal language use.

The inverse relationship between productivity and storage is a crucial insight; however, modularity as a theory of productivity has a serious flaw. As we noted earlier, productivity is a property that cuts across different kinds of linguistic units and levels of linguistic structure. Researchers have not been able to identify independent properties of different linguistic units, representations, or mechanisms that can be used to strictly segregate productive and unproductive processes into separate modules of grammar. As Jackendoff (1975) pointed out, idiomatic syntactic constructions suggest that syntactic rules sometimes operate in a storage mode, while creative word formation (e.g., *pine-scentedness*) shows that morphological rules sometimes do not. This latter point was brought to the forefront of morphological research by Aronoff (1976), who discussed the disparity between formations in productive *-ness* and unproductive *-ity* and reached the conclusion that *-ity* forms are stored in the lexicon, while *-ness* forms are not. Exceptional patterns like this rob the modularity-based approach of its explanatory power, since they require additional mechanisms to explain variation in productivity within a module.

The clearest statement of this perspective is offered by Di Sciullo and Williams (1987) who give extensive arguments against the use of productivity in arguments for modularity (p. 2). In their words:

> The distinction between words and phrases is a distinction in the theory of grammar. The listed/unlisted distinction has nothing to do with grammar. Syntax and morphology are both recursive definitions of sets of objects—but of different sets, with different

atoms and different rules of combination. These are the only differences the grammarian need acknowledge.

The most immediate consequence of this view is that productivity and listedness are not grammatical concepts. [...]

Di Sciullo and Williams (1987) conclude that storage can vary from form to form, cutting across levels of linguistic structure, and that productivity can vary from rule to rule, also cutting across different rule types.

In this book, we will adopt the insights of the literature we have just briefly reviewed. Any module of grammar, rule, or linguistic unit can, in principle, be involved in both storage and productive computation. However, there is an inverse relationship between the two. The more that a process results in storage the less likely it will be to generate novel forms, and, conversely, the more that a process generates novel forms, the less likely it is to store its outputs.

1.2.2 The Idiosyncrasy-Based Approach

A second approach to productivity that has been frequently invoked in linguistic and psycholinguistic theories is the *idiosyncrasy-based* approach (e.g., Bloomfield, 1933; Di Sciullo and Williams, 1987; Pinker and Prince, 1988). This approach holds that only idiosyncratic structures are stored; all structures that are computable by regular rules are productive. The appeal of idiosyncrasy-based accounts of productivity is that they build on a logical truism: purely idiosyncratic structures must be stored, while novel structures, produced by productive rules, cannot be stored prior to the first time they were produced. Since idiosyncratic structures must be stored, it may be reasonable to hypothesize that that only idiosyncratic structures are stored.

Note that modularity-based proposals (i.e., only words or morphemes or the like are stored) and idiosyncrasy-based proposals (i.e., only idiosyncratic structures are stored) are logically distinct. It is not the case that there must be only one kind of linguistic representation that allows for idiosyncrasies, nor is it the case that all structures of a single kind must be idiosyncratic. Nevertheless, there has been a tendency for theorists to tacitly adopt a mixture of the modularity-based and idiosyncrasy-based positions.

Despite their widespread adoption, idiosyncrasy-based proposals face a challenge similar to modularity-based approaches. Idiosyncrasy and regularity are not necessary and sufficient conditions for storage and productivity. Although idiosyncratic forms must be stored under any theory, idiosyncratic patterns are sometimes partially generalizable (e.g., *-ity* after *-able*). Cases of semi-productivity like these call into question the clean divide between idiosyncratic stored forms and productive rules.

The most important evidence against a idiosyncrasy-based model of productivity and storage comes from psycholinguistic studies of word structure. Over the past forty years, such studies have demonstrated that even fully regular, compositional forms are often stored. We review this evidence in the next section.

1.2.3 Evidence from Psycholinguistic Studies of Word Structure

As discussed above, Di Sciullo and Williams (1987) call into question modularity-based approaches to productivity and reuse and propose that productivity and reuse are properties that can vary independently across different linguistic modules, structures, and rules. Since the early 1970s, a long sequence of psycholinguistic studies of word processing have supported this view. However, these studies also provide evidence—*contra* Di Sciullo and Williams (1987)—that a idiosyncrasy-based theory must also be rejected. In this section, we give a brief overview of this work. See Chapters 4 and 6 for more detailed discussions.

The first studies that we will review examine the question of whether and when words are retrieved whole from memory or composed on the fly during lexical access. The most common paradigm for asking such questions has been the *lexical decision* experiment. In these experiments, subjects are asked to classify strings of letters (or, occasionally, sequences of sounds) as words or non-words, and their response time is measured. Whether words have been processed compositionally or retrieved whole from memory is assessed in one of two ways. First, some studies vary the degree of morphological overlap between target items (words or non-words) and existing words. For example, Taft and Forster (1975) compared how long it took subjects to reject real stems (e.g., *juvenate* derived from *rejuvenate*) with how long it took them to reject pseudo-stems (e.g., *luvenate* which does not occur in an existing English word). They found that the former took longer to reject, suggesting that the stem *juvenate* is represented independently of the prefix *re-*. Many similar experiments have found that shared substructure affects processing time, thus providing evidence for decomposition (e.g., Andrews, 1986; Caramazza et al., 1988; Taft and Forster, 1975, 1976). However, for existing words, the effect of shared structure is mediated by the frequency of the form (Colé et al., 1997);

shared structure affects the processing time of low-frequency words more than it affects the processing time of high-frequency words. The latter result suggests that high-frequency forms are more likely to be stored as wholes than low-frequency forms.

The second approach to studying (de)composition in lexical decision paradigms exploits the fact that people respond more quickly to items that are more frequent. Such **frequency effects** are observed in a wide variety of cognitive domains and at multiple levels of linguistic analysis, and are assumed to reflect some property of storage whereby more frequent stored representations are easier to access. A number of such studies have found whole-form frequency effects in morphologically complex words, strongly suggesting that (some of) these words are stored as wholes. Importantly, such whole-form frequency effects are often present even for words formed by perfectly regular morphological processes (Bertram et al., 2000c; Manelis and Tharp, 1977; Rubenstein et al., 1970; Sereno and Jongman, 1997), seriously calling into question a idiosyncrasy-based account where only idiosyncratic structures are stored.

However, frequency effect data also often provides evidence for decomposition. For example, in some studies the frequency of bases (e.g., *agree* in *agreeable*) predicts processing latency, even when whole-form frequencies are controlled. In other studies, whole-form frequency effects are present for high-frequency words while base frequency effects are present for lower-frequency words (see, Alegre and Gordon, 1999; Baayen et al., 1997b, 2003; Bradley, 1980; Burani and Caramazza, 1987; Burani et al., 1984; Colé et al., 1989; Gordon and Alegre, 1999; Moscoso del Prado Martín et al., 2005; Taft, 1979; Vannest and Boland, 1999).

Evidence for both decomposition and retrieval of both idiosyncratic and regular morphology has also been found using **primed lexical decision**. In primed lexical decision experiments, the lexical decision task described above is manipulated by the inclusion of prime trials. Prime items are manipulated to share different amounts of structure with target items. For example, if the target item is *agreeability*, a prime trial might consist of a word that shared the same stem (e.g., *agreement*); a word that shared a suffix, or combination of suffixes (e.g., *curiosity/drinkability/watchable*); the word itself (known as **identity priming**); or a word that shared no structure with the target (e.g., *drink*). If any of these primes facilitates the recognition of the target, this can be taken as evidence that whatever is shared between the prime and target is represented independently in the lexicon.

Some studies have found evidence for identity priming (e.g., Katz et al., 1991), suggesting whole-form retrieval during processing. Other studies have found evidence that identity priming and priming by shared structures, such as bases, are equally strong, suggesting decomposition (Forster and Azuma, 2000; Fowler et al., 1985). In some studies, both kinds of priming have been discovered, but identity priming is larger in magnitude than priming by shared structures (e.g., Stanners et al., 1979b). Furthermore, the difference in the size of the priming effect can be modulated by other factors. For instance, more phonologically regular or semantically transparent complex words are primed by shared structures to a greater degree than less regular or transparent words (Kielar et al., 2008; Marslen-Wilson et al., 1994; Sandra, 1990; Stanners et al., 1979a). Priming is also modulated by frequency. Higher-frequency items show smaller effects of priming by shared structures than lower-frequency items (de Vaan et al., 2007; Meunier and Segui, 1999). Overall, these results suggest that word processing is characterized by a mixture of decomposition and retrieval.

To summarize, psycholinguistic evidence calls into question both the modularity-based and idiosyncrasy-based accounts of productivity. Productivity and storage appear to be partially independent of other aspects of linguistic representations and can vary from word to word and rule to rule. Furthermore, even non-idiosyncratic structure can sometimes be stored.

1.2.3.1 Learning Synergies

In the preceding sections, our goal has been to review the evidence in favor a view of productivity and reuse where these properties cut across different levels of linguistic structure and different kinds of representations, and cannot be uniquely identified with (non)idiosyncrasy. However, while phonological regularity, semantic transparency, and grain-size of linguistic unit cannot be necessary and sufficient conditions for productivity, there is an obvious empirical correlation between productivity and these other properties. In the case of morphology, productive affixes like *-ness* tend to be more regular, more transparent, and tend to occur in a greater number of low-frequency forms. Less productive affixes like *-th* tend to be less transparent, less regular, and occur more often in a small number of high-frequency forms.

The correlation between these properties is an example of what might be called a *learning synergy*—several different kinds of evidence available to the language learner all make similar, redundant predictions about the productivity of word-formation processes (Johnson, 2008b). We will review further evidence for such synergies in Chapters 4 and 6. However, in this book we will only speculate on the consequences of including additional phonological and semantic sources of evidence to the learner, focusing instead on those aspects of productivity which can be predicted just from the distribution of words and word parts in the input.

1.2.4 Converging Evidence from Syntax

Our discussion so far has focused on the need for a theory that allows arbitrary mixtures of productivity and reuse in morphology. Converging evidence from syntax suggests that such a theory is necessary in that domain as well. The grammatical system must be able to store linguistic units larger than individual words in order to account for syntactic non-compositionality like that exhibited by idioms such as *kick the bucket*. Some accounts of idiom structure have tried to preserve some version of the modularity-based approach. For example, Emonds (1970) and Chomsky (1981) proposed that idioms like *kick the bucket* are not verb phrases, but rather that they are a special kind of complex verb. This theoretical move preserves the distinction between a fully compositional syntactic component and a lexicon that is the repository for stored, idiosyncratic items. However, as Di Sciullo and Williams (1987) observed, these accounts suffer from the fact that they require a potential replication of theoretical machinery. *Kick the bucket* behaves like other, fully compositional verb phrases in many respects. For example, it can be inflected for tense—*kicked the bucket*, *will kick the bucket*. If such an item is just a special kind of verb, no different from other simple verbs such as *eat* or *walk*, then accounting for the ways in which it patterns like other compositional verb phrases requires replication of the rules of syntax in the lexicon. Moreover, such an account poses a formidable learning problem for the child. How can the child recognize that this structure, which appears to be like any other verb phrase in many ways, is actually a special kinds of lexical verb (see also Jackendoff, 2002a; Section 6.9)?

The problem is pervasive. Idiomaticity, partial compositionality, and, in general, constructions that in some ways generalize productively and in other ways are idiosyncratic are all frequent phenomena in syntax. Nunberg et al. (1994) examined a large range of idioms, and argued that these occupy a cline of compositionality and productivity. At one end are idioms, such as *shoot the breeze*, which are almost completely non-compositional, and share very little with more productive structures (e.g., they cannot be passivized **The breeze was shot by me and John*). On the other end are idioms, such as *leave no stone unturned*, which are semi-compositional and can enter into a variety of productive syntactic alternations and constructions (e.g., *no stone was left unturned, We left no legal stone unturned*). Recent experimental work has also provided evidence that idiom processing has both compositional and non-compositional aspects (Peterson et al., 2001; Sprenger et al., 2006).

The need for a theory of variable productivity and reuse in syntax is not limited to idioms which consist only of strings of words. Fillmore et al. (1988) highlighted the important phenomenon of productivity within the context of an idiomatic frame, showing that the non-compositional construction [*let alone X*] generalizes regularly to a variety of novel expressions (e.g., *let alone John, et alone leaving work early today, let alone too young to drink*). Culicover (1999) identified a large number of problematic syntactic constructions that are compositional, but exhibit some exceptionality in their distribution. For example, he looked at a number of prepositions in English that vary in properties such as whether they can appear before or after a noun phrase (e.g., *in two years* v. *two years ago*), arguing that the pervasive exceptionality in the system must be learned on a case-by-case basis. Although there have been attempts to deal with some of these issues by positing more abstract or complex underlying representations (e.g., Coppock, 2007), many of the problems have resisted theoretical treatment. In addition there is accumulating experimental evidence from psycholinguistics that even fully compositional and regular syntactic structures larger than words might also sometimes be stored (e.g., Arnon and Snider, 2010; Bannard and Matthews, 2008; Berant et al., 2008; Bod, 2001; Conklin and Schmitt, 2008; Gurevich and Goldberg, 2010; Tremblay and Baayen, 2010).

1.2.5 Heterogeneous Lexica

In the preceding sections, we have outlined prior attempts to reduce productivity and storage to a distinction between linguistic modules (e.g., phrase structure versus word structure) or a side effect of idiosyncrasy. In both cases, the proposals fail to provide necessary and sufficient conditions on storage and computation. Instead, productivity appears to be a property that varies from form to form and rule to rule, cutting across levels of representation and degrees of regularity.

This conclusion has been widely accepted in the field, and a new class of linguistic and psycholinguistic theories have been developed to account for these phenomena. Jackendoff uses the term ***heterogeneous lexicon*** theories to refer to these approaches because they allow storage and computation for a variety of different kinds of units, including morphemes, words produced by rules (regular or otherwise), and larger chunks of syntactic structure (see, e.g., Jackendoff, 2002a). Heterogenous approaches to the lexicon have been widely adopted in both morphology (see, e.g., Aronoff and Fudeman, 2005; Bauer, 2001; Pinker, 1999; Plag, 2003, 2004), and syntax (see, e.g., Bybee and Hopper, 2001; Croft, 2001; Culicover and Jackendoff, 2005; De Marcken, 1996b; Goldberg, 2006; Kay, 2002; Kay and Fillmore, 1999; Nooteboom et al., 2002; Sag et al., 2003; Wood, 2010).

Heterogenous lexicon theories assume that the grammar and the lexicon are a single system consisting of a small number of structure-building operations (e.g., UNIFY, MERGE, APPLY, ADJOIN, SUBSTITUTE, etc.) together with a large collection of heterogenous fragments of structure. These fragments vary from the abstract, regular units traditionally identified with productive rules, to specific, idiosyncratic units traditionally understood as stored lexical primitives.

While heterogenous lexicon theories appear to be needed to account for adult knowledge of language, they raise serious questions about language learning and use. Heterogenous theories assume that a variety of structures—large and small, corresponding to words or phrases—can be stored. Furthermore, they assume that productivity can vary from process to process and be localized to specific contexts and constructions (e.g., *-ability*). As a result, under such theories, the problems of acquiring an inventory of stored items and of comprehending or producing an utterance are massively ambiguous. For example, to generate or analyze a word like *Googleability* a heterogenous system has many options: whole-form storage (i.e., *Google⁀able⁀ity*), full composition (i.e., *Google+able+ity*), or various kinds of partial storage/composition (e.g., *Google⁀able+ity* or *Google+able⁀ity*). The learner must determine the basic combinatorial units of their language, and the adult language user must be able to access this knowledge to determine which structure-building process can give rise to novelty and which cannot (see also Jackendoff, 2002a; Section 6.9).

What kinds of evidence might a learner use to determine which units are stored and which units are computed? Some theorists have proposed that learners use information such as the phonological regularity and semantic transparency of word-formation processes as cues to productivity (e.g., Marcus et al., 1995, 1992). As we discussed above, these properties are correlated with productivity and reuse, and are, therefore, likely important cues for the learner. However, in this book we focus primarily on another kind of evidence available to the language learner: the distribution of forms in the linguistic input. We ask how the learner can infer the correct pattern of productivity and reuse for his language using the frequency and pattern of co-occurence of forms and subforms within and across the words he encounters.

In the next section, we briefly discuss several distributional proposals for learning the pattern of productivity and reuse. However, we defer a detailed discussion and comparison of individual models to later chapters.

1.2.6 Distributional Theories of Productivity and Reuse

Distributional theories of productivity are based on the frequencies of forms and subforms in the linguistic input that the learner receives. When large corpora of words are examined, suffixes such as *-ness*, *-ity*, and *-th* show important differences in their frequency distributions. These differences are the direct consequence of productivity. Highly productive suffixes such as *-ness* frequently give rise to new words. Because of this, the distributions associated with these suffixes tend to have large numbers of low-frequency words and often exhibit high *type frequency*—the number of distinct word forms using the suffix. By contrast, less productive affixes such as *-ity* and, especially, *-th* are mostly reused in existing forms and, therefore, tend to appear in words across a variety of token frequencies, often with a bias towards more frequent items (Baayen, 1992, 1993, 1994, 2001, 2006; Baayen and Lieber, 1991; Baayen and Renouf, 1996; Barnwell, 2010; Bauer, 1992, 2001, 2005; De Marcken, 1996a,b; Du and Zhang, 2010; Fernández-Domínguez et al., 2007; Gaeta and Ricca, 2006; Hay and Baayen, 2002; Plag, 2004; van Marle, 1992; Yang, 2005).

A useful tool for visualizing frequency distributions is the *frequency spectrum*. Figure 1.2.6 shows the frequency spectra for *-ness*, *-ity*, and *-th*, as computed against the training data set used in the study of derivational morphology presented in Chapter 7. The x-axis in each plot represents the different frequencies at which words can appear (on a log scale). The height of each bar in the plot represents the density of words in the frequency range spanned by the width of the bar.[7] The red lines are a smoothed kernel-density estimate of the densities of words at various frequencies.

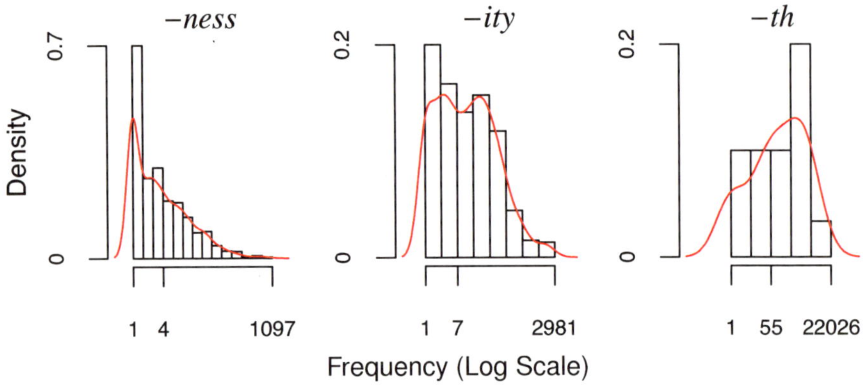

Figure 1.1
Frequency Spectra for *–ness*, *–ity*, and *–th*

The far left of Figure 1.2.6 shows the frequency spectrum for *-ness*. Because *-ness* is highly productive, the spectrum is steeply peaked at low frequencies indicating the large number of new coinages using this suffix. The distribution falls off quickly towards the higher frequency ranges, although there are still a few very frequent words using *-ness*. Although there has been some discussion in the literature about the exact mathematical form of word-frequency distributions (see Baayen, 2001; for a comprehensive review), productive affixes have distributions that fall into a class known as *large number of rare event* distributions (Baayen, 2001; Khmaladze, 1987), indicating their fundamental property: productivity gives rise to new forms (i.e., rare events). The spectra for *-ity* and *-th* are quite different. These suffixes have a smaller proportion of low-frequency words and a higher proportion of high-frequency words. Some authors have claimed that unproductive affixes such as these tend to to be normally distributed in their log frequencies (e.g., Baayen and Lieber, 1996). Regardless of the exact mathematical form of these distributions, they are characterized by a less sharp peak at low-frequency ranges and a greater spread of word forms throughout higher-frequency ranges.

7. In other words, the sum of the heights of each bar multiplied by their width is equal to one.

There have been a number of proposals in the literature for how such distributional facts can be used to infer levels of productivity for an affix. The simplest kind of distributional cues are those based purely on type frequencies of morphemes. Bybee (1995), for example, proposed that the type frequency of an affix (also known as the morphological family size) predicts its productivity. However, this measure systematically overestimates productivity when a morpheme has both high type frequency and high token frequency. Baayen (2006) pointed out, for example, that the Dutch verbal prefix *ver-* has greater type frequency than the Dutch agentive suffix *-ster*, but that the former is judged less productive than the latter by speakers. He argues that this is because most types using *ver-* have a high token frequency.

Crucial to understanding this insight is the fact that the type and token frequencies of morphologically complex words represent a trade-off. Productive suffixes give rise to new forms, increasing type frequency, but repeated use of individual forms using a suffix encourages storage, decreasing productivity. Thus productivity is predicted by the shape of the distribution of tokens over types, and, therefore, any measure of productivity based only on type or token frequencies of words (or parts of words) will be inaccurate in some cases. For this reason, many researchers have developed distributional models that combine type and token frequencies. These fall into two main classes: statistics designed to estimate productivity from a corpus sample, and cognitive models, intended to capture the mental processes by which people acquire and use their grammatical knowledge. We discuss each of these in turn.

A number of researchers have proposed statistics intended to estimate the probability that a given affix (e.g., *-ness*) will give rise to novel forms. An early example of such a statistical estimator was the proposal of Aronoff (1976) that the (non-)productivity of a process (such as *-ness* affixation) can be measured by taking the proportion of all existing word types that use the process and dividing by the number of potential word types that could use the process. The greater proportion of words which could exist that do, the less productive the suffix. In a more recent proposal, Barnwell (2010) defined an index of productivity using the product of the log frequency of an affix and the entropy of the set of bases using that affix. The first term in this formula takes into account the token frequency of words using an affix, while the second term takes into account the variability in how the affix is used.

However, by far the most well-known measures of productivity are Baayen's hapax-based statistical estimators, $\mathcal{P}$ and $\mathcal{P}^*$, which respectively estimate the conditional probability that an affix will be used to form a new word, and the conditional probability that a new word will use a particular affix (Baayen, 1992, 1993, 1994, 2001, 2003, 2006; Baayen and Lieber, 1991; Baayen and Renouf, 1996; Du and Zhang, 2010; Hay and Baayen, 2002; van Marle, 1992). We will discuss these measures in much greater detail in Chapters 6 and 7.

Most cognitive theories of productivity have seen the problem as a matter of optimizing a trade-off between storage and computation. There are two ways of construing this trade-off. Under one view, what is being optimized is the use of two resources: the space used in memory by stored items, and the amount of computation (measured in number of computational steps, or time) needed to produce and comprehend linguistic forms (e.g., Baayen and Schreuder, 2000; Baayen et al., 2000; Bertram et al., 2000c; Frauenfelder and Schreuder, 1991; Kuperman et al., 2010; Prasada and Pinker, 1993; Sandra, 1994; Stanners et al., 1979a; Yang, 2009, 2010). In general, reducing the burden on one of these resources will increase the burden on the other. For example, if we reduce the memory load by storing only the smallest, most abstract units, then building expressions will require more computational steps because each word will have to be built from scratch. In contrast, if we minimize computational load by storing large complex structures, the demands on memory will increase.

A second view, espoused in this book, analyzes this trade-off in terms of the ability to succinctly represent the input data and thereby accurately predict new forms (Cohn et al., 2010, 2009; De Marcken, 1996a,b; O'Donnell, 2011; O'Donnell et al., 2009, 2011; Post and Gildea, 2009; Zuidema, 2007). This general approach was first suggested by De Marcken (1996a) using the ***minimum description length*** framework which finds an optimally succinct code for representing a set of outcomes drawn from some probability distribution (Grünwald, 2007; Rissanen, 1978, 2007; Vitányi and Li, 2008).[8] The distributional approach to productivity was also anticipated by work in the data-oriented parsing framework (e.g., Bod, 1998; Bod et al., 2003; Scha, 1990, 1992; see Section 2.4.3). prediction-based models use probability (or, relatedly, code-succinctness) as the currency of optimization, balancing the amount of probability assigned to reusing existing forms against the the possibility of generating novel forms in the future. The resource-based and prediction-based perspectives are clearly related. To predict that a structure will be reused, it must be stored in memory, and this requires space. In contrast, novel forms require

8. De Marcken (1996a) developed a formal system very similar to the fragment grammars model, but did not examine the issue of morphological productivity and only published simulation results for a special case, more closely related to adaptor grammars. See Section 2.4.4.1.

computation to produce and comprehend, and this takes time. Consequently, some models in the literature have combined aspects of both approaches (e.g., Taatgen, 2001; Taatgen and Anderson, 2002; Taatgen and Dijkstra, 2003; Taatgen and Lee, 2003).

The framework proposed in this book adopts a strictly prediction-based view on the matter. In the next sections of this chapter, we will discuss the psychological assumptions underlying the model, and give an informal overview of its workings. Later in the book, we will examine two morphological systems: the English past tense and English derivational morphology. These two domains are of special interest in evaluating models of productivity and reuse because they exhibit very different distributions of these properties. Derivational morphology is characterized by a broad array of affixes of differing levels of productivity. This kind of system is exactly what one might expect probabilistic models, such as the one advocated in the book, to excel at capturing (see, e.g., Hay and Baayen, 2005). By contrast, the English past tense's regular +/*d*/ rule (e.g., *walk/walked*) is highly productive, while the various irregular form classes (e.g., *sing/sang*, *sleep/slept*, etc.) generalize only very rarely (Prasada and Pinker, 1993; see, also Chapter 4). Thus the English past tense provides an important counterpoint to the variable structure of derivational morphology. A model of productivity and reuse must be able to handle both kinds of linguistic systems, those with widely varying mixtures of productive computation and reuse, and those with sharp, nearly deterministic dichotomies between the two.

1.3 The Proposal

In this section we describe our proposal and define a formal model, fragment grammars, which instantiates this theory. We begin by enumerating the psychological claims of the theory, and then discuss the trade-off between productivity and reuse implicit in these claims. In addition to our proposal that productivity and reuse should be considered a probabilistic inference, we also consider three other approaches, which we review below. Each of these models has been chosen to formalize a specific historical proposal about productivity and reuse, while otherwise keeping the models maximally similar to one another.

1.3.1 Psychological Claims

Our model is built on four assumptions about the nature of computation in the linguistic system. First, the ability to comprehend and produce novel forms is the result of an underlying system of hierarchical, recursive, discrete computation. We treat this underlying system as an input to all of the models studied in this book. Second, frequently reused sequences of computations can be stored and treated as a single unit in future computations. In principle, any (contiguous) combination of computations is a candidate for storage. Third, the fact that any combination of computations can be stored gives rise to an inference problem. Given the observed data, the learner must determine which parts of structures in the linguistic input should be stored and retrieved as wholes and which should be computed on the fly. We quantify uncertainty about storage and computation using probability, and assume that the learner uses probabilistic conditioning to find an approximately optimal solution given the data. Fourth, this inference involves two competing prior expectations on the part of the learner: that the learner prefers to store fewer, simpler items, and that the learner prefers to minimize the number of independent computational steps in the derivation of any particular form. Note that these biases compete, and therefore our system approximately optimizes the trade-off between them. We discuss each of these assumptions in more detail in the following sections. Note that we assume that the same principles apply to both comprehension and production in our model. Therefore, when we discuss the construction of a form, we have in mind both processing during production and parsing during comprehension.

1.3.1.1 The Underlying Computational System

All approaches to linguistic structure must satisfy a basic principle called the *productivity constraint* by Frauenfelder and Schreuder (1991), which states that any theory must provide a mechanism for generating novel expressions. We satisfy the productivity constraint by adopting the classic linguistic and psycholinguistic solution to the problem of generalization, assuming an underlying system of hierarchical, recursive, and discrete computation. Using such a system as an underlying model of computation allows for a potentially unbounded number of forms to be described by a finite system. In what follows, we make use of Probabilistic Context-Free phrase structure Grammars (PCFGs) to formalize this underlying system (see, Autebert et al., 1997; Manning and Schütze, 1999; for reviews).

A PCFG consists of several elements. The first is the set of *terminal* symbols—written with lowercase letters—which are the most primitive, atomic elements out of which larger constituent categories can be built. In what follows, terminals will represent morphemes such as *-ness* or *happy*. The second element is the set of ***nonterminals***. By convention, nonterminals are written with capital letters and represent categories of constituents such as ADJECTIVE (`Adj`), or NOUN (`N`). One nonterminal is known as the start symbol, and represents the category of complete structures. In the examples below, it will correspond to the category of complete words (`W`). Finally, there is the set of production rules (or just rules). The production rules are written $\mathtt{A} \longrightarrow \gamma$, where γ is some sequence of terminals and nonterminals, and `A` is a nonterminal. These rules define the set of possible `consists-of` relations between categories of constituents. For example the rule $\mathtt{N} \longrightarrow \mathtt{Adj}$ `-ness` says that a noun (`N`) can consist of the suffix *-ness* appended to an adjective (`Adj`). The list of symbols to the right of the arrow is referred to as the ***right-hand side*** of that rule. The nonterminal to the left of the arrow is the rule's ***left-hand side***. Starting with the start symbol `W` and applying the rules until only terminals remain, it is possible to derive sequences of terminals such as *happy -ness* (*happiness*).

As an example, consider the PCFG in Figure 1.2. The rules of the PCFG are shown on the right and a probability associated with each rule is shown on the left. Suppose that we wish to construct a noun (`N`) using this PCFG. The grammar shows five rules with `N` on their left-hand side. The probabilities associated with these five rules sum to 1 (i.e., $\sum_i p_{\mathrm{N}_i} = 1$), and, thus, specify a distribution that can be used to sample one of the rules. To generate a noun using this PCFG, one of these five rules is chosen according to this distribution. Once a particular rule has been chosen, sampling continues recursively on each nonterminal of that rule's right-hand side. For example, if the rule $\mathtt{N} \longrightarrow \mathtt{Adj}$ *-ity* was chosen, the process would continue by sampling an adjective (`Adj`) from the distribution over rules with `Adj` on their left-hand sides.

The particular set of choices associated with building a linguistic expression can be represented as a tree, called a ***derivation*** or ***parse tree***. Figure 1.3 shows the tree associated with the process of generating the word *agreeability*. In this figure, and in the derivation trees shown later, we have suppressed the top-level `W` category.

There are several reasons that we selected PCFGs as the underlying model of computation for the models in this book. First, they are simple, well-understood, widely used in computational linguistics, and have well-known inference algorithms (see, Manning and Schütze, 1999). Second, fragment grammars build on the adaptor grammars model (Johnson et al., 2007a), which employs PCFGs as its underlying model of computation. Third, and most

$p_{\mathtt{W}_1} = \frac{1}{4}$	W	$\longrightarrow$	N	
$p_{\mathtt{W}_2} = \frac{1}{4}$	W	$\longrightarrow$	V	
$p_{\mathtt{W}_3} = \frac{1}{4}$	W	$\longrightarrow$	Adj	
$p_{\mathtt{W}_4} = \frac{1}{4}$	W	$\longrightarrow$	Adv	
$p_{\mathtt{N}_1} = \frac{1}{10}$	N	$\longrightarrow$	Adj	*-ness*
$p_{\mathtt{N}_2} = \frac{1}{10}$	N	$\longrightarrow$	Adj	*-ity*
$p_{\mathtt{N}_3} = \frac{1}{100}$	N	$\longrightarrow$	*electro-*	N
$p_{\mathtt{N}_4} = \frac{1}{230}$	N	$\longrightarrow$	*magnet*	
$p_{\mathtt{N}_5} = \frac{1}{50}$	N	$\longrightarrow$	*dog*	
	...			
$p_{\mathtt{V}_1} = \frac{1}{300}$	V	$\longrightarrow$	N	*-ify*
$p_{\mathtt{V}_2} = \frac{1}{100}$	V	$\longrightarrow$	Adj	*-ize*
$p_{\mathtt{V}_3} = \frac{1}{40}$	V	$\longrightarrow$	*re-*	V
$p_{\mathtt{V}_4} = \frac{1}{50}$	V	$\longrightarrow$	*agree*	
$p_{\mathtt{V}_5} = \frac{1}{300}$	V	$\longrightarrow$	*count*	
	...			
$p_{\mathtt{Adj}_1} = \frac{1}{500}$	Adj	$\longrightarrow$	*dis-*	Adj
$p_{\mathtt{Adj}_2} = \frac{1}{300}$	Adj	$\longrightarrow$	V	*-able*
$p_{\mathtt{Adj}_3} = \frac{1}{25}$	Adj	$\longrightarrow$	N	*-ic*
$p_{\mathtt{Adj}_4} = \frac{1}{1000}$	Adj	$\longrightarrow$	N	*-al*
$p_{\mathtt{Adj}_5} = \frac{1}{35}$	Adj	$\longrightarrow$	*tall*	
	...			
$p_{\mathtt{Adv}_1} = \frac{1}{2}$	Adv	$\longrightarrow$	Adj	*-ly*
$p_{\mathtt{Adv}_2} = \frac{1}{100}$	Adv	$\longrightarrow$	*today*	
	...			

Figure 1.2

Probabilistic Context-free Grammar: A fragment of a PCFG for English derivational morphology.

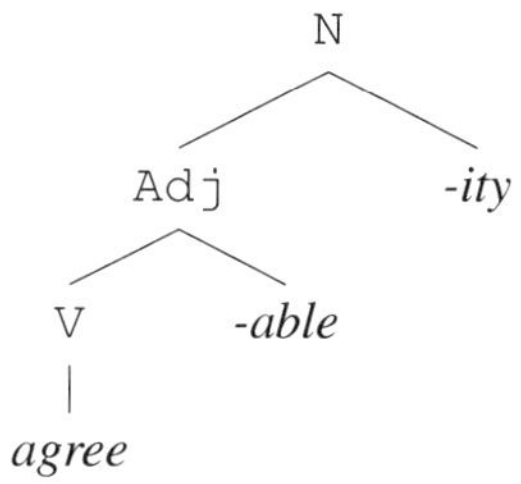

Figure 1.3

Derivation Tree: A tree showing the set of random choices needed to sample the word *agreeability*. Note that we have suppressed the top W node.

important, PCFGs allow us to describe simple hierarchical (and potentially recursive) structures of the kind needed to describe morphologically complex words. Although there is considerable debate in morphology about the appropriate representations for word structure, the kinds of hierarchical constituent structures that can be described by PCFGs are sufficient to study the phenomena of productivity and reuse that we focus on in this book. In particular, there is wide-spread consensus that the string sets characterizing natural language morphology fall into a class contained within the context-free string-sets (indeed, the weaker class of finite-state systems may suffice for morphology and phonology. See e.g., Beesley and Karttunen, 2003; Mohri, 1997; Roark and Sproat, 2007).[9]

Nevertheless, the adoption of PCFGs is a convenience. There are a wide variety of systems that could be used in place of this formalism to describe our underlying theory of computation (see the next chapter for generalizations that apply to a wider range of generative models).

1.3.1.2 Arbitrary Storage and Computation

The second assumption of our model is that any (contiguous) sequence of computations can be stored and reused as a single unit. Thus, our theory posits a heterogeneous lexicon like those discussed above.

9. The current consensus is that CFGs are inadequately expressive for syntax, however. From the perspective of computational linguistics see, for example, discussions in Jurafsky and Martin (2000); from a linguistic perspective see Chomsky (1956); Huybregts (1985); Pullum and Gazdar (1982); Shieber (1985); and for a psycholinguistic perspective see Levy (2008).

When the underlying model of computation is a PCFG, the storage of (partial) computations corresponds to the storage of tree fragments such as those shown in Figure 1.4.

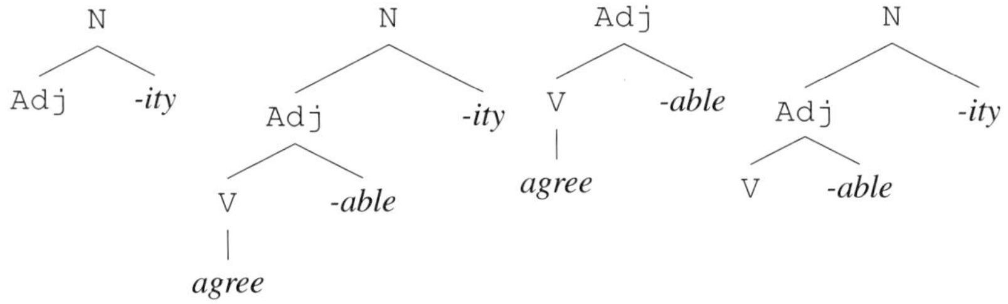

Figure 1.4

PCFG Sub-computations: Some sub-computations of the PCFG derivation tree shown in Figure 1.3.

The first subtree in Figure 1.4 corresponds to a minimal computational step, which results from choosing the underlying PCFG rule: `N` ⟶ `Adj` *-ity*. The second subtree represents the entire computation of *agreeability*. The third and fourth fragments show intermediate subcomputations of *agreeability*. Crucially, these tree fragments can represent arbitrary mixtures of computation and storage. For example, the right-most tree in Figure 1.4 allows the combination *-ability* to be generated in a single step but leaves the choice of stem `V` as a *variable*, requiring further computation on the fly.

1.3.1.3 Productivity and Reuse as an Inference

In a heterogeneous system allowing arbitrary mixtures of storage and productive computation, the combination of stored units and on-the-fly computation giving rise to any observed form (or set of forms) is necessarily highly ambiguous. Figure 1.5 illustrates this for the word *agreeability* by showing three ways in the which the word could be computed via different combinations of stored items. The far left panel shows a computation in which only minimal-sized stored units are used, the middle shows a computation in which the entire form has been stored and retrieved as a whole from memory, and the far right shows one of the possible intermediate cases.

As mentioned above, our theory assumes that previous choices about computation and storage affect the future degree to which a morpheme tends to be

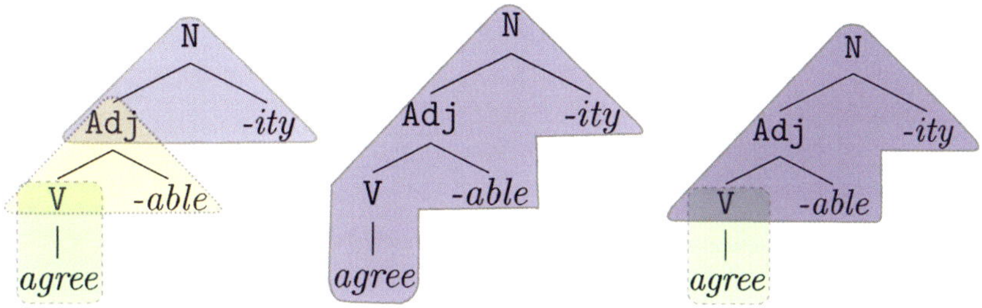

Figure 1.5

Ambiguity in the Computation of *agreeability*: This figure shows three ways in which the word *agreeability* can be derived using different combinations of storage and productive computation.

stored or computed on the fly. However, because any form is highly ambiguous with regard to computation and storage (as just illustrated), the learner must have mechanisms for quantifying and reasoning about uncertainty in the pattern of computation and storage in their language.

In this book, we propose a theory in which uncertainty is quantified using probabilities and the language user reasons about the pattern of storage and computation in their language using probabilistic conditioning. By assigning probabilities to each possible combination of storage and computation for a set of forms, the model provides a measure of how well particular hypotheses predict the distribution of data in the input. By using probabilistic conditioning, the model can update these hypotheses in the face of new evidence, balancing the ability to accurately predict the data seen so far against the prior biases discussed in the next section.

1.3.1.4 Prior Expectations of the Learner

The mathematical model that defines this inference process implicitly encodes two competing assumptions about the nature of storage and computation in the linguistic system. First, the model penalizes overly large and complex lexica. In particular, it assigns higher probability to smaller lexica with less complex stored items. Second, the model penalizes derivations of individual forms requiring many independent computational steps, assigning higher probability to derivations that require fewer independent choices to generate. Note that these are opposing biases. All else being equal, a larger and more complex lexicon will require fewer computations to generate or analyze individual forms because larger subparts of these forms will be stored. By contrast, a smaller and more compact lexicon will require more computation on average since complex forms will have to be generated from smaller parts (De Marcken, 1996b). Because our model simultaneously prefers both small, simple lexica, and fewer computational steps per form, this leads to a trade-off between computation and storage, or productivity and reuse. By using probabilistic conditioning, our model finds a coherent explanation of the pattern of morphemes in the input which optimally balances these competing forces.

1.4 Productivity and Reuse as a Trade-off: Four Approaches

To illustrate the trade-off between productivity and reuse in more detail, we consider four schematic approaches to the problem of what to store and what to compute: *full parsing*, *full listing*, *exemplar based*, and *inference based*. These four approaches will form the basis of the five implemented models we study in the remainder of the book. All of these approaches start from a underlying system that defines the space of possible computations. Here we assume that this starting system provides the ability to generate a set of tree-shaped computations; that is, it is a context-free grammar (CFG).

These approaches have been chosen to encode theoretical proposals in the literature, including several of the classical linguistic accounts discussed earlier in the chapter.[10] As we decribe in detail in Chapters 2 and 3, the approaches have been implemented using state-of-the-art probabilistic models, which have been chosen to be maximally similar to one another, differing only in the strategies that they use to determine which subcomputations to compute or reuse. In this section, we give an informal overview of these strategies.

1.4.1 Full Parsing

Under this approach, all structures are the result of computation using minimal-sized units. No larger items are stored in memory (Figure 1.6). In such a setting, each lexical primitive is highly reusable. However, the generation or analysis of any particular form will involve many small, abstract primitives. A full-parsing model predicts the maximal degree of combinatoriality, variability, and novelty

10. The three-way distinction between full listing, full parsing, and models mixing computation and storage (often known as models in the literature) is common in discussions of word processing in psycholinguistics (See, e.g., Baayen et al., 1997b; Domínguez et al., 2000).

in the data, and, therefore, can serve as a baseline against which to compare models which provide other mechanisms for constraining productivity.

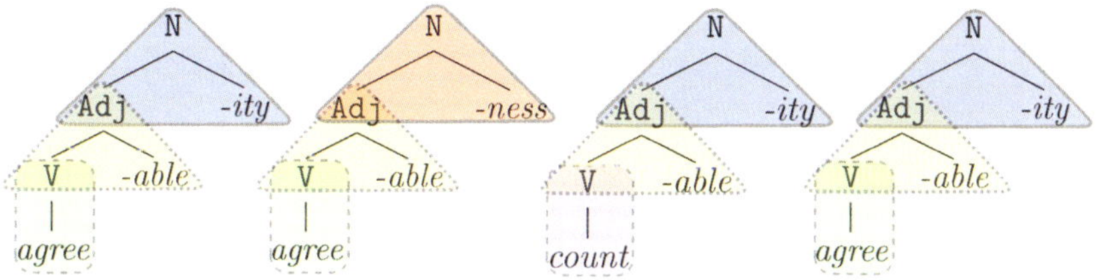

Figure 1.6

Full parsing: In this model, small, reusable subcomputations can be shared across many forms, as shown by the highlighting in the same color. However, generating each form requires many random choices.

In later chapters, the full-parsing approach will be formalized using a mathematical model known as Dirichlet-Multinomial Probabilistic Context-Free Grammars (DMPCFG; Johnson et al., 2007b; Kurihara and Sato, 2006).

1.4.2 Full Listing

Under this approach, the first time a structure is built, it is stored in its entirety. This system can account for productive generalization (through the rules from the underlying computational system), but is very conservative, preferring to reuse previously built structures whenever possible (Figure 1.7). In this storage regime, maximal units (i.e., full words) are stored and thus each item is highly specific and limited in the contexts where it can be reused. Such an approach to reuse will be most useful in domains consisting of a number of specific, but frequently reused forms.

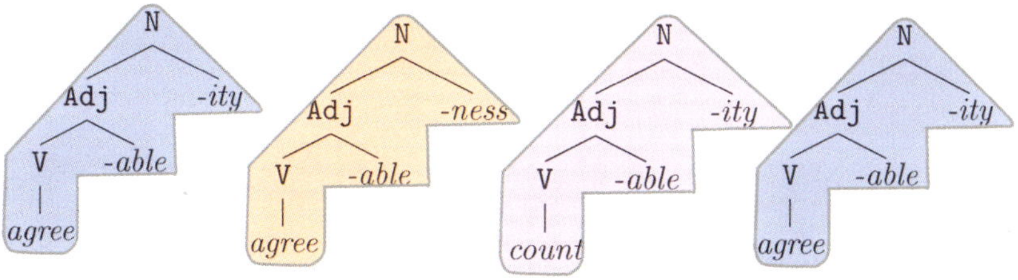

Figure 1.7

Full listing: This figure shows the consequences of the full-listing model. Stored subcomputations are very specific. Their substructures cannot be shared with other forms; however, they can be reused in their entirety with high probability.

Full-listing models like the one studied in this book can be seen as implementing a probabilistic version of classical lexicalist theories. Under such theories, lexical rules stored their outputs, and were used to express generalizations inherent in stored items, thereby aiding learning and being useful in cases of analogy, but not operating in typical language use (e.g., Bresnan, 1982; Jackendoff, 1975; Lieber, 1980).

Importantly, however, the variant of the full-listing approach used in this book goes beyond from classical linguistic proposals in two important ways. First, it is probabilistic. This means that the tools of probabilistic inference can be employed to provide a precise theory of learning (i.e., acquisition) and inference (i.e., word generation and processing). Second, the full-listing model used in this book is fully recursive. If the generation of a word requires multiple ordered rule applications, then intermediate steps of computation are also stored in memory and can be reused to generate other words.

The full-listing approach is formalized using a mathematical model known as Maximum a posteriori Adaptor Grammars (MAG; Johnson et al., 2007a).

1.4.3 Exemplar Based

This strategy stores all structures consistent with the data, both the small and abstract, the large and specific, and those in between (Figure 1.8). This model differs from the model proposed in this book in that it does not commit to a single analysis of each observed word, but rather spreads probability across many different, mutually inconsistent abstractions.

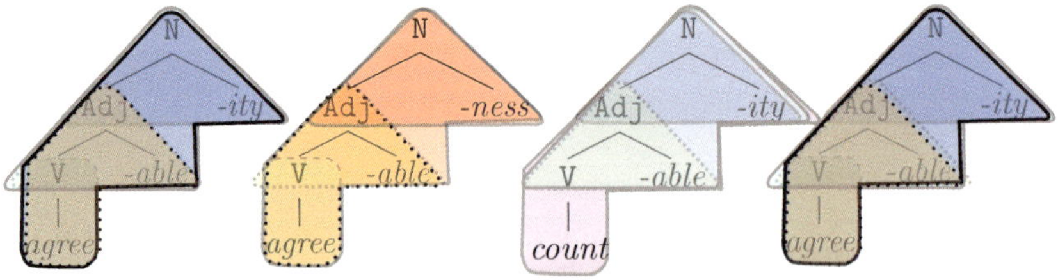

Figure 1.8

Exemplar-based Productivity and Reuse: This figure shows exemplar-based inference. Here every possible subtree consistent with the data is stored. As the figure illustrates, this leads to many overlapping analyses for each item.

There are a large variety of exemplar models in psychology and linguistics (e.g., Daelemans and van den Bosch, 2005; Murphy, 2004; Pierrehumbert, 2001). However, one unifying characteristic of such models is that they operate by storing some representation of every data point they encounter. Generalizations to new data points are handled by computing the similarity of new observations with this database of stored exemplars. Exemplar models are often contrasted with theories that store summary representations of categories of input data (e.g., prototype theories). In the case of the models in this book, data are represented in terms of their derivation trees. Each tree represents a hypothesis about how a word was constructed. Unlike the other models, the exemplar-based model does not assign a single hypothesis to each data point, but, rather represents all possible derivations.

The exemplar-based approach is formalized using two estimators for the probabilities of derivation tree fragments drawn from the literature on *data-oriented parsing* (DOP; Bod, 1998; Bod et al., 2003; Scha, 1990, 1992) The first, the *Data-Oriented Parsing 1* (`DOP1`; Bod, 1998; Sangati and Zuidema, 2011; Scha, 1990, 1992), weights all tree fragments proportional to their frequency in the input corpus. The second, the *Data-Oriented Parsing: Equal-Node estimator* (`ENDOP`; Bod, 2003; Goodman, 2003; Sangati and Zuidema, 2011), weights equally each tree fragment and each derivation tree node.

1.4.4 Inference Based

The model advocated in this book, Fragment Grammars (`FG`), treats the question of which subcomputations should be stored as an inference problem. Like the full-parsing approach, fragment grammars can store abstract structures; like the full-listing approach, they can store specific structures; and, like the exemplar-based approach, they can store any intermediate structure. However, unlike the previous approaches, the particular solutions a fragment grammar finds to the productivity and reuse problem are determined by using probabilistic inference to optimize the trade-off between storage and computation with respect to the data and the prior biases of the model. For each data point to which it is exposed, the model determines whether it is better to account for the structure using reusable stored items, productive computation, or some mixture of both (Figure 1.9).

1.5 Remarks on Modeling Assumptions

Before moving on to a more detailed discussion of the models, we make some further remarks on the assumptions made in this book. In later chapters, we

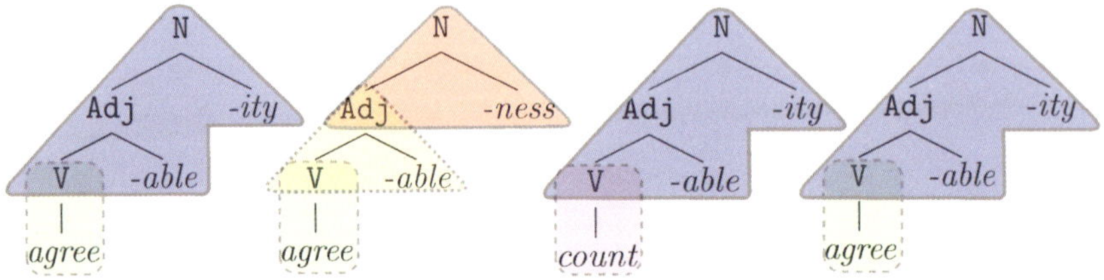

Figure 1.9

Productivity and Reuse as an Inference: This figure shows the consequences of inferring the set of stored sub-computations that best account for the data. In this example, more sharing is allowed on average than full parsing with less computation on average than full listing.

will apply the models studied in the book to two domains of morphology: the English past tense, and English derivational suffixation. For both domains we make use of an underlying computational system (i.e., a context-free grammar) that is already quite specialized to the domain, and we provide morphological analyses as inputs to the models according to these grammars. For example, in Chapter 5, where we examine the English past tense, the underlying context-free grammar defines computations similar to those shown in Figure 1.10.

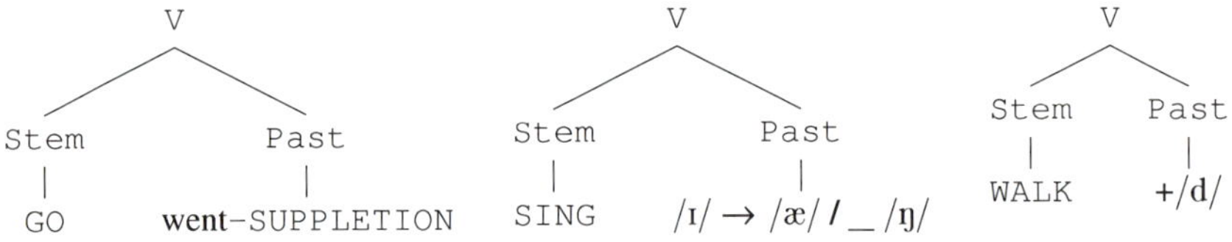

Figure 1.10

Past Tense Computations: Trees representing the computation of some English past tense forms.

These representations are specific to the English verbal system and quite abstract. Stems are segmented from their inflections. Abstract stems such as GO are identified when they underlie irregular past forms such as *went*. The space of possible verb tenses is fixed in advance and provided as part of the grammar and tenses are labeled on individual forms. At the same time, the system does not encode any information about the semantic or phonological selectional restrictions of various morphemes. It allows, for example, a rule like /ɪ/ → /æ/ to apply to a stem like EAT, even though this is a phonological impossibility. In some ways, when compared with the starting state of the child learner, these modeling assumptions provide far too much information. The child does not have access a priori to morpheme segmentations, to the fact that irregular morphemes like *go* and *went* are part of the same verbal paradigm, or to the categories of different morphemes. In other ways, however, these representations provide much less information than is available to the child learner. For instance, they do not encode any phonological information and do not provide any grounding in meaning.

We wish to emphasize, therefore, that our simulations are not intended to correspond to a particular stage of development or learning procedure used by children. Instead, the question addressed by all of the simulations is: what does the distribution of morphemes in each empirical domain imply for productivity and reuse under the assumptions encoded by each model? Providing segmentations, categorial information, and the abstract identity of morphemes allows us to focus on just the patterns of reuse and novelty in this distribution. While a more complete model would simultaneously learn these other kinds of structure, it would also make the underlying representations themselves a moving target, complicating interpretation.

By not modeling semantic and phonological selectional restrictions, just those aspects of productivity and reuse that depend only on the distribution of morphemes can be isolated. This choice can sometimes be adversarial to learning the correct patterns of productivity and reuse. As we mentioned above, all of the empirical domains examined in this book exhibit learning synergies. Productivity is correlated with phonological regularity and semantic transparency, while reuse is correlated with irregularity and non-transparency. If phonology and semantics were modeled, this might provide a valuable additional source of evidence about the productivity of various word-formation processes. For example, cases like the past tense allomorph /ɪ/ → /æ/ applying to EAT would be ruled out a priori. As we will show, fragment grammars are able to rule out incorrect combinations like this despite the lack of phonological and semantic structure. This strongly supports the idea that the distribution of morphemes alone provides robust evidence of the correct pattern of productivity. Furthermore, it suggests that the construction of a more complete model integrating these other kinds of structure would lead to improved performance.

Finally, note that all of the assumptions about the starting context-free grammar, the set of abstract morphemes, and the phonological and semantic selectional restrictions are shared identically by all of the models explored in the book. Therefore, to the degree that they affect the performance of any of the models, they affect the performance of all of the models.

2 The Framework

Timothy J. O'Donnell and Noah D. Goodman

This chapter consists of four parts. In Section 2.1, we discuss the modeling idiom adopted in this book—structured generative modeling. In Section 2.2, we discuss the psychological interpretation of this approach to cognitive modeling, focusing on the broad ideas behind the framework. In Section 2.3, we develop the modeling framework from the perspective of probabilistic programming using the Church programming language (Goodman et al., 2008). This formalization makes explicit the relationship between the model and two important technical ideas from computer science—*memoization* and *order of evaluation* (Abelson and Sussman, 1996). By adopting stochastic generalizations of these existing techniques, we show how adaptor grammars and fragment grammars can be formalized as generalizations of probabilistic context-free grammars. We then show how to generalize these methods further so that they apply to arbitrary generative models. One theme that runs through this part of the book is that techniques that have consequences only for the efficiency of traditional deterministic programs often have effects on the meaning of probabilistic programs. Finally, in Section 2.4, we provide more discussion of the five formal models evaluated in this book and connections to earlier literature not reviewed elsewhere.

2.1 Structured Generative Modeling

The modeling framework used in this book is known as *structured probabilistic generative modeling*. This paradigm makes use of probabilistic inference over structured representations. We discuss each part of the framework in more detail below.

2.1.1 Probabilistic Inference

Inference is the process of inferring hidden structure, discovering latent causes, and generalizing from noisy, incomplete, and ambiguous data. In this book, we use a fully probabilistic formulation of the inference problem, in particular, representing our data and hypothesis spaces in terms of *random variables* and reasoning about hypotheses using *probabilistic conditioning*.[1]

1. Adopting a fully probabilistic model of inference using conditioning is often referred to as the Bayesian approach (Bernardo and Smith, 1994; de Finetti, 1974; Jaynes, 2003; Laplace, 1814;

Intuitively, a random variable can be thought of as a placeholder that can take on particular values, from some specified set X according to a given probability distribution P. Sampling from a random variable produces a single value x (known as a ***draw***) from X, with probability $P(x)$. We say that X has taken on the value x and write $X = x$.

A probability distribution that is defined for a collection of random variables $V = \{X, Y, Z\}$: $P(X = x, Y = y, Z = z)$ is known as the ***joint distribution*** of V. Note that any set of random variables that has a joint distribution can be considered as a single (complex) random variable, that is, $V = v$ is just shorthand for $\{X = x, Y = y, Z = z\}$. When considering the random variable representing an entire distribution (i.e., the random variable that ranges over all possible combinations of random choices in a model), we sometimes use the term *state*.

A typical probabilistic model has many random variables of interest. The models used in this book define random variables that range over objects like words, morphemes, parses, rules, and grammars. To illustrate this, consider the probabilistic context-free grammar for a fragment of English morphology shown in Figure 2.1.

As we discussed in the last chapter, this grammar defines a probability distribution over derivations of English word forms. In such a setting, there are many random variables of potential interest. For example, if this grammar is used to generate N words, then there is a random variable D (for data), which ranges over all generable sets of N English words (according to this grammar). If we are interested in the possible values for one specific word in this set, say the sixth word W_6, we can consider this a random variable as well. The set of derivation trees that records the process by which each word $W_i = w_i$ was generated is also a random variable. In this book, we study systems that can learn to store and reuse fragments of derivation structure. The set of such fragments can be thought of as a lexicon of stored structures and also constitutes a random variable L, which ranges over different sets of stored derivation fragments given a base grammar and set of observed forms.

Joint distributions over multiple random variables allow inference problems to be represented as problems of probabilistic conditioning. Conditioning is a form of hypothetical reasoning that asks: how likely are various outcomes when we force some of the random variables to take on certain values? For

Sivia, 1996), although this term also carries other connotations and meanings about which we are agnostic (e.g., about the subjective versus objective interpretation of probabilities).

$p_{\mathtt{W}_1} = \frac{1}{4}$	W	$\longrightarrow$	N	
$p_{\mathtt{W}_2} = \frac{1}{4}$	W	$\longrightarrow$	V	
$p_{\mathtt{W}_3} = \frac{1}{4}$	W	$\longrightarrow$	Adj	
$p_{\mathtt{W}_4} = \frac{1}{4}$	W	$\longrightarrow$	Adv	
$p_{\mathtt{N}_1} = \frac{1}{10}$	N	$\longrightarrow$	Adj	*-ness*
$p_{\mathtt{N}_2} = \frac{1}{10}$	N	$\longrightarrow$	Adj	*-ity*
$p_{\mathtt{N}_3} = \frac{1}{100}$	N	$\longrightarrow$	*electro-*	N
$p_{\mathtt{N}_4} = \frac{1}{230}$	N	$\longrightarrow$	*magnet*	
$p_{\mathtt{N}_5} = \frac{1}{50}$	N	$\longrightarrow$	*dog*	
	...			
$p_{\mathtt{V}_1} = \frac{1}{300}$	V	$\longrightarrow$	N	*-ify*
$p_{\mathtt{V}_2} = \frac{1}{100}$	V	$\longrightarrow$	Adj	*-ize*
$p_{\mathtt{V}_3} = \frac{1}{40}$	V	$\longrightarrow$	*re-*	V
$p_{\mathtt{V}_4} = \frac{1}{50}$	V	$\longrightarrow$	*agree*	
$p_{\mathtt{V}_5} = \frac{1}{300}$	V	$\longrightarrow$	*count*	
	...			
$p_{\mathtt{Adj}_1} = \frac{1}{500}$	Adj	$\longrightarrow$	*dis-*	Adj
$p_{\mathtt{Adj}_2} = \frac{1}{300}$	Adj	$\longrightarrow$	V	*-able*
$p_{\mathtt{Adj}_3} = \frac{1}{25}$	Adj	$\longrightarrow$	N	*-ic*
$p_{\mathtt{Adj}_4} = \frac{1}{1000}$	Adj	$\longrightarrow$	N	*-al*
$p_{\mathtt{Adj}_5} = \frac{1}{35}$	Adj	$\longrightarrow$	*tall*	
	...			
$p_{\mathtt{Adv}_1} = \frac{1}{2}$	Adv	$\longrightarrow$	Adj	*-ly*
$p_{\mathtt{Adv}_2} = \frac{1}{100}$	Adv	$\longrightarrow$	*today*	

Figure 2.1

Probabilistic Context-free Grammar: A probabilistic context-free grammar (PCFG) for English derivational morphology.

example, the conditional probability $P(L = l|D = d)$ is the probability distribution over possible lexica given that a particular set of words has been generated. Conditional probability is given by Equation 2.1.

$$P(L = l|D = d) = \frac{P(D = d, L = l)}{\sum_{\forall l' \in L} P(D = d, L = l')} \tag{2.1}$$

Conditioning can be understood in terms of a simple two-step procedure. The procedure is given two inputs: a joint distribution, and a set of particular values for a subset of the random variables in the distribution (called the *conditioner*). In general, the conditioner can encode any assumption, hypothesis, observation of data, or other fact about the random variables in the joint distribution, although in practice it is often used to represent observed data. The procedure first checks every combination of values in the joint distribution and throws away those for which the conditioner does not hold. This is represented in the equation by the numerator $P(D = d, L = l)$. This term is non-zero only when the conditioner is true (i.e., $D = d$).

After combinations of variables inconsistent with the conditioner have been thrown away, the procedure takes the probability mass from the rejected outcomes and redistributes it over the remaining possibilities such that their probability remains proportional to their probability before conditioning. This is represented by the denominator of the right-hand side of the equation $\sum_{\forall l' \in L} P(D = d, L = l')$. This denominator sums over all of the joint probabilities for each $L = l'$, and divides the numerator by this total. The conditional distribution over L given $D = d$ is just the relative degree of belief in each outcome in L, restricting attention to just those situation where $D = d$ is true. Note that in a certain sense, conditioning represents the most conservative, multiplicative update to the joint distribution that one can make while ensuring that the conditioned variables have the specified values. Because all probability distributions sum to one (a convention sometimes called the *law of conservation of belief*), the relative probabilities of the possible outcomes do not change after impossible outcomes are removed.

2.1.1.1 Bayes' Rule

The *chain rule*, given in Equation 2.2, states that the joint probability of two random variables can be rewritten as the product of the conditional probability of one random variable given the other, multiplied by the prior probability of the other random variable.

$$P(D=d, L=l) = P(D=d|L=l)P(L=l) \tag{2.2}$$

The chain rule allows us to rewrite the definition of conditional probability as follows.

$$P(L=l|D=d) = \frac{P(D=d|L=l)P(L=l)}{\sum_{\forall l' \in L} P(D=d|L=l')P(L=l')}$$

This version of the formula for conditional probability is known as ***Bayes' rule***. The left-hand side of the equation is known as the ***posterior probability*** of the lexicon $L=l$ given the data $D=d$. The first term in the numerator on the right-hand side is known as the ***probability of the data*** $D=d$, given the lexicon $L=l$, or alternatively as the ***likelihood*** of the lexicon l given the data d.[2] The second term is known as the ***prior probability*** of the lexicon $L=l$. The denominator of the right-hand side of Bayes' rule is known as the ***marginal likelihood*** of the data. Note that the marginal likelihood is just the probability of the particular dataset $D=d$ summed, or ***marginalized***, over all possible hypotheses. The marginal likelihood can be thought of as the prior probability of the data. If a lexicon were chosen at random from the prior over possible lexica, and then a random dataset were sampled from this lexicon, the marginal likelihood gives the probability that this dataset equals the observed data.

Bayes' rule is important because it makes explicit a structure that is common in probabilistic models. Often models can be stated in terms of a set of unobservable hypotheses (in this case, grammars) and some prior distribution over those hypotheses. After observing data drawn from an unknown hypothesis in this space, Bayes' rule can be used to update beliefs about which hypotheses were likely to have generated that data.

While Bayes' rule is conceptually important, complex models frequently involve large (often infinite) numbers of random variables in complicated dependency relationships. In such settings, there are many potential conditional distributions which are of interest, and it is sometimes less useful to identify particular subsets of random variables as corresponding to likelihood or prior components of the model. It is often most clarifying to think of a model

2. This terminology is sometimes confusing. From the point of view of inference, the likelihood is considered to be a function of the lexicon for a fixed dataset. From this perspective, the likelihood is not a probability. The likelihoods of different lexica do not sum to one. Rather, the likelihood can be thought of as a score assigned to each lexicon by a fixed dataset. This is the perspective taken by maximum-likelihood based approaches to inference, which try to find the grammar that maximizes this score.

in terms of its full joint distribution with some random variables conditioned to take on particular values.

2.1.2 Hierarchical Generative Modeling

There are different approaches to building probabilistic models. In this book, we adopt a *hierarchical generative* approach to model-building—also known as *analysis by synthesis*.

Both the language scientist and the language learner are interested in the hidden causal processes (e.g., choosing a word from the lexicon) that generate observed data. That is, they are interested in $P(L = l|D = d)$. Generative approaches start from the observation that, because of Bayes' rule, the probability of interest, $P(L = l|D = d)$, is proportional to $P(D = d|L = l)P(L = l)$; the factor relating these two expressions is just the marginal likelihood of the data (the denominator in Bayes' rule). Instead of giving a recipe to directly compute the probability of hypotheses given the data, $P(L = l|D = d)$, generative models specify $P(D = d|L = l)P(L = l)$ and then use conditional probability to reason about the former quantity.

Hierarchical modeling refers to the construction of generative models in which the random variables are arranged in a hierarchical order meant to reflect the causal processes by which data is generated. For example, from the perspective of language production, the lexicon and the data have a natural ordering in terms of their part in the causal process of generating an utterance. The lexicon consists of an inventory of stored items that are reused in producing utterances. We can imagine different potential processes for generating this inventory, for example, the full-listing, exemplar-based, or inference-based approaches discussed in the previous chapter. If we define a distribution over these four approaches to lexicon acquisition, represented by a random variable S (for storage strategy), we can extend our hierarchical generative process by another level. First, a storage strategy $S = s$ is chosen, then a particular lexicon that uses that strategy is sampled $L = l|S = s$, and finally a dataset is sampled using that lexicon $D = d|L = l$. The joint distribution over all three random variables can be written as in Equation 2.3, where the ordering of the variables represents the dependencies in the causal chain.

$$P(D = d, L = l, S = s) = P(D = d|L = l)P(L = l|S = s)P(S = s) \tag{2.3}$$

2.1.3 Structured Models

The final element of the modeling framework in this book is the use of structured probabilistic models. The probabilistic generative approach uses probabilities to measure uncertainty over the different ways in which data might have been generated. However, it does not presuppose anything about the types of objects that can be specified. There has been a tendency in the history of cognitive science to treat statistical approaches and more traditional symbolic approaches as fundamentally distinct.[3] However, probability theory presupposes no such distinction. ***Structured modeling*** refers to the application of probability theory to representations from classical cognitive science, linguistics, logic, artificial intelligence, and programming.

2.2 Probability Theory, Optimality, and Cognitive Science

Structured generative approaches approaches in cognitive science are often associated with the idea that models should be optimal. The search for optimal models in cognitive science can be interpreted in one of two ways: either as a hypothesis or alternatively as a methodological claim. The hypothesis has been stated most clearly by John Anderson in his ***principle of rationality*** (Anderson, 1990, 1991). In his words:

Principle of Rationality: The cognitive system optimizes the adaptation of the behavior of the organism.

In other words, the hypothesis states that cognitive systems attempt to find optimal solutions to the problems an organism faces in its environment. Anderson (1990) argues that this principle can be motivated from an evolutionary perspective. Since the brain has evolved to guide behavior, it might be expected that there has been pressure over evolutionary time for it to select approximately optimal solutions to behavioral problems, within the constraints provided by biological endowment of the organism.

Crucially, the principle of rationality is a hypothesis, not a conclusion that can be drawn a priori from evolutionary considerations. Evolution does not always optimize. While natural selection tends to favor behaviors with greater fitness, there are other mechanisms of evolutionary change that do not, such as

3. This has led to unfortunate use of terminology, such as the use of the general term *statistical learning* to refer to a class of Markov models (e.g., Saffran, 2002, 2003).

genetic drift (Gillespie, 2004). Even when evolution does optimize, it does so only approximately, and furthermore evolutionary change is subject to strong constraints because natural selection can only operate with the material at hand.

It is sometimes argued that even if the principle of rationality turns out to be wrong as a hypothesis about cognition, it is still useful to develop models from an optimality perspective. Cognitive systems are computational information-processing systems. If the task of cognitive science is to reverse engineer the computational systems underlying cognition, the methodological perspective proposes that the first step is to understand the exact information processing task of each system. This has been stated most clearly by Marr (1982):

> An algorithm is likely to be understood more readily by understanding the nature of the problem being solved than by examining the mechanism (and the hardware) in which it is solved.

The intuition underlying this proposal is that understanding the function of a system, and which solutions provide better or worse solutions to the information-processing problem, strongly constrains on the space of possible models of the system. John Anderson has used the term ***rational analysis*** to refer to this methodological approach (e.g., Anderson, 1991). A rational analysis starts by precisely specifying the goals of the system, formalizing the space of computational mechanisms that can subserve these goals, and deriving the optimal or approximately optimal behavior of the system with respect to these goals and mechanisms. If the actual performance of the system matches the predictions of the rational analysis, then this can be taken as evidence for its purported function, and for the hypothesis that the system is optimal. On the other hand, when predictions are not borne out, it is often possible to isolate which assumptions about the information-processing problem and computational structure of the system were flawed (see, Barton et al., 1987; for similar suggestions with respect to complexity-theoretic analyses of grammatical systems).

There are many different notions of optimality that can be relevant in a rational analysis: rewards in terms of resources for certain behaviors, computational efficiency, memory succinctness, etc. In a probabilistic setting, one particular dimension is often emphasized: prediction accuracy.[4] An optimal system under

4. But see Lewis et al. (2014) and Griffiths et al. (2015) for approaches to integrating other computational considerations into rational probabilistic models.

this criterion is the one which makes the best predictions about the domain in which the system operates. An important reason that probability theory is often used to formulate prediction accuracy is the idea that it is a consistent or coherent theory of uncertain reasoning in certain formal senses of these terms. Note that the claim is that probability theory makes consistent predictions from starting premises. If the premises themselves are flawed, the system can make perfectly coherent, but wholly inaccurate predictions.

There have been two main classes of argument that the laws of probability lead to coherent reasoning. In the first, the ***axiomatic approach***, the laws of probability are derived from a set of axioms that are intended to characterize consistent reasoning. The first such axiomatic result was proven by Cox (1946) who proposed the following three axioms.[5] Let $B(x)$ represent the degree of belief in some statement x.

1. Degrees of belief can be ordered. If $B(x) > B(y)$ and $B(y) > B(z)$ then $B(x) > B(z)$.
2. The degree of belief in a statement is related to the degree of belief in the negation of that statement. That is, there exists a function f such that $B(\neg x) = f\,[B(x)]$
3. The degree of belief in the conjunction of statements $B(x \wedge y)$ is related to the degree of belief in the conditional statement $x|y$ and the degree of belief in y. That is, there exists a function g such that $B(x \wedge y) = g\,[B(x|y), B(y)]$.

Cox showed that probability theory is the unique theory, up to isomorphism, that satisfies these three axioms. In subsequent decades, the axiomatic approach has been both extended and challenged in various ways (see e.g., Bernardo and Smith, 1994; Halpern, 1999). Nevertheless, the approach remains an important theoretical justification for the use of probability theory to model coherent reasoning.

A second class of arguments for the optimality of probability theory are so-called ***Dutch book*** arguments (de Finetti, 1974). Dutch book arguments start from the assumption that probabilities can be interpreted as degrees of belief in gambling outcomes. An individual's degree of belief, $B(x)$, in a statement, x, is interpreted as the proportion of a wager that he or she would be willing to bet on x being true. The term Dutch book refers to a set of bets that guarantees a net loss for one of the participants. In de Finetti (1974) it was shown that a

5. This discussion follows MacKay (2003).

necessary and sufficient condition for guaranteeing that no Dutch book could be made against an individual was that all of his or her beliefs follow the laws of probability theory.

Axiomatic and Dutch book arguments have both been used to argue that using the laws of probability to reason about uncertain beliefs is coherent. In this book, we adopt the rational analysis methodology using probability theory. In later chapters, we show that the model proposed here explains a large number of phenomena in English morphology. This lends support to both the computational machinery that we propose, as well as to the idea that individuals use (approximate) probabilistic inference to reason about this domain.

2.3 The Modeling Framework from a Probabilistic Programming Perspective

In this section, we use the framework known as probabilistic programming, and the programming language Church (Goodman et al., 2008), to explain the main ideas behind the models in the book.[6] Church, a LISP dialect and stochastic variant of the λ-calculus, is especially well-suited for expressing the kinds of recursive functions needed to generate linguistic structure, and we use it to develop the critical ideas underlying the models in a compositional and stepwise fashion. We first define the structure-building recursion underlying Probabilistic Context-Free Grammars (PCFGs). We then extend this definition, using stochastic generalizations of ideas from the theory of programming languages. Stochastic memoization is used to extend probabilistic context-free grammars to adaptor grammars, and stochastically lazy evaluation is used to extend adaptor grammars to fragment grammars. In the final part of this section, we demonstrate how arbitrary stochastic λ-defined processes can be "fragmentized," allowing for a mixture of productivity and reuse in any generative model.

2.3.1 Distributions as Programs and Church

Probabilistic generative models can be used to define distributions over many different mathematical structures: strings, trees, graphs, logical expressions, continuous values, and functions, to name a few. One way of unifying these

6. See the next chapter for a more complete, mathematical discussion of the models and inference algorithms.

disparate kinds of distributions and probabilistic processes is by viewing them as programs that can be executed on a natively probabilistic model of computation. This perspective is known as the probabilistic programming approach (Freer et al., 2012; Goodman et al., 2008; Mansinghka, 2009; Pfeffer, 2001; Roy, 2011).

The idea of programmable machine models of computation is familiar from computer science where many such models have been studied. Some well-known examples include Turing machines, Post canonical systems, register machines, Markov algorithms, μ-recursive functions, and the λ-calculus. A natively probabilistic model of computation can be defined by taking such a model and adding a source of randomness and the ability to make decisions based on this random source during computation. Building on the tradition of functional programming languages, many probabilistic programming approaches have chosen the λ-calculus as an underlying model of computation (Erwig and Kollmansberger, 2006; Goodman et al., 2008; Kiselyov and Shan, 2009; Pfeffer, 2001). The λ-calculus was introduced in Church (1932) to formalize the notion of a ***computable function***. McCarthy (1960) showed that λ-defined functions over lists could be the basis for a practical programming language he called LISP (for LISt Processor). In this book, we use a probabilistic variant of LISP called Church, after Alonzo Church (Goodman et al., 2008).

In functional programming languages, such as Church, the basic units used to implement programs are ***procedures***—recipes for computing various mathematical functions. Church uses a LISP-like syntax where the application of an operator to its operands is in Polish notation, meaning that the operator (i.e., function) comes first, followed by its operands (i.e., arguments). Thus, in Church, the addition procedure + applied to two arguments `2` and `3`, is written `(+ 2 3)`, rather than `2 + 3`.

The most important operator in the λ-calculus is λ (which is written `lambda` in Church). This is the basic abstraction operator, which is used to define procedures. In Church, this has the syntactic form `(lambda arguments body)` where `arguments` is a list of input arguments to the procedure and `body` defines the computation that the procedure will carry out on those arguments. Figure 2.2 shows a simple example of a procedure definition, which takes three arguments `x`, `y`, and `z`, and returns the result of adding them together. Figure 2.2 also shows the use of `define` which allows values to be named in Church and is used here to name the procedure `add3`.

```
(define add3
  (lambda (x y z)
    (+ x y z)))
```

Figure 2.2

A Simple Procedure in Church: A definition for a simple procedure. This procedure adds the values of its three inputs.

In the λ-calculus every expression has a ***value***. The value is the outcome that results from performing the computation defined by the expression, a process known as ***evaluation***. For example, the expression `(+ 1 2 3)` evaluates to `6`. The expression `(lambda (x y z) (+ x y z))` evaluates to a procedure of three arguments. This last example illustrates an important point. In functional programming languages like Church, procedures are first-class objects just like other, more familiar datatypes such as integers, strings, or lists. Procedures can be the result of evaluating an expression, they can be passed as arguments to other procedures, they can be returned by other procedures, and they can be stored in memory just like other values.

Probabilistic programming languages like Church add probabilistic primitives, called ***elementary random procedures*** (ERPs), to the language. A deterministic procedure always returns the same value when applied to the same arguments. A random procedure, on the other hand, defines a distribution over outputs given inputs. An ERP that may be familiar from many existing programming languages is the `rand` or `random` function, which returns a value in the interval 0, 1 with uniform probability. Another example of a random procedure is `flip`, which flips an unbiased coin, returning `true` or `false` with equal probability. A different version of flip takes an argument that specifies the weight of a biased coin, for example, `(flip 0.2)`.

An inventory of such basic ERPs can be used to construct more complex random procedures. Figure 2.3 shows a simple Church procedure, which defines a function called `noisy-or`. With probability `epsilon`, `noisy-or` returns the result of applying `or` to its arguments. However, with probability (1−`epsilon`), it returns the opposite of that result (i.e., if `(or arguments)` is `true`, `noisy-or` returns `false`, and vice versa). Note that the syntactic form of an `if` statement in Church is `(if condition do-if-true do-if-false)` where the first expression (`condition`) is a logical condition, the second expression (`do-if-true`) is evaluated if this condition is true, and, otherwise, the third expression (`do-if-false`) is evaluated.

```
(define noisy-or
  (lambda (epsilon boolean1 boolean2)
    (if (flip epsilon)
        (not (or boolean1 boolean2))
        (or boolean1 boolean2))))
```

Figure 2.3

Church Code for `noisy-or` Function: Church code defining a noisy version of the logical operator `or`. This operator is identical to logical `or` except that with some probability `epsilon` it reverses its output.

The probability of some output of a complex random procedure is the product of the probabilities of each output of an ERP that is called during its evaluation. For example, if `noisy-or` is applied to the arguments `false` and `true` (i.e., `(noisy-or 0.01 false true))` then the probability of the outcome `true` is 0.99 and the probability of `false` is 0.01. In more complex examples, such as `(noisy-or 0.01 (flip 0.7) (flip 0.7))`, the probability of `true` depends both on the evaluations of the calls to `flip` that determine the two arguments, as well as the `flip` inside the procedure body.

A procedure in Church can be understood as a *sampler* for outputs given its input arguments, and therefore defines a family of probability distributions, one for each possible combination of input arguments.[7] When a Church procedure is applied to some arguments, it will be evaluated to produce a sample from the distribution over outputs (given those inputs). For example, `noisy-or` can be understood as a sampler for a distribution over {`true, false`} determined by the arguments passed to the procedure.

Probabilistic generative models for complex, structured representations can be constructed by using stochastic procedures defined in a probabilistic language like Church. An important technique for doing this is the use of *recursive procedure definitions*. In a recursive procedure definition, a procedure is defined in terms of itself; that is, a recursive procedure invokes itself somewhere in its own body. A well-known recursive function from mathematics

7. Families of probability distributions are often informally called "distributions." For instance, the Gaussian distribution is actually an infinite family of related distributions each parameterized by a mean, μ, and a variance, σ^2.

is the Fibonacci function, which maps from an integer n to the nth number in the Fibonacci sequence, F_n, defined inductively as: $F_0 = 0$, $F_1 = 1$, and $F_n = F_{n-1} + F_{n-2}$ (Equation 2.4).

$$F_n = \begin{cases} 0 & n = 0 \\ 1 & n = 1 \\ F_{n-1} + F_{n-2} & \forall n > 1 \end{cases} \tag{2.4}$$

Figure 2.4 shows a recursive procedure that computes the nth Fibonacci number in Church. The computation of a Fibonacci number by this procedure can be visualized as a tree as is shown in Figure 2.5. Each node in this tree represents an application of the `fib` function and the hierarchical organization reflects the recursive nature of the process.

```
(define fib
  (lambda (n)
    (case n
      (0 0)
      (1 1)
      (else
       (+ (fib (- n 1)) (fib (- n 2)))))))
```

Figure 2.4
Chuch Code for Fibonacci Function: Procedure to compute the nth Fibonacci number.

Recursive procedures can be used to define distributions over infinite numbers of objects. For example, the ***geometric distribution*** is distribution over all non-negative integers. The probability of each integer i is given by $p(1-p)^i$ where p is a probability: $p \in [0, 1]$. The geometric distribution can be described by the following generative process: start at 0; flip a coin with weight p; if it comes up heads, return 0 and stop; if it comes up tails, move on to the next integer (in this case 1) and repeat. The term $(1-p)^i$ in the expression $p(1-p)^i$ corresponds to the number of integers that were passed over by this process (i.e., where the coin flip returned tails). The term p in the expression corresponds to the choice that led to the procedure returning a value. In Church, this distribution can be represented by the simple recursive procedure shown in Figure 2.6.

This procedure samples from the integers, returning integer i with probability $p(1-p)^i$. The use of recursion allows this infinite distribution to be expressed in only a few lines of code. As discussed in the preceding chapter,

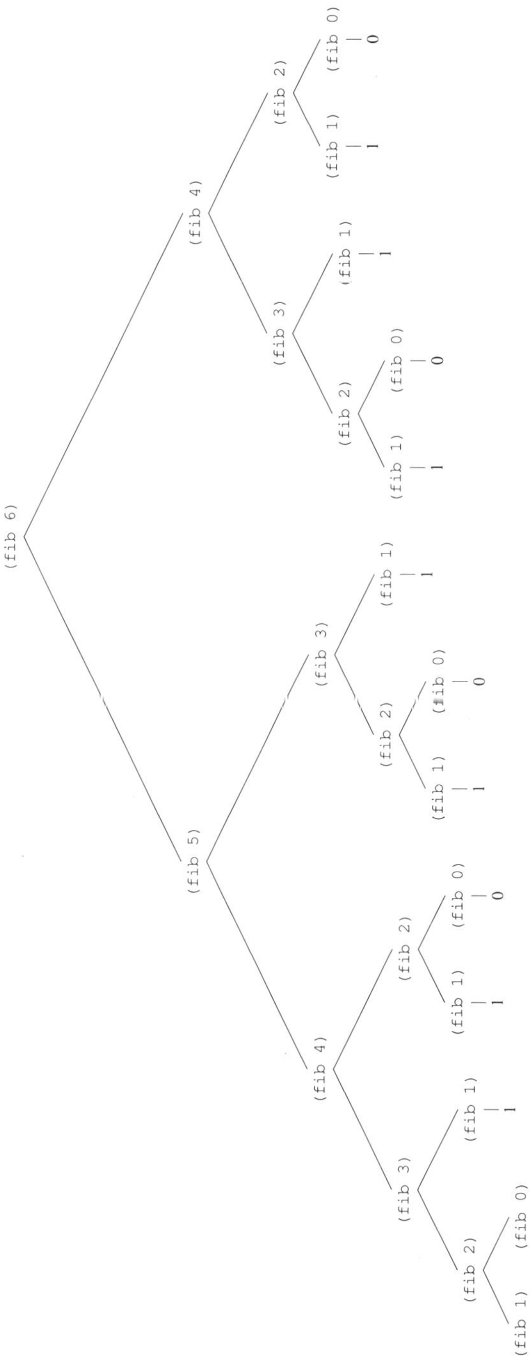

Figure 2.5

Computation of F_6: This figure shows the computation of the 6th Fibonacci number by the procedure shown in Figure 2.4. When applied to 0 or 1 `fib` returns its input. When applied to values $n > 1$ it calls itself recursively on $n - 1$ and $n - 2$.

models of linguistic structure often make use of recursion. We will exploit the ability of Church to succinctly represent these processes in the next section.

```
(define geometric
  (lambda (p)
    (if (flip p)
        0
        (+ 1 (geometric p)))))
```

Figure 2.6

Church Code for the Geometric Distribution: Probabilistic procedure that samples from a geometric distribution with parameter *p*.

2.3.2 Structure Building with `unfold`

In the last chapter, we introduced a simple model of computation, the PCFG (Probabilistic Context-Free [Phrase-structure] Grammar). PCFGs consist of a set of rules, like `N` ⟶ `Adj` *-ness*, which can be understood as declarative statements about the possible composition of constituent types. Implementing a PCFG as a computational model requires specifying a procedure, which we call `unfold`, that takes a (terminal or nonterminal) symbol as an input argument and uses the declarative knowledge in the rule set to recursively build expressions. Recall that there may be many rules associated with a nonterminal in a PCFG. When `unfold` is passed a nonterminal as an argument, it samples one of the corresponding rules (i.e., a rule with this nonterminal on its left-hand side). It then recursively calls itself on each symbol on the right-hand side of the chosen rule. Alternatively, if passed a terminal symbol (i.e., a word or morpheme), `unfold` simply returns the symbol, ending the recursion. This function can be defined by the Church procedure shown Figure 2.7.

```
(define unfold
  (lambda (symbol)
    (if (terminal? symbol)
        symbol
        (map unfold (sample-rhs symbol)))))
```

Figure 2.7

`unfold`: The `unfold` procedure implements the basic recursion underlying computation in a PCFG.

In Figure 2.7, the procedure application `(terminal? symbol)` returns `true` if `symbol` is a terminal and `false` otherwise. The procedure `sample-rhs` encodes the database of PCFG rules. It takes a nonterminal as an argument, and, when applied to this nonterminal, samples one of the rules that share this nonterminal on their left-hand side. One simple implementation of `sample-rhs`, which encodes the PCFG shown in Figure 1.2 of Chapter 1, is shown in Figure 2.8.

The procedure `map` takes two arguments. The first argument is a procedure and the second is a list of values: `(map procedure list)`. It applies `procedure` to each element of `list` and returns the new list that results. For example, if given a simple procedure that adds one to each input value, `(define add1 (lambda (n) (+ n 1)))`, the procedure application `(map add1 (list 1 2 3))` returns the list `(2 3 4)`. In the body of `unfold`, `map` recursively applies `unfold` to each symbol on the right-hand side of the rule that is chosen by `sample-rhs`. For example, if `(sample-rhs 'N)` were to return `(Adj` *-ness*`)`, `(map unfold '(Adj` *-ness*`))` would result in the recursive calls `(unfold 'Adj)` and `(unfold` *-ness*`)`.[8]

A particular evaluation of `unfold` can be visualized with a tree like the one in Figure 2.9. This figure shows one possible evaluation of `unfold` when applied to the nonterminal `N`, which has resulted in the sampling of the word *agreeability*. Note that because `unfold` is a stochastic procedure, this is only one of many words that might have been sampled by this application of the function to some arguments.

Many linguistic generative models can be formalized as variants of `unfold`.[9] The importance of `unfold` is that it expresses the fundamental recursive computation used to build expressions. It separates the specific knowledge of a particular grammar (encoded here by the procedure

8. Note that in LISP the quote symbol, `'`, tells the interpreter not to evaluate an expression. `'N` is just the nonterminal symbol `N`. Without a quote, `N` is treated as the name of a variable referring to some other value. Likewise `'(A B C)` is a list with three symbols. The unquoted list `(A B C)` is treated as the procedure `A` applied to input arguments `B` and `C`.

9. The name `unfold` derives from the fact that this procedure is the dual of the `fold` procedure (also sometimes called `reduce` or `inject`), which is common in functional programming. Fold-type procedures take a structure and reduce them to a single value by recursively applying a function which combines substructures. Unfold-type procedures take a value and build a structure by recursively applying a function, like `sample-rhs`, that expands the input at each step. These operations can be generalized and formalized as category-theoretic constructs known as *catamorphisms* (`folds`) and *anamorphisms* (`unfolds`; Meijer et al., 1991).

```
(define sample-rhs
  (lambda (nonterminal)
    (case nonterminal
      (('W) (multinomial (list (list 'N) (list 'V) (list 'Adj) (list 'Adv) ... )
                         (list p_W1 p_W2 p_W3 p_W4 ...)))
      (('N) (multinomial (list (list 'Adj 'ness) (list 'Adj 'ity)
                               (list 'electro 'N) (list 'magnet) (list 'dog) ...)
                         (list p_N1 p_N2 p_N3 p_N4 p_N5 ...)))
      (('V) (multinomial (list (list 'N 'ify) (list 'Adj 'ize) (list 're 'V)
                               (list 'agree) (list 'count) ...)
                         (list p_V1 p_V2 p_V3 p_V4 p_V5 ...)))
      (('Adj) (multinomial (list (list 'dis 'Adj) (list 'V 'able) (list 'N 'ic)
                                 (list 'N 'al) (list 'tall) ...)
                           (list p_Adj1 p_Adj2 p_Adj3 p_Adj4 p_Adj5 ...)))
      (('Adv) (multinomial (list (list 'Adj 'ly) (list 'today) ...)
                           (list p_W1
                           p_W2 ...)))))))
```

Figure 2.8

Procedure for Sampling the Right-Hand Side of Rules: `sample-rhs` takes a nonterminal symbol as input and samples one of the possible RHSs associated with that nonterminal by the PCFG.

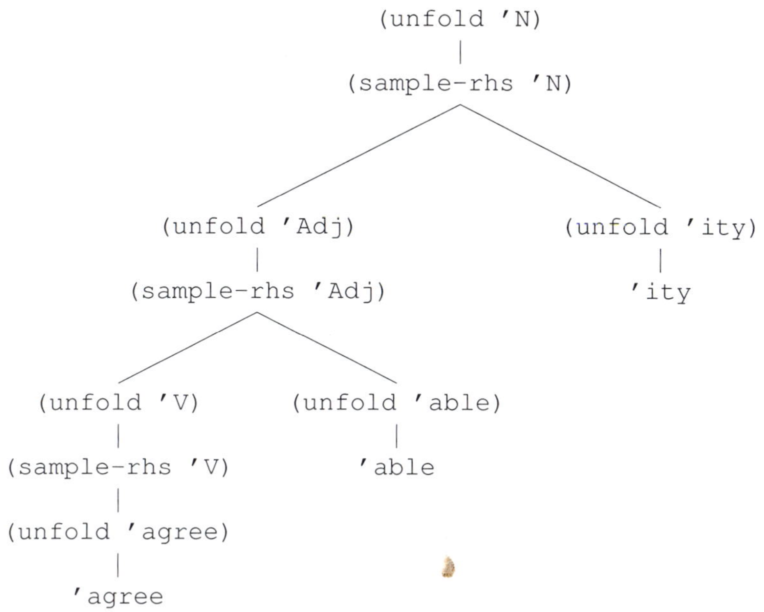

Figure 2.9

One Possible Evaluation of `unfold`: This figure shows a possible computation resulting from applying `unfold` to the nonterminal `N`. When applied to a nonterminal `unfold` calls `sample-rhs` to sample an expansion for that nonterminal. It then recursively applies itself to each symbol of that expansion. When called with a terminal, `unfold` returns that terminal, stopping the recursion.

`sample-rhs`) from the general recursive computation needed to generate expressions.

In this book, the full-parsing model described in Section 1.4.1, is formalized by Dirichlet-Multinomial Probabilistic Context-Free Grammars (`DMPCFG`s). This model corresponds to a modified version of the `unfold` procedure described here with slightly modified probability distributions over the right-hand side of rules.[10]

2.3.3 Remembering Earlier Work with Memoization

Memoization refers to a widely-used technique in computer science where the results of computations are stored for later reuse. In functional programming,

10. The `DMPCFG` model places a Dirichlet prior on the rule weights in the underlying grammar.

memoization is often applied directly to procedures using a higher-order function known as `mem`. To memoize a procedure, each call to the procedure is intercepted. If the procedure has been called before with the current arguments, the result is looked up in a table called the ***memotable*** and returned. Otherwise, the procedure is first evaluated on the arguments, the result is stored in the memotable, and, finally, it is returned to the caller.

Because computations—especially those involving recursion—often consist of frequently reoccurring sub-computations, memoization can sometimes lead to a significant improvement in efficiency. This idea can be illustrated with an example. As we discussed in Section 2.3.1, the Fibonacci function can be defined using a recursive procedure (see Figure 2.4). Because of its recursive structure, a naive implementation of `fib` can be unnecessarily inefficient. Consider Figure 2.5 of Section 2.3.1. In the pictured tree `(fib 2)` is evaluated 5 times, and `(fib 4)` is evaluated twice. If, instead of recomputing these values each time they are needed, the values are stored after their first evaluation and then subsequently reused, the time needed to compute a value will be greatly reduced. Figure 2.10 shows the reduction in work that is possible by use of memoization. The naive implementation takes a number of recursive calls that is exponential in the input, n, while the memoized version is linear in n.

In many programming languages, including Church, the higher-order procedure `mem` is provided as a primitive in the language. In the case of `mem`, a memoized version of the input procedure is returned. Thus `(mem fib)` will return a memoized version of `fib`, which intercepts calls and checks if they are in the memotable before repeating unnecessary work.

Because of its potential to improve efficiency, memoization has been applied widely in the design of algorithms, especially ***dynamic programming*** algorithms, which figure prominently in linguistic applications such as parsing (see, e.g., Jurafsky and Martin, 2000; Manning and Schütze, 1999).[11] It has also played a prominent role in the implementation of lazy functional programming languages (see below).

11. Also see Norvig (1991) and Johnson (1995) for a discussion of the relationship between memoization in functional programming languages and CFG parsing algorithms, especially Earley parsing.

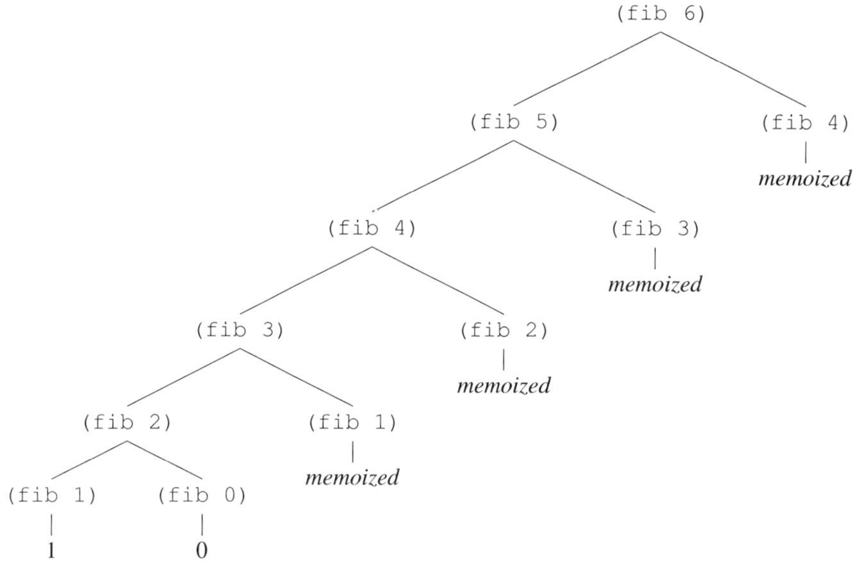

Figure 2.10

Memoized Fibonacci: This figure shows the computation of the memoized Fibonacci function applied to input 6: `((mem fib) 6)`.

2.3.3.1 Stochastic Memoization

In a deterministic setting, every function that is applied to some arguments has a unique output value (when it is defined at those arguments). Therefore, memoizing a deterministic procedure cannot change its meaning; the mathematical function which it computes remains the same. By contrast, in a probabilistic setting, a procedure samples from a distribution over values. Therefore, memoization fundamentally changes the distribution over return values and, hence, the meaning of the memoized procedure. The first value sampled from a memoized stochastic procedure applied to some particular arguments will be drawn from the distribution defined by the procedure without memoization. However, subsequent calls to the procedure with the same arguments will deterministically return the same value. In this book, we use memoization to formalize lexicon learning. Therefore, we require a notion of memoization which can model probabilistic reuse. In order to do this, the notion of memoization must be generalized to *stochastic memoization*.

A deterministic memoizer builds a memotable associating each input with a particular output. A stochastic memoizer associates each input with a distribution over output values (Goodman et al., 2008). This distribution, called the ***memoization distribution***, specifies probability that each previously computed value will be reused on the next call to the procedure (when applied again to the same input arguments). Note that in order to probabilistically reuse earlier work, the memoization distribution must assign some probability to the possibility of sampling a value from the underlying procedure and storing it in the memotable.

Although memoization is usually understood as a technique for improving program efficiency, the technique was originally introduced by Michie (1968) as a method for learning. Michie explicitly highlighted how memoization allows systems to mix rule-based computation and retrieval from memory in a single domain.

> A function is a function, and the means chosen to find its value in a given case is independent of the function's intrinsic meaning. This is no more than a restatement of a mathematical truism, but it is one which has been lost sight of by the designers of our programming languages. By resurrecting this truism we become free to assert: (1) that the apparatus associated with any given function shall consist of a "rule part" (computational procedure) and a "rote part" (look-up table); (2) that evaluation in the computer shall on each occasion proceed either by rule or rote, or by a blend of the two, solely as dictated by the expediency of the moment; (3) that the rule versus rote decisions shall be handled by the machine behind the scenes; and (4) that various kinds of interactions be permitted to occur between the rule and the rote part. Thus each evaluation by rule adds a fresh entry to the rote.

Although Michie applied memoization to problems of learning and inference in artificial intelligence, his notion was not probabilistic, and thus he introduced various other techniques to model uncertainty. The technique of stochastic memoization proposed in Goodman et al. (2008) and used here can be understood as extending Michie's proposal to the probabilistic setting.

Stochastic memoization of probabilistic generative models, when combined with probabilistic conditioning, provides a recipe for doing inference over a set of stored computations. By conditioning on a set of outputs from a memoized procedure, we ask the question: which computations were likely to have been stored and reused in generating this data?

2.3.3.2 Balancing Novelty and Reuse with Pitman-Yor Processes

There are many possible distributions one could use to implement a stochastic memoizer. In this section, we describe the one used in this book: the *Pitman-Yor process*, a distribution from non-parametric Bayesian statistics, which can be used to model processes mixing reuse of previously computed items with creative sampling of new items (Pitman and Yor, 1995).[12] Following Johnson et al. (2007a) and Goodman et al. (2008) we will use it as a memoization distribution.

A Pitman-Yor process can be described in terms of a sequential sampling scheme using the metaphor of a restaurant.[13] Imagine a restaurant with an infinite number of tables. The first customer enters the restaurant and sits at the first unoccupied table. The $(N+1)$th customer enters the restaurant and sits at either (i) an already occupied table with probability $\frac{y_i - a}{N+b}$, where N is the total number of customers, y_i is the number of customers already seated at table i, and a and b are parameters, or (ii) at a new table with probability $\frac{Ka+b}{N+b}$, where K is the total number of occupied tables. Note that the tables are indexed by $i \in \{1,..,K\}$, the parameter $a \in 0, 1$ is known as the *discount parameter* of the model, and the parameter $b \geq -a$ is known as the *concentration parameter* of the model.

According to the process just defined, customers sit at an already-occupied table with probability proportional to the number of individuals at that table minus a constant discount factor a which lies between 0 and 1, or at a new table with probability proportional to b, plus the total number of previously occupied tables multiplied by the discount factor a. This is illustrated in Figure 2.11.

To use a Pitman-Yor process as a memoization distribution for some procedure `proc` applied to some argument `arg`, we associate each table in the restaurant with a value called a *dish* that is sampled from `(proc arg)`. In the literature, the distribution defined by evaluating `(proc arg)` is known as the *base distribution* of the Pitman-Yor process. When a Pitman-Yor process is

12. The term *non-parametric* refers to statistical models whose complexity can grow with the data, rather than being specified in advance.

13. This construction of the Pitman-Yor process is called the Chinese restaurant process in Pitman (2002). However, in the literature the term Chinese restaurant process is now generally reserved for a Pitman-Yor process where the parameter a is set to 0.

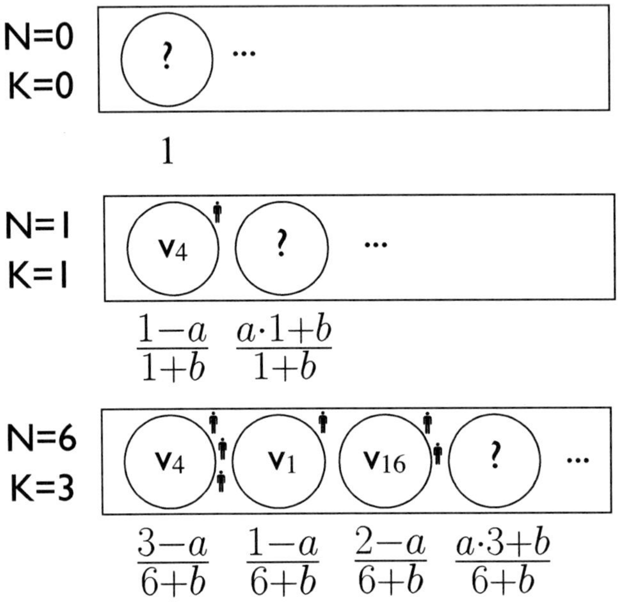

Figure 2.11

The Pitman-Yor Process: This figure shows one possible series of distributions that might result from the sequential sampling scheme for the Pitman-Yor process. Shown is the distribution over the next customer after N customers have been seated at K tables. Associated with each table is a value $\mathtt{v}_i$, which is drawn from the underlying function that has been stochastically memoized using the Pitman-Yor process as a memoization distribution.

used as a memoization distribution, each table represents a reusable computation sampled from the procedure that has been memoized.[14] The probability of sampling the new table, and thus re-evaluating `(proc arg)`, depends on the number of times that a new table was sampled in the past (K in the expression above).

To model productivity and reuse in English morphology, we might use a Pitman-Yor process in the following way: we use the probabilistic procedure `proc` to define the space of possible words, we let dishes that are sampled from `(proc arg)` and associated with tables in the PYP represent observed word types, and we let customers seated at tables represent observed word tokens.

14. Note that multiple tables can share the same label. This is a necessary due to the *exchangeability* of the PYP. However, the a priori highest probability assignments of customers and dishes to tables will be those which assign a single unique dish to each table.

In the Church definitions that follow, we will assume a higher-order procedure `PYmem` which implements a Pitman-Yor-based stochastic memoizer. Thus, the procedure `proc` can be memoized with the syntax `(PYmem a b proc)`.

2.3.3.3 Properties of the Pitman-Yor Process

The usefulness of the Pitman-Yor process as a model of productivity and reuse lies in the way that it balances competiton between existing and new values. The first thing to note is that the PYP implements a simplicity bias on the set of stored items. It assigns a higher probability to restaurants that (i) have fewer customers, (ii) have fewer tables (iii) for a fixed number of customers N assign the customers to the smallest number of tables. Furthermore, the PYP implements a rich-get-richer scheme. The probability of reusing an existing table is proportional to the number of customers seated at that table. As the number of customers at a table increases, so does the probability that the next customer will be seated at that table. Because of these biases, when used as a model of lexical storage, the PYP will tend to prefer smaller, simpler lexica, and will be able to capture differences in word frequency.

The PYP also encodes another, opposing bias. The probability of a new table $\frac{aK+b}{N+b}$ depends on K, the number of times a new table has been generated in the past. Thus, in a Pitman-Yor process, the probability of generating a novel table is a function of the number of novel tables that have been generated in the past. The more often that novelty has occurred in the past, the more likely it will be in the future.

2.3.4 Stochastic Memoization of `unfold`: Adaptor Grammars

In the preceding sections, we introduced the notion of stochastic memoization and defined and discussed the Pitman-Yor process as a memoization distribution, stressing the way in which it balances productivity and reuse. In this section, stochastic memoization is applied to the probabilistic `unfold` recursion that was discussed in Section 2.3.2. The resulting model, known as adaptor grammars, was introduced in Johnson et al. (2007a).

Adaptor grammars can be defined in Church as shown in Figure 2.12. Each time `adapted-unfold` is invoked, the call will be intercepted by the memoizer and either (i) a previously sampled expression will be returned, or (ii) a new expression will be sampled, stored in the memotable, and then returned. Although this model is very simple to express, it has a number of subtleties and

```
(define adapted-unfold
  (PYmem a b
         (lambda (symbol)
           (if (terminal? symbol)
               symbol
               (map adapted-unfold (sample-rhs symbol))))))
```

Figure 2.12

Adaptor Grammars: By stochastically memoizing `unfold`, the results of earlier context-free derivations can be reused. The resulting model is known as adaptor grammars.

can result in surprisingly complex behaviors. Because `adapted-unfold` is recursive, when recursing on the right-hand side of a rule like `N` ⟶ `Adj` *-ity*, the expansion of the nonterminal `Adj` will also be drawn from a memoized distribution, and therefore may correspond to a previously built complex adjective already stored in memory, such as the following tree:

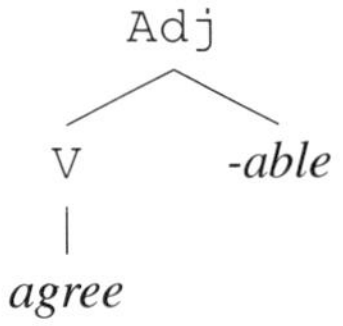

A second important thing to note about adaptor grammars it that they only store complete expressions. For example, if `adapted-unfold` is applied to the symbol `N`, it can either choose to retrieve a complete noun from the stored sets of nouns, or it can build a new noun, but it cannot ever store and retrieve a partial structure like the following tree:

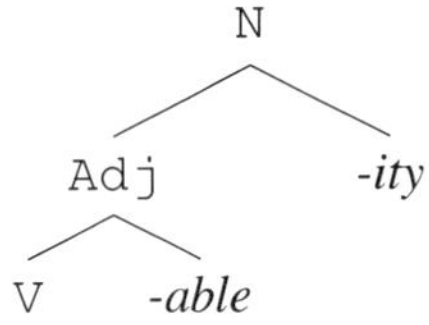

A slightly modified variant of adaptor grammars, known as Maximum a posteriori Adaptor Grammars (MAGs), are used in this book to model the full-listing approach to lexical storage,[15] and can be interpreted as a modern update of the idea of the lexical redundancy rule from classical lexicalist theories of grammar (e.g., Bresnan, 1982; Jackendoff, 1975; Lieber, 1980). Under such theories, lexical redundancy rules stored their outputs, and were used to express generalizations inherent in stored items. Adaptor grammars go beyond these classical rules in several ways. In an adaptor grammar, storage is fully recursive so that stored items can be composed of other stored items; furthermore, adaptor grammars provide a fully probabilistic formulation of structure building, and, therefore, can can capture aspects of learning and use not addressed by classical theories. Nevertheless, they still preserve the key insight that some rules store their output forms after the first time they are created.

2.3.5 Lazy and Eager Evaluation Strategies

In order to turn the λ-calculus into a practical model of computation, one must fix an *evaluation strategy* for programs. The idea of an evaluation strategy can be illustrated with an example (See Abelson and Sussman, 1996; for a detailed introduction to evaluation strategies). Consider the procedure `add3` discussed above, and listed again in Figure 2.13.

```
(define add3
  (lambda (x y z)
    (+ x y z)))
```

Figure 2.13
add3: This procedure adds the values of its three inputs.

Imagine that this procedure is applied to three complex expressions as arguments, for example, `(add3 (+ 1 2 3) (* 2 4) (- 3 1))`. There are several orders in which the final value of this expression can be computed. Two are shown in Figure 2.14.

In the top row of the figure, each argument expression is first evaluated, and then these values are substituted into the body of the `add3` procedure. Only after this is done is the body of the procedure evaluated to produce a final output for the whole computation. In functional programming, an evaluation

15. The version of adaptor grammars used in this book and developed originally in Johnson et al. (2007a) uses an underlying PCFG in which rule weights have a Dirichlet prior. See the next chapter for details.

	Eager Evaluation	Lazy Evaluation
1	`(add3 (+ 1 2 3) (* 2 4) (- 3 1))`	`(add3 (+ 1 2 3) (* 2 4) (- 3 1))`
2	`(add3 6 8 2)`	`(+ (+ 1 2 3) (* 2 4) (- 3 1))`
3	`(+ 6 8 2)`	`(+ 6 8 2)`
4	`16`	`16`

Figure 2.14

Evaluation Strategies: The eager evaluation strategy first evaluates all arguments to a procedure, and then applies the procedure to these arguments. The lazy evaluation strategy waits until an argument value is needed before evaluating the argument expression.

strategy like that shown in the first row of Figure 2.14 is known as *applicative order* or *eager* evaluation.

By contrast, in the bottom row, the expressions passed as arguments to the procedure are not evaluated immediately; rather, they are substituted into the body of `add3` first. Only when their values are needed by the primitive procedure + (i.e., the sum procedure) are they evaluated. In functional programming, this evaluation strategy is known as *normal order* or *lazy* evaluation. Under lazy evaluation, arguments to a procedure are *delayed* until they are required in order to proceed to a final value for some expression. These delayed objects are known as *promises* since they "promise" to return a value when needed. When the value of a delayed promise is needed, it is *forced*. Programming languages that make use of lazy evaluation force the value of delayed objects at two places: the top-level of the entire program when a value must be returned, and whenever delayed values are passed into primitive procedures, such as + in the example above.

The Church-Rosser theorem for the deterministic λ-calculus states that the order in which expressions are evaluated does not affect the final result of the computation (see, Hindley and Seldin, 2008). Nevertheless, the choice of evaluation strategy can still have impact on the efficiency of computation. To see this consider the procedure in Figure 2.15.

```
(define silly-procedure
  (lambda (first-argument second-argument)
    (if true
        (do-something first-argument)
        (do-something-else second-argument))))
```

Figure 2.15

Advantages of Being Lazy: This figure shows a procedure where there is an advantage to a lazy evaluation strategy. In this example, the value of the second argument is never needed and therefore an eager evaluator wastes work computing it.

This procedure takes two arguments; it only evaluates the second argument when the condition in the `if` statement is `false`. However, in this example, the `if` statement is always `true` and therefore the second argument will never be used. Suppose that this procedure was applied to two complex expressions, such as in `(silly-procedure expression1 expression2)`.

Under eager evaluation, both expressions will be evaluated before they are passed to `silly-procedure`. The work involved in evaluating `expression2` will be wasted. By contrast, a lazy evaluator will not do this extra work.

While this example is admittedly contrived, there are many real-life cases in complex programs where the value of a particular argument expression turns out not be needed. This has led to the development of lazy functional programming languages such as Haskell. For certain kinds of computations, lazy languages can lead to significant efficiency gains.

```
(define silly-procedure2
  (lambda (argument)
    (do-something argument argument argument)))
```

Figure 2.16

Advantages of Being Eager: This procedure shows a situation where it is advantageous to be eager. A lazy evaluator evaluates the input argument expression three times, while an eager evaluator only does so once.

However, lazy evaluation can also sometimes have disadvantages, as illustrated in Figure 2.16. Suppose that this procedure was applied to a single very complex expression, such as in the following expression.

```
(silly-procedure2 complex-expression).
```

Under an eager evaluation strategy, `complex-expression` would be evaluated before it was passed to `silly-procedure2`, and thus the computation involved would be performed only once and shared between the three uses in the body of the procedure. However, the lazy evaluator would repeat this work three separate times, unnecessarily.[16]

2.3.5.1 Laziness and Eagerness in a Probabilistic Language

Like memoization, the choice of lazy versus eager evaluation does not change the meaning of deterministic programs, but does have consequences for the meaning of probabilistic programs. This can be illustrated with the simple example shown in Figure 2.17.

This figure shows a procedure called `check-equality`, which takes a single argument and branches conditional on whether or not the argument

16. In real-world lazy languages, this problem is solved by memoizing the result of computations so that they are only computed once. This strategy is known as *call-by-need* evaluation. In a stochastic setting, a call-by-need strategy can restore applicative order semantics to a lazy evaluator (see the next section).

```
(define check-equality
  (lambda (argument)
    (if (= argument argument)
        (do-if-true)
        (do-if-false))))
```

Figure 2.17

A Procedure with Different Meanings under Stochastic Lazy and Eager Strategies: This procedure takes a single argument and checks if it is equal to itself.

is equal to itself. Consider what happens when this procedure is applied to the expression `(flip)`, as in `(check-equality (flip))`. Under an eager evaluation strategy, the expression `(flip)` is evaluated once and a particular value is passed into `check-equality`. This value will always be equal to itself, and, therefore, the procedure will always execute the true branch of the `if` statement. However, with a lazy evaluation strategy the expression `(flip)` will be evaluated twice in the expression `(= argument argument)`. Because `(flip)` evaluates to a fair coin toss, the condition will be true only half the time—when both evaluations come up either `true` or `false`.

This example illustrates the difference in meaning between eager and lazy evaluation strategies in a probabilistic program. Under a lazy evaluation strategy, if an argument expression is used multiple times by a procedure, then each time that expression is needed, it will be re-evaluated. Particular values will only be sampled at the last possible moment, and, thus, the value of an argument expression can vary over the course of the evaluation of a procedure. On the other hand, under an eager evaluation strategy, a particular value for an argument expression will be sampled as early as possible, and that particular value will be shared and reused over the course of the evaluation. A natural idea is to allow a probabilistic mixture of lazy and eager evaluation strategies and thus be able to learn which computations should be carried out ahead of time and reused, and which should be delayed and computed on the fly when needed. In the next section, we explore this idea in the context of the `unfold` procedure.

```
(define delay-or-unfold
  (lambda (symbol)
    (if (terminal? symbol)
      symbol
      (if (flip)
        (delay (stochastic-lazy-unfold symbol))
        (stochastic-lazy-unfold symbol)))))

(define stochastic-lazy-unfold
  (lambda (symbol)
    (if (terminal? symbol)
        symbol
        (map delay-or-unfold (sample-rhs symbol)))))
```

Figure 2.18

Stochastic Lazy Unfold: When passed a nonterminal as an argument, this version of `unfold` first samples a right-hand side. For each symbol on that right-hand side, it calls `delay-or-unfold` which randomly chooses to continue or delay the recursion.

2.3.5.2 Stochastic Lazy `unfold`

In this section, we propose a version of `unfold` that mixes lazy and eager evaluation strategies probabilistically. By stochastically mixing lazy and eager evaluation, this procedure will be able to return arbitrary, partial computations, and, in combination with stochastic memoization, will result in the fragment grammars model.

To develop a stochastically lazy version of `unfold`, we assume a version of Church that uses an eager evaluation strategy, but comes equipped with a `delay` statement, which delays the evaluation of an expression until its value is needed.[17] In this version of Church, it is possible to write the stochastically lazy version of `unfold` shown in Figure 2.18.

Figure 2.19 shows a possible evaluation of `stochastic-lazy-unfold` when applied to the nonterminal `N`. The procedure first samples a right-hand side for `N`, in this case `N` ⟶ `Adj` *-ity*. Once this rule has been sampled, the

17. We assume that the interpreter recursively forces delayed expressions at the top-level and when they are needed as the arguments to primitive procedures. `stochastic-lazy-unfold` could also be implemented in a version of Church that was fundamentally lazy, but contained a primitive procedure, `non-recursive-force`, which forces a single delayed value (i.e., that forces computation a single step further, then stops). The code would be identical to Figure 2.18 with `delay` replaced by `non-recursive-force`.

procedure recursively applies `delay-or-unfold` to each element of the right-hand side. Because *-ity* is a terminal, the recursion stops. For the nonterminal `Adj`, the procedure `delay-or-unfold` decides whether to continue the recursion or to delay sampling a new adjective until a later time. In Figure 2.19, the recursion is continued by sampling the rule `Adj` $\longrightarrow$ `V` *-able*. Once again, `stochastically-lazy-unfold` recursively applies `delay-or-unfold` to the right-hand side. The terminal *-able* halts the recursion on that branch. In the case of `V`, `delay-or-unfold` delays the recursive call and returns the delayed object. At this point there is no more work to be done, and the procedure returns.

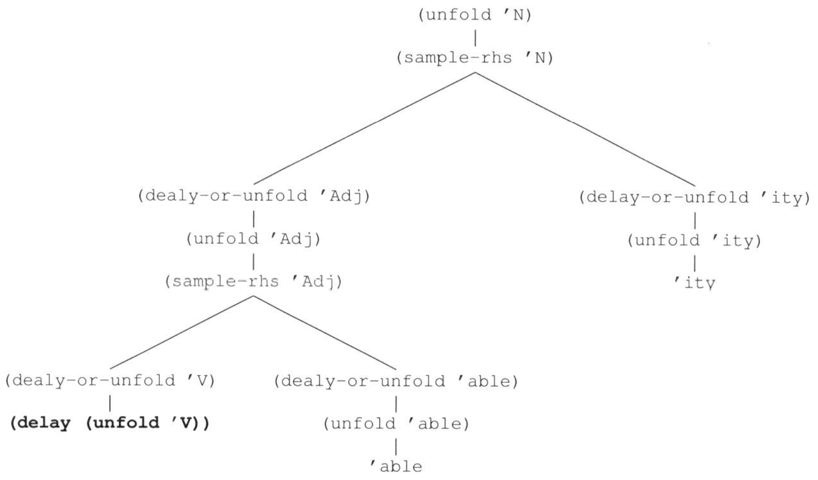

Figure 2.19

Possible Computation of `stochastic-lazy-unfold`: `stochastic-lazy-unfold` chooses whether or not to delay each recursion on the right-hand side of sampled rules. Here it has expanded a noun with the affixes *-able* and *-ity* but delayed the sampling of `V`. The delayed branch of execution is highlighted in bold.

The result of sampling from `stochastically-lazy-unfold` is an object in which some of the potential recursive branches of computation have been completed, but others have been delayed. Such partially delayed objects correspond to derivation tree fragments in a PCFG. For example, the delayed object in Figure 2.19 corresponds to the partially evaluated PCFG computations discussed in the last chapter.[18]

18. The delayed call to `(unfold 'V)` will eventually be forced by the interpreter at the top-level, and a full expression will be sampled.

The procedure `stochastic-lazy-unfold` demonstrates how mixing lazy and eager evaluation strategies can lead to the sampling of partial computations during the evaluation of a probabilistic program. The return value of `stochastic-lazy-unfold` is a complex structure including delayed subparts. However, under the definitions above, the delayed objects are ephemeral. They cease to exist at the end of the particular evaluation that created them. What is required is a way of storing such partial computations for later reuse. In the next section, we show how this can be done with stochastic memoization.

2.3.6 Fragment Grammars

```
(define fragment-unfold
  (PYmem a b
        (lambda (symbol)
          (if (terminal? symbol)
              symbol
              (map delay-or-unfold (sample-rhs symbol))))))
```

Figure 2.20

Fragment Grammars: The fragment grammars model results from stochastically memoizing `stochastic-lazy-unfold`. The resulting procedure can reuse partial computations across multiple calls.

Having outlined the techniques of stochastic memoization and stochastic lazy evaluation, we can now define the model proposed in this book: fragment grammars.[19] By memoizing `stochastic-lazy-unfold`, partial sub-computations can be shared across multiple calls to the procedure shown in Figure 2.20 (note that the procedure `delay-or-unfold` is defined as in Figure 2.18.)

Each time `fragment-unfold` is called, one of two things will happen: the procedure will return a previously sampled partial computation from the memotable, or it will sample a new partial subcomputation from the underlying procedure. In either case, delayed subparts of this computation will ultimately be forced, and a complete expression will result. Like adaptor grammars, the

19. The version of fragment grammars used in the rest of this book differs slightly from the presentation in this section. In particular it (i) places a Dirichlet prior on the weights of underlying PCFG rules, and (ii) places a beta-distributed prior on the probability of delaying each branch of subcomputation. See Section 3.1.8 for details.

sampled computations are defined recursively, and therefore when a novel partial computation is sampled, it may be built of previously stored parts.

2.3.7 A Generalization: `fragment-lambda`

In the preceding sections, we developed the fragment grammars model by stochastically memoizing a stochastically lazy version of the `unfold` recursion. In this section, we show how this technique can be generalized in a way that allows it to be applied to any procedure in Church and therefore allows the *fragmentization* of any probabilistic generative model.

We propose a novel language construct called `fragment-lambda`, which—like λ in the λ-calculus—can be used to define procedures. It differs from λ in that the procedures it defines automatically stochastically reuse partial subcomputations. The operator `fragment-lambda` can be defined via a program transformation as shown in Figure 2.21.

```
                                              (PYmem a b
                                                 (lambda arguments
(fragment-lambda arguments body)      ⟶            (if (delay? arguments)
                                                         (delay body)
                                                         body)))
```

Figure 2.21

Fragment Lambda: Shown here is a program transformation implementing a new form of abstraction operator called `fragment-lambda`, which allows arbitrarily defined procedures to automatically reuse subcomputations.

Figure 2.22 shows this program transformation applied to the `unfold` recursion. Under this transformation, when the `unfold` recursion is applied to each symbol on the right-hand side of a sampled PCFG rule, it will sometimes return a delayed object representing the application itself, and other times it will recurse. The resulting model is very similar to the definition of fragment grammars given above.[20]

The technique encapsulated by the `fragment-lambda` operator opens up the possibility of developing fragmentized versions of diverse models across language. For example, in this book we focus on fragmentized variants of context-free grammars. However, with `fragment-lambda` it is possible to define fragmentized variants of more sophisticated grammatical formalisms,

20. In the `fragment-lambda`-defined model, it is possible for a category like `N` to be stored as a partial computation at a table in the memoizer. By contrast, fragment grammars only store partially expanded subtrees of at least depth 1 on tables in their memoizer.

```
(define unfold
  (fragment-lambda (symbol)
                   (if (terminal? symbol)
                       symbol
                       (map unfold (sample-rhs symbol)))))
```

⟶

```
(define unfold
  (PYmem a b
         (lambda (symbol)
           (if (flip)
               (delay (if (terminal? symbol)
                          symbol
                          (map unfold (sample-rhs symbol))))
               (if (terminal? symbol)
                   symbol
                   (map unfold (sample-rhs symbol)))))))
```

Figure 2.22

Defining Unfold with `fragment-lambda`: Defining the recursive `unfold` procedure with `fragment-lambda` results in a model similar to fragment grammars.

such as those on the so-called *mildly context-sensitive* hierarchy (e.g., tree-adjoining grammars, derivational minimalist grammars, multiple context-free grammars, etc. Joshi and Schabes, 1997; Joshi et al., 1991; Seki, 1991; Stabler, 1997). Fragmentization could also be applied to other domains of cognition. For instance, an influential theory in the realm of human image understanding is the recognition-by-components theory (Biederman, 1987). In this theory, it is hypothesized that objects are recognized as assemblages of simple geometric components such as blocks and cylinders. An implementation of such a model with `fragment-lambda` would allow increasingly complex, reusable components to be built out of simpler ones, leaving certain subcomponents underspecified as variables.

There are, of course, many open questions about modeling and inference using `fragment-lambda` that we leave for future research. However, `fragment-lambda` illustrates how the idea of reuse of partial subcomputations provides a domain-general way of thinking about the interaction of domain-specific structure-building processes and memory (see Taatgen and Lee, 2003; for a similar proposal).

2.4 The Models Examined in the Book and Related Work

In this section, we discuss some properties of the five formal models that we introduced in the last chapter. A full formal definition of these models is given in the next chapter. Here, we will provide some intuitions necessary to understand their performance in later chapters. This section can be referred back to for readers wishing to clarify the reasons for the results in later chapters.

2.4.1 Full Parsing: Dirichlet-Multinomial PCFG (DMPCFG)

Recall from Chapter 1 that the full-parsing approach to productivity and reuse stores only minimal-sized fragments equivalent to rules in the underlying CFG. We formalize this approach using Dirichlet-Multinomial Probabilistic Context-Free Grammars, which we will abbreviate as DMPCFGs throughout the book. A DMPCFG is a probabilistic context-free grammar where rule probabilities have a Dirichlet prior distribution (Johnson et al., 2007b; Kurihara and Sato, 2006). Putting a prior on rule weights allows us to compute the conditional distribution over rule weights given the input, that is, the posterior distribution over weights, thus learning the weights from the data. In this book, we will always work with the set of rule weights with the maximum probability under the posterior distribution.

One intuition that is helpful for understanding the behavior of the DMPCFG model later in the book is that an DMPCFG learns rule weights that are related to the token frequency of each rule in the training corpus. In particular, high posterior-probability rule weights will tend to be proportional to the token count of a rule in the training set. For the maximum a posteriori grammars we use in later chapters, the rule weights are directly proportional to the frequency of the depth-one trees found in the training data set.

2.4.2 Full Listing: Maximum A Posteriori Adaptor Grammars (MAG)

In Chapter 1, we discussed the full-listing approach to productivity and reuse, which stores all computations in their entirety after the first time they are computed. This model is formalized by a variant of the adaptor grammars model of Johnson et al. (2007a) where we use the adaptor grammar with the maximum score under the posterior distribution given the training data—Maximum a posteriori Adaptor Grammars (MAGs).[21]

MAGs have a number of subtle properties that are important for understanding their behavior in the simulations reported in later chapters. Although a MAG always stores each computation (and subcomputation) in its entirety when it is first computed, it reuses these stored computations probabilistically. In particular, a (sub)computation will be reused at a rate that is proportional to the token frequency of that subtree in the training set. Because reuse of subtrees is probabilistic, it is always possible for a MAG to build a new subtree to account for an input form, even when that exact form is already stored. Nevertheless, in general, a MAG will prefer to reuse previously computed forms.

The degree to which an adaptor grammar prefers to reuse previously stored computations can be controlled by adjusting the a and b parameters of the Pitman-Yor memoization distributions associated with each nonterminal. Johnson and Goldwater (2009) show that the performance of adaptor grammars can be improved by learning these hyperparameters. This technique allows the system to learn that some categories of constituent store their computations, while others are more likely to productively compute fresh forms. Note that this approach differs from the fragment grammar model in this book because

21. Note that this contrasts with many applications in the literature (e.g., Johnson, 2008a,b; Johnson and Goldwater, 2009) where an approximation to the full posterior distribution over adaptor grammars is considered.

it allows levels of productivity to be inferred only on a per-nonterminal basis, rather than at the granularity of individual words and rules.

Another important aspect of the MAG model is the way in which it estimates the probability of the rules in the underlying PCFG defining the set of possible trees. These rules receive a probability proportional to the number of stored computations in which they take part, that is, they are estimated based on type frequency of the rule across stored trees. As we will explain in subsequent chapters, because type frequency is sometimes a good approximate estimator of productivity, a MAG can exhibit good generalization performance in some domains. For example, we will see that a MAG can learn that novel forms ending in *-ness* (e.g., *wugness*) are more probable than novel forms ending in *-ity* (e.g., *wugity*), due to the higher type frequency of *-ness*. However, as we will also discuss, type frequency is not always a good predictor of productivity (Baayen, 2006), and this aspect of MAG generalization performance is limited to single rules in the underlying PCFG, such as N ⟶ Adj *-ness*. A MAG can never learn that a combination of morphemes such as *-ability* can also generalize productively.

2.4.3 Exemplar-Based Productivity and Reuse: Data Oriented Parsing (DOP)

In section 1.4.2, we discussed the exemplar-based approach to productivity and reuse, under which all possible generalizations consistent with the data are stored. In this book, we formalize the exemplar-based approach using the data-oriented parsing grammar-estimation framework (Bod et al., 2003)

There are many different techniques that fall under the DOP umbrella, differing primarily in how they estimate the probability of stored tree fragments. We use two variants of DOP. The first, known as Data Oriented Parsing 1 (DOP1), sets the probability of a subtree proportional to its token frequency in the training corpus (Bod, 1998; Scha, 1990).

There a number of well-known problems with DOP1. For example, Johnson (2002) shows that the DOP1 estimator is both biased and inconsistent (in the statistical senses of these two terms). Another problem with DOP1 is that it tends to overweight training data nodes that appear higher and in larger trees, and, thus, produces a pronounced bias towards larger stored items (see Goodman, 2003). For this reason, we also explore a second DOP estimator, called the Equal-Node Estimator (ENDOP; Bod, 2003; Goodman, 2003), which assigns equal weight to each training data node and each training data item. The ENDOP variant of DOP has been shown to significantly improve performance in syntactic parsing compared to DOP1 (Bod, 2003).

Although there are many other variants, in its original form, DOP was committed to the storage of *all* subtrees, and it is this feature that corresponds to the exemplar-based approach.[22] Some versions of DOP have relaxed the all-subtrees assumption by introducing techniques penalizing either large subtrees or long derivations (see various chapters in Bod et al., 2003).[23] We believe that these approaches can be understood as moving toward the use of optimal sets of subtrees, like the fragment grammars model advocated in this book. DOP1 and ENDOP on the other hand, straightforwardly encode the idea of storing *all* generalizations consistent with the data.

Unlike the full-parsing, full-listing, and inference-based approaches, DOP models spread their probability mass across all possible derivations consistent with the input data. This fact is important for understanding the behavior of the DOP models later in the book. DOP does not commit to a single hypothesis that a form was derived productively or that it was stored. Instead, it hedges across both possibilities, and thus tends to favor grammars that blend computation and reuse, rather than sharply distinguishing between them.

The two DOP models differ from one another in a simple way: DOP1 assigns more probability mass to larger subtrees, and therefore tends to behave more like a full-listing model; ENDOP assigns more probability mass to smaller subtrees, and therefore tends to behave more like a full-parsing model.

2.4.4 Inference-Based Productivity and Reuse: Fragment Grammars (FG)

The inference-based model advocated in this book, fragment grammars (FG), differs from the earlier models in that it tries to find the set of stored items and points of productivity that best predicts the training data. It inherits from MAG

22. Another class of all-fragment or all-subtree approaches is based on the use of *kernel methods*. In such methods, a function, called a kernel, is defined that measures similarity between pairs of inputs (e.g., pairs of trees). Using the kernel, an input training data set can be transformed into an inner product space in which discriminative machine learning techniques can be applied (e.g., support vector machines or perceptrons). Collins and Duffy (2002) show how to define tree kernels that compute the similarity between two trees by considering all subtrees, but in a way that sidesteps many of the efficiency problems inherent in DOP.

23. Some other versions of DOP are aimed at different problems entirely. For example, unsupervised data-oriented parsing (Bod, 2006) is aimed at the problem of unsupervised grammar induction, and is therefore not strictly comparable to the present models.

the ability to store any sub-computation. However, it generalizes MAG by allowing stored sub-computations to include variables that require productive computation. Whereas MAG can only productively compute forms using the underlying PCFG, FG can productively compute forms by reusing stored structures with variables. Like MAG, FG estimates the probability of stored items based on their token frequency and the probability of underlying PCFG rules based on their type frequency. Unlike with MAGs, it is not possible to efficiently compute the maximum a posteriori fragment grammar directly from the training data, and so we approximate this most probable grammar by using stochastic search.

Like the full-parsing (DMPCFG) and full-listing (MAG) models, fragment grammars commit to a single analysis for each form in the data. However, fragment grammars differ in a more fundamental way from all of the other models studied in this book. The other models all follow a fixed strategy for what is stored. The full-parsing model (DMPCFG) stores any and all depth-one subtrees occurring in the analysis of a form, the full-listing model (MAG) stores all complete subtrees occurring in an analysis,[24] and the exemplar-based (DOP) models store all full and partial subtrees consistent with a form. By contrast, for a FG, the set of subtrees stored for a given form depends on earlier observed data. A FG chooses an analysis of an input tree in which structure is shared in an optimal way with earlier analyses.

One consequence of this is that FGs exhibit a kind of attractor behavior. In particular, they prefer that all instances of individual word types be analyzed in the same way. For example, if a corpus contains many copies of a form such as *goodness*, the FG model prefers to treat the copies as all retrieved whole or all composed. The system also prefers to treat all instances of a particular subform, such as *-ness*, in a consistent manner. If it decides that *-ness* is productive, then there is pressure to treat all forms using it as composed; if it decides that *-th* is unproductive, then there is pressure to treat all forms using this suffix as retrieved whole. Although these are only prior preferences—there can be exceptions—they are an important part of the explanation for why an FG is able to account for the linguistic data.

24. Note that MAG storage is deterministic only for tree fragments stored in the top level of the system. Token uses of CFG base rules may or may not be stored depending on whether a top level tree was composed anew or retrieved from memory during computation.

2.4.4.1 Other Inference-Based Models of Productivity and Reuse

As mentioned in Section 1.2.6, De Marcken (1996a,b) introduced a model of lexicon learning which treats the problem as an inference in the minimum-description length framework (Grünwald, 2007; Rissanen, 1978; Vitányi and Li, 2008), jointly minimizing the sum of lexicon complexity and the complexity of all derivations given the lexicon. While De Marcken discussed a version of his framework very similar to the model proposed in this book, he did not pursue this in published work, instead extensively investigating a simpler system that can be viewed as a forerunner to adaptor grammars. Most interestingly, De Marcken combines his storage proposal with a notion of *lexical perturbation* whereby stored items can accumulate phonological, semantic, or other idiosyncrasies. This idea may present a promising avenue for integrating more realistic phonological and semantic assumptions into the fragment grammars model.

Zuidema (2007) introduced Parsimonious Data-Oriented Parsing (PDOP), a version of DOP that explicitly eschews the all-subtrees approach in favor of finding a set of subtrees that best explains the data. PDOP uses a heuristic algorithm, similar to the inside-outside algorithm (see Manning and Schütze, 1999), to search for good sets of subtrees. Two other models, developed simultaneously with the current framework, make use of similar Bayesian non-parametrics to define generative models over sets of subtrees (Cohn et al., 2010, 2009; Post and Gildea, 2009). These two approaches differ from fragment grammars in that they do not define the distribution over stored structures recursively—a new stored item cannot be built out of other stored items. Nevertheless, we predict that these models would perform similarly to fragment grammars on the datasets studied in this book.

3 Formalization of the Models and Inference

Timothy J. O'Donnell and Noah D. Goodman

In this chapter, we present the mathematical details of the models studied in this book. We also discuss the inference algorithms used for each of the models and various other issues of practical concern for the simulations that we report later.

3.1 Formal Details of the Models

The mathematical formalization of the models we present in this section follows the order of their development from the perspective of probabilistic programming in the previous chapter. We start by defining context-free grammars, their probabilistic generalization, and prior distributions on rule weights. We then introduce data-oriented parsing estimators for probabilistic tree-substitution grammars, and, finally, generalize context-free grammars to adaptor grammars and fragment grammars using the techniques of stochastic memoization and stochastically lazy evaluation introduced in the last chapter.

3.1.1 Context-Free Grammars

All of the models studied in this book use Context-Free Grammars (CFGs) as their underlying model of computation. CFGs are a simple, widely-known, and well-studied formalism for modeling hierarchically structured computation (Autebert et al., 1997). Formally, a CFG is a 4-tuple, $\mathcal{G} = \langle V_\mathcal{G}, T_\mathcal{G}, R_\mathcal{G}, \mathtt{W} \rangle$.

- $V_\mathcal{G}$ is a finite set of *nonterminal* symbols.
- $T_\mathcal{G}$ is a finite set of *terminal* symbols (distinct from $V_\mathcal{G}$).
- $R_\mathcal{G} \subseteq V_\mathcal{G} \times (V_\mathcal{G} \cup T_\mathcal{G})^*$ is a finite set of *production rules.*
- $\mathtt{W} \in V_\mathcal{G}$ is a unique, distinguished *start symbol.*

By convention, nonterminals are written with capital letters, and, when used to model linguistic structure, they represent categories of constituents such as, "verb" (V) or "noun" (N). The unique, distinguished nonterminal known as the start symbol is often written S (for "sentence"). However, because this book focuses on the domain of morphology, we will use W instead to refer to the category of "word." This symbol represents the category of complete derivations: sentences for syntax, words for morphology. Terminal symbols, written here in italics, typically represent atomic words or morphemes (e.g., *agree*, or *-ness*).

Production rules are written A $\longrightarrow \gamma$, where γ is some sequence of terminals and nonterminals, and A is a nonterminal. The set of production rules defines the possible computations for a CFG. For example, the rule, N $\longrightarrow$ Adj *-ness*, says that a constituent of type N can be constructed by first constructing an adjective Adj, and then concatenating this adjective with the morpheme *-ness*. The list of symbols to the right of the arrow is referred to as the right-hand side of that rule. The nonterminal to the left of the arrow is the rule's left-hand side. The functions lhs(r) and rhs(r) return the left-hand side and right-hand side of rule r, respectively. The ***language*** $\mathcal{L}_{\mathtt{A}}$ associated with nonterminal A is the set of expressions that can be derived from that nonterminal by recursive application of the production rules of the grammar. The language associated with a CFG $\mathcal{G}$ is $\mathcal{L}_{\mathtt{W}}$.

A CFG encodes the possible choices that can be made in computing an expression, but does not specify how to make the choices. It chooses amongst the production rules which expand a nonterminal ***non-deterministically***. Like many other non-deterministic generative systems, CFGs also have a number of probabilistic variants where each choice is instead specified by sampling from a probability distribution over possible production rules. In the next section, the simplest kind of Probabilistic Context-Free Grammar (PCFG), the multinomial PCFG, is introduced.

3.1.2 Multinomial Context-Free Grammars

A multinomial distribution is the simplest way of specifying a distribution over a finite number of discrete choices. If there are K possible choices, we simply specify a vector of probabilities of length K (i.e., a vector of real numbers $\vec{\theta}$, such that $\theta_i \geq 0$ and $\sum_i \theta_i = 1$).[1] A multinomial PCFG is the simplest version of a probabilistic CFG, where the distributions over rule choices are given by multinomial distributions.

A ***derivation*** or ***parse tree*** is a tree representing the trace of the computation of some expression $\vec{w}$, beginning from some nonterminal category A. A derivational tree is ***complete*** if all of its leaves are terminal symbols. A derivation tree

1. This is a slight abuse of terminology. What we refer to here as a multinomial distribution is technically a *categorical distribution*, the multinomial distribution being the distribution over counts of occurrences of each of the K possible outcomes in N draws. However, it is common to refer to categorical distributions as multinomial distributions in the literature, and we follow this convention here.

fragment is a derivation tree whose leaves may be some combination of terminals and nonterminals. Given a derivation d, define the function $\texttt{yield}(d)$ to be the function returning the leaves of d (complete or fragmentary) as a list. Define the function $\texttt{root}(d)$ to be the function returning the nonterminal at the root of the d. Define the function $\texttt{top}(d)$ to be the function returning the topmost production rule (i.e., depth-one tree) in d. A nonterminal $\texttt{A}$ derives some expression $\vec{w} \in T^*$ if there is a complete derivation d such that $\texttt{root}(d) = \texttt{A}$ and $\texttt{yield}(t) = \vec{w}$.

Formally, a multinomial PCFG $\langle \mathcal{G}, \{\vec{\theta}^{\texttt{A}}\}_{\texttt{A} \in V_{\mathcal{G}}} \rangle$ is a CFG $\mathcal{G}$ together with a set of probability vectors $\{\vec{\theta}^{\texttt{A}}\}_{\texttt{A} \in V_{\mathcal{G}}}$. Each vector $\vec{\theta}^{\texttt{A}}$ represents the parameters of a multinomial distribution over the set of rules that share $\texttt{A}$ on their left-hand sides. We write $\theta_r^{\texttt{A}}$ or θ_r to mean the component of vector $\vec{\theta}^{\texttt{A}}$ associated with rule r. Because each vector $\vec{\theta}^{\texttt{A}}$ represents a probability distribution, they satisfy Equation 3.1.

$$\sum_{r \in R^{\texttt{A}}} \theta_r^{\texttt{A}} = 1 \tag{3.1}$$

By convention, we will label the k immediate children of a root node of some derivation d as $\hat{d}_1, \cdots, \hat{d}_k$, that is $\texttt{yield}(\texttt{top}(d)) = \langle \texttt{root}(\hat{d}_1), \cdots, \texttt{root}(\hat{d}_k) \rangle$. The basic structure-building recursion for PCFGs, described in Section 2.3.2, can be expressed mathematically by the following recursive stochastic equation.

$$G^{\texttt{a}}_{\texttt{pcfg}}(d) = \begin{cases} \displaystyle\sum_{r \in R_{\mathcal{G}}\, \texttt{a} \rightarrow \texttt{root}(\hat{d}_i), \cdots, \texttt{root}(\hat{d}_k)} \theta_r \prod_{i=1}^{k} G^{\texttt{root}(\hat{d}_i)}_{\texttt{pcfg}}(\hat{d}_i) & \texttt{root}(d) = \texttt{a} \in V_{\mathcal{G}} \\ 1 & \texttt{root}(d) = \texttt{a} \in T_{\mathcal{G}} \end{cases}$$

This equation states that the probability of a tree d given by the multinomial PCFG $\mathcal{G}$ is the product of the probability of the rules used to build that tree from depth-one subtrees, summed over all rules whose left-hand side matches the root of d, and whose right-hand side matches the immediate children of the root of d.[2] Note that the distribution over full, derived structures is given by $G^{\texttt{W}}$.

2. Typically, for a PCFG this set of rules will be singleton.

Multinomial PCFGs make two strong conditional independence assumptions. First, all decisions about expanding a nonterminal are local to that nonterminal and cannot make reference to any information other than the identity of the nonterminal and the multinomial distribution associated with it. Second, as a consequence of the first independence assumption, expressions themselves are generated independently of one another. These assumptions together mean that, given a multinomial PCFG $\mathcal{G}$ the probability of a set of computations (i.e., parse trees) can be computed directly from the counts of the rules used in the set without knowing precisely where or how the individual rules were used. The probability of a derivation d is just the product of the probabilities of the individual rules it contains, as shown in Equation 3.2.

$$P(d|\mathcal{G}) = \prod_{r \in d} \theta_r \tag{3.2}$$

The probability that a PCFG assigns to a particular expression (i.e., sequence of terminals) $\vec{w}$ is computed by marginalizing (i.e., summing) over all derivation trees sharing that expression as their yield.

$$P(\vec{w}|\mathcal{G}) = \sum_{d \,|\, \texttt{yield}(d)=\vec{w}} P(d|\mathcal{G}) \tag{3.3}$$

Given an expression $\vec{w}$ and a rule r, we define the ***inside probability*** of $\vec{w}$ given r as:

$$P(\vec{w}|r,\mathcal{G}) = \sum_{d \,|\, \texttt{yield}(d)=\vec{w} \,\wedge\, \texttt{top}(d)=r} P(d|\mathcal{G}) \tag{3.4}$$

The inside probability of a sequence of terminals $\vec{w}$, given rule r, is the probability that the sequence is the yield of some complete tree for $\vec{w}$ whose topmost depth-one subtree corresponds to r.[3]

Define a ***corpus*** of derivation trees $\mathbf{D}$ of size $N_{\mathbf{D}}$ with respect to a grammar to be the trees resulting from deriving $N_{\mathbf{D}}$ expressions from the start symbol `W`. Let $\mathbf{X} = \{\vec{x}^{\mathtt{A}}\}$ be the set of count vectors for each rule used in $\mathbf{D}$. The function `counts` takes a set of derivation trees and returns the corresponding vectors

3. Note that we have defined inside probabilities with respect to rules; in most work, they are instead defined with respect to nonterminal categories. The inside probability of a sequence of terminals $\vec{w}$ given nonterminal `A` is computed by additionally marginalizing over rules that share `A` on their left-hand side: $P(\vec{w}|\mathtt{A},\mathcal{G}) = \sum_{r \,|\, \texttt{lhs}(r)=\mathtt{A}} \left[\sum_{d \,|\, \texttt{yield}(d)=\vec{w} \,\wedge\, \texttt{root}(d)=r} P(d|\mathcal{G}) \right]$.

of rule counts: $\texttt{counts}(\mathbf{D}) = \mathbf{X}$. We will occasionally abuse notation and use counts to mean the function that returns the rule counts associated with some particular derivation tree: $\texttt{counts}(d^{(i)}) = \texttt{counts}(\{d^{(i)}\}) = \mathbf{x}_{d^{(i)}}$. The probability of a set of derivation trees $\mathbf{D}$ under a PCFG can be computed from the count vectors $\texttt{counts}(\mathbf{D}) = \mathbf{X}$ according to the function **pcfg** below.

$$\begin{aligned} P(\mathbf{D}|\mathcal{G}) &= \mathbf{pcfg}(\mathbf{X}; \mathcal{G}) \\ &= \prod_{\mathtt{A} \in V_{\mathcal{G}}} \prod_{r \in R_{\mathcal{G}}^{\mathtt{A}}} [\theta_r^{\mathtt{A}}]^{x_r^{\mathtt{A}}}, \end{aligned} \tag{3.5}$$

where $x_r^{\mathtt{A}}$ is the number of times that rule r with left-hand side A was used in the corpus. That is, the probability of a corpus of derivations under a multinomial PCFG is given by the product of the probabilities of all the rules used in the corpus.

3.1.3 Dirichlet-Multinomial Distributions

In this section, we define what is known as the ***Pólya-urn*** representation of the Dirichlet-multinomial distribution. A Dirichlet-multinomial distribution places a Dirichlet prior on the vector $\vec{\theta}$ of probabilities, which parameterizes a multinomial distribution over K elements.[4]

The Dirichlet-multinomial distribution is usually described via a two-step sampling process. First, draw a probability vector $\vec{\theta}$ from a Dirichlet distribution; second, draw N observations from the multinomial distribution parameterized by $\vec{\theta}$. However, the Dirichlet-multinomial distribution also has another, equivalent representation, which is the finite analog of the sequential sampling scheme described for the Pitman-Yor process in the previous chapter.[5] It is this representation which we use in this book.

4. The Dirichlet is commonly used as the prior because it is *conjugate* to the multinomial. The notion of conjugacy is an important one in Bayesian statistics (see, e.g., Gelman et al., 2003); it refers to pairs of distributions—such as the multinomial and Dirichlet distributions—where the posterior distribution over the parameters is in the same family of distributions as the prior. For example, the posterior distribution over $\vec{\theta}$, when the prior is Dirichlet and the likelihood is multinomial, is also a Dirichlet distribution.

5. The fact that this hierarchical process can be represented as a sequential sampling scheme is the result of *de Finetti's Theorem* (de Finetti, 1974; Freer and Roy, 2009), a fundamental result in Bayesian statistics, which states that all *exchangeable* sequences of random variables can also be defined in terms of a hierarchical process where observations are drawn i.i.d. (i.e., independently and with identical distribution) from a random measure.

Define the following sequential process: the first observation, $v^{(1)}$, is sampled according to expression 3.6.

$$v^{(1)}|\pi_1,\cdots,\pi_K \sim \frac{\pi_1}{\sum_{i=1}^K \pi_i}\delta_{[v^{(1)}=v_1]} + \cdots + \frac{\pi_K}{\sum_{i=1}^K \pi_i}\delta_{[v^{(1)}=v_K]}, \tag{3.6}$$

where δ_X is a δ-distribution that returns 1 if the proposition X is true and 0 otherwise, and the π s are ***pseudo-counts***, which can be thought of as counts of imaginary prior observations of each of the K possible outcomes. Thus, the first observation is sampled with probability proportional to the pseudocounts of the K possible outcomes.

After N observations have been sampled, the $(N+1)$th observation is sampled according to the following distribution,

$$\begin{aligned} v^{(N+1)}|v^{(1)},\cdots,v^{(N)},\pi_1,\cdots,\pi_K \sim & \frac{\pi_1+x_1}{\sum_{i=1}^K[\pi_i+x_i]}\delta_{[v^{(N+1)}=v_1]} + \cdots \\ & + \frac{\pi_K+x_K}{\sum_{i=1}^K[\pi_i+x_i]}\delta_{[v^{(N+1)}=v_K]}, \end{aligned}$$

where x_i is the number of previous draws of value v_i. In other words, the probability of subsequent draws of each of the possible K outcomes is proportional to the number of previous draws of that outcome plus the pseudo-count associated with that outcome.

This process implements a simple rich-get-richer dynamic, which increases the probability of each outcome as more values of that outcome are sampled. Note that, unlike a multinomial, the draws are not independent, rather the probability of the next draw is dependent on the history of previous samples. Nevertheless, the draws are ***exchangeable***. The joint probability over a fixed set of N observations does not depend on the order of the draws; the counts of each of the possible K outcomes are sufficient to compute the probability of the sequence (whatever order the outcomes were drawn).

Due to the exchangeability of the Pólya-urn process, we can interpret the Dirichlet-multinomial distribution as assigning probabilities to different partitions of N objects amongst K bins. Let $\vec{x}$ be a vector of length K that counts the number of observations for each of the K possible outcomes. The probability of a particular partition (i.e., count vector of draws of the K outcomes) can be given by the following closed-form equation.

$$P(\vec{x}|\vec{\pi}) = \frac{\prod_{i=1}^K \Gamma(\pi_i+x_i)}{\Gamma(\sum_{i=1}^K \pi_i+x_i)}\frac{\Gamma(\sum_{i=1}^K \pi_i)}{\prod_{i=1}^K \Gamma(\pi_i)}, \tag{3.7}$$

where $\Gamma(\cdot)$ is the ***gamma function***—a continuous generalization of the factorial function; in particular, $m! = \Gamma(m+1)$ for discrete m. This expression corresponds to the result of integrating out $\vec{\theta}$ from the product of a multinomial likelihood and Dirichlet prior.

3.1.4 Dirichlet-Multinomial Context-Free Grammars

For a multinomial PCFG, the set of weight vectors $\{\theta^{\mathrm{A}}\}_{\mathrm{A} \in V_{\mathcal{G}}}$ is pre-specified as a model parameter. It is possible instead to draw these weight vectors from a Dirichlet prior (Johnson et al., 2007b; Kurihara and Sato, 2006). This variant, known as Multinomial-Dirichlet Probabilistic Context-Free Grammar (DMPCFG), will serve as the formalization of the full-parsing theory throughout this book.

An DMPCFG associates a Pólya-urn process with each nonterminal in its underlying CFG, parameterizing each with a vector of pseudo-counts. Thus, a DMPCFG $\mathcal{M} = \langle \mathcal{G}, \{\vec{\pi}^{\mathrm{A}}\}_{\mathrm{A} \in V_{\mathcal{G}}} \rangle$ is a context-free grammar $\mathcal{G}$ together with a set $\{\vec{\pi}^{\mathrm{A}}\}_{\mathrm{A} \in V_{\mathcal{G}}}$ of vectors of pseudo-counts.

If instead of the Pólya-urn representation of the Dirichlet-multinomial distribution we use the standard hierarchical representation (i.e., we explicitly draw the probability vectors $\vec{\theta}^{\mathrm{A}}$ from a Dirichlet distribution), then the distribution over derivations described by an DMPCFG can be represented by the following recursive stochastic equations.

$$G^{\mathrm{a}}_{\mathtt{dmpcfg}}(d) = \begin{cases} \displaystyle\sum_{r \in R_{\mathcal{G}} \mathrm{a} \to \mathtt{root}(\hat{d}_i), \cdots, \mathtt{root}(\hat{d}_k)} \theta_r \prod_{i=1}^{k} G^{\mathtt{root}(\hat{d}_i)}_{\mathtt{dmpcfg}}(\hat{d}_i) & \mathtt{root}(d) = \mathrm{a} \in V_{\mathcal{G}} \\ 1 & \mathtt{root}(d) = \mathrm{a} \in T_{\mathcal{G}} \end{cases}$$

$$\vec{\theta}^{\mathrm{A}} \sim \mathrm{DIRICHLET}(\vec{\pi}^{\mathrm{A}})$$

Note that these equations are identical to those for the multinomial PCFG except that the probability vectors $\vec{\theta}^{\mathrm{A}}$ are now drawn hierarchically from $\mathrm{DIRICHLET}(\vec{\pi}^{\mathrm{A}})$.

Given a corpus of derivation trees **D** the counts of individual rule uses $\mathbf{X} = \{\vec{x}^{\mathrm{A}}\}$ again provides sufficient information to compute the probability of the corpus. The probability of a corpus of derivations **D** under an DMPCFG, $\mathcal{M}$, can be given in terms of these counts by the function **dmpcfg** below using the closed-form probability of a partition for the Pólya-urn scheme.

$$P(\mathbf{D}|\mathcal{M}) = \mathbf{dmpcfg}(\mathbf{X};\mathcal{M}) \tag{3.8}$$

$$= \prod_{\mathtt{A}\in V}\left[\frac{\prod_{i=1}^{K^{\mathtt{A}}}\Gamma(\pi_i^{\mathtt{A}}+x_i^{\mathtt{A}})}{\Gamma(\sum_{i=1}^{K^{\mathtt{A}}}\pi_i^{\mathtt{A}}+x_i^{\mathtt{A}})}\frac{\Gamma(\sum_{i=1}^{K^{\mathtt{A}}}\pi_i^{\mathtt{A}})}{\prod_{i=1}^{K^{\mathtt{A}}}\Gamma(\pi_i^{\mathtt{A}})}\right] \tag{3.9}$$

3.1.5 Data Oriented Parsing

As we described in Chapter 1, in this book we formalize exemplar-based productivity and reuse with two variants of the Data Oriented Parsing (DOP) formalism for tree-substitution grammar estimation (Bod et al., 2003). A Tree-Substitution Grammar (TSG) is a generalization of context-free grammars where the basic units of combination can be arbitrary tree fragments, rather than rules. A derivation in a TSG proceeds via the ***substitution operation*** which replaces a nonterminal leaf node in a tree fragment with label $a \in V_{\mathcal{G}}$ with another tree fragment which has a labeling its root. A CFG can be seen as a special case of a TSG where all fragments are restricted to be depth-one trees. A Probabilistic Tree-Substitution Grammar (PTSG) is a TSG where each subtree is equipped with a probability.

Define the function $\mathtt{prefix}(d)$ to enumerate all prefixes of a given derivation tree, that is, all fully connected fragments of the given derivation tree that include the root of the entire tree. Given a tree prefix $s \in \mathtt{prefix}(d)$ write the n subtrees of d at the leaves of s as $s_i', \cdots, s_n'$, that is, $\mathtt{yield}(s) = \langle \mathtt{root}(s_i'), \cdots, \mathtt{root}(s_n')\rangle$. The probability of a derivation d under a PTSG is defined by the following recursive stochastic equations.

$$G_{DOP}^{\mathtt{a}}(d) = \begin{cases} \sum\limits_{s\in\mathtt{prefix}(d)} \mathrm{PROB}(s)\prod\limits_{i=1}^{n} G_{DOP}^{\mathtt{root}(s_i')}(s_i') & \mathtt{root}(d) = \mathtt{a} \in V_{\mathcal{G}} \\ 1 & \mathtt{root}(d) = \mathtt{a} \in T_{\mathcal{G}} \end{cases}$$

This equation is essentially the same as the recursive stochastic equation for PCFGs except that tree prefixes have been substituted for rules. The probability of a particular derivation is the sum over all possible derivation tree fragments that can be combined to give rise to the target derivation d.

DOP is a family of estimation techniques for PTSGs. DOP provides a way of estimating the probability of stored tree fragments, $\mathrm{PROB}(s)$, from a corpus of examples. There are many techniques that fall under the DOP umbrella;

however, most DOP estimators for $\text{PROB}(s)$, including the two used in this book, are based directly on corpus counts.

Note that we are assuming that the training corpus consists of labeled derivation trees. The first variant of DOP, known as Data-Oriented Parsing 1 (DOP1) (Bod, 1998; Scha, 1990), sets the probability of each derivation tree fragment proportional to its token frequency in the training corpus. Let $F_{\mathbf{D}}(s)$ be the frequency of derivation subtree s in a set of derivations $\mathbf{D}$. DOP1 uses the following estimator.

$$\text{PROB}_{\texttt{DOP1}}(s) \propto F_{\mathbf{D}}(s) \tag{3.10}$$

A well-known problem with DOP1 is that it tends to overweight training data nodes that appear higher and in larger trees (see, e.g., Goodman, 2003). For this reason, we explore a second DOP estimator in this book. This estimator, which we call the Equal-Node Estimator (ENDOP), assigns equal weight to each training data node and each training data item, effectively down-weighting larger derivation tree fragments (Bod, 2003; Goodman, 2003). If $d \in \mathbf{D}$ is some derivation tree in a corpus of derivations, then let $F_d(s)$ be the number of times that a subtree (complete or fragmentary) s appears in d. Let $\texttt{subtrees}(d)$ be the function returning the set of all subtrees of derivation tree d, and $|\texttt{subtrees}(d)|$ be the cardinality of that set. The equal-node estimator is given by the following expression (Sangati and Zuidema, 2011).

$$\text{PROB}_{\texttt{ENDOP}}(s) \propto \sum_{d \in \mathbf{D}} \frac{F_d(s)}{|\texttt{subtrees}(d)|} \tag{3.11}$$

In other words, the equal-node estimator re-weights each derivation tree fragment in the corpus by the total number of trees which appear in its derivation. As a result, nodes appearing higher in derivation trees in the training set will receive the same weight as nodes appearing lower in trees in the training set.

3.1.5.1 PCFG Reductions of DOP Grammars

The number of subtrees for a given derivation tree is exponential in the number of nodes in the tree. TSGs that result from DOP are enormous—far too large to allow for efficient parsing. Early versions of DOP dealt with this problem by working with approximations (e.g., Bod, 1995; randomly excludes 95% of potential fragments from the TSG). However, Goodman (1998, 2003) showed that it was possible to give a PCFG that is linear in the number of nodes in the training corpus and gives exactly the same distribution over derivations as the corresponding DOP TSG.

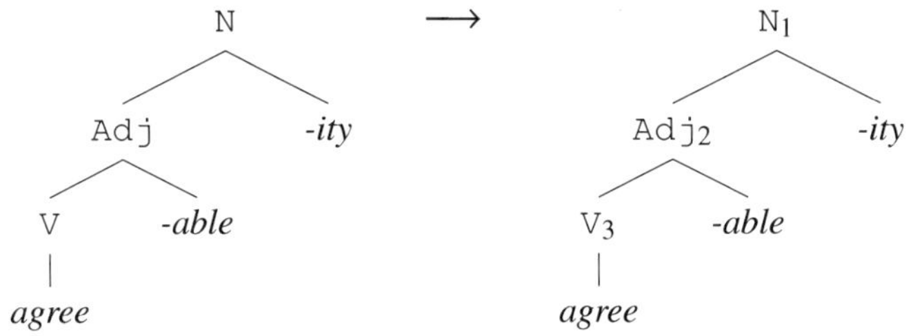

Figure 3.1

Node Addresses: The DOP PCFG reductions defined in Goodman (2003) require annotating training data nodes with unique node addresses like those shown on the right of the figure.

For simplicity, assume that all derivation trees in the set **D** are in Chomsky normal form (i.e., all nodes are either binary branching to two nonterminals, or are unary branching to a single terminal). Furthermore, assume that each node in the set of training derivations is labeled by an integer address to distinguish it from other nodes with the same label. This is shown in Figure 3.1, where each node has been labelled with an integer address so that, for example, $\mathtt{N}_1$ refers specifically to a particular node in a particular training data token.

		DOP1					ENDOP		
$1/a_j$	$\mathtt{A}_j$	$\longrightarrow$	$\mathtt{B}$	$\mathtt{C}$	$1/a_j$	$\mathtt{A}_j$	$\longrightarrow$	$\mathtt{B}$	$\mathtt{C}$
b_k/a_j	$\mathtt{A}_j$	$\longrightarrow$	$\mathtt{B}_k$	$\mathtt{C}$	b_k/a_j	$\mathtt{A}_j$	$\longrightarrow$	$\mathtt{B}_k$	$\mathtt{C}$
c_l/a_j	$\mathtt{A}_j$	$\longrightarrow$	$\mathtt{B}$	$\mathtt{C}_l$	c_l/a_j	$\mathtt{A}_j$	$\longrightarrow$	$\mathtt{B}$	$\mathtt{C}_l$
b_kc_l/a_j	$\mathtt{A}_j$	$\longrightarrow$	$\mathtt{B}_k$	$\mathtt{C}_l$	b_kc_l/a_j	$\mathtt{A}_j$	$\longrightarrow$	$\mathtt{B}_k$	$\mathtt{C}_l$
$1/a$	$\mathtt{A}$	$\longrightarrow$	$\mathtt{B}$	$\mathtt{C}$	$1/a\bar{a}$	$\mathtt{A}$	$\longrightarrow$	$\mathtt{B}$	$\mathtt{C}$
b_k/a	$\mathtt{A}$	$\longrightarrow$	$\mathtt{B}_k$	$\mathtt{C}$	$b_k/a\bar{a}$	$\mathtt{A}$	$\longrightarrow$	$\mathtt{B}_k$	$\mathtt{C}$
c_l/a	$\mathtt{A}$	$\longrightarrow$	$\mathtt{B}$	$\mathtt{C}_l$	$c_l/a\bar{a}$	$\mathtt{A}$	$\longrightarrow$	$\mathtt{B}$	$\mathtt{C}_l$
b_kc_l/a	$\mathtt{A}$	$\longrightarrow$	$\mathtt{B}_k$	$\mathtt{C}_l$	$b_kc_l/a\bar{a}$	$\mathtt{A}$	$\longrightarrow$	$\mathtt{B}_k$	$\mathtt{C}_l$

Figure 3.2

PCFG Reductions for DOP1 (left) and ENDOP (right): This figure shows the PCFG reductions given in Goodman (2003) for the DOP1 estimator (left) and ENDOP estimator (right).

Let a_j represent the number of subtrees occurring under node $\mathtt{A}_j$, let a be the total number of subtrees occurring under nodes labeled $\mathtt{A}$ in the whole corpus, that is, $a = \sum_j a_j$, and let $\bar{a}$ be the total number of nodes of category $\mathtt{A}$ in the training corpus. Consider a depth-one subtree, such as

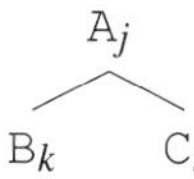

For each such depth-one subtree in the training corpus, add the rules shown in Figure 3.2 to the PCFG reduction. Goodman (2003) shows that the resulting PCFG defines exactly the same distribution as the `DOP1` estimator (left-hand reduction) and `ENDOP` estimator (right-hand reduction). For the results reported in this book, we use a generalized version of these reductions allowing non-binary input trees.

3.1.6 The Pitman-Yor Process

In the last chapter, we introduced the notion of stochastic memoization and defined the Pitman-Yor process (PYP) for use as a memoization distribution. The PYP is typically described using the metaphor of a restaurant. Imagine a restaurant with an infinite number of tables. The first customer enters the restaurant and sits at the first table. The $(N+1)$th customer enters the restaurant and sits at either an already occupied table or a new, unoccupied table, according to the following distribution:

$$\tau^{(N+1)}|\tau^{(1)},\cdots,\tau^{(N)},a,b \sim \sum_{i=1}^{K}\frac{y_i-a}{N+b}\delta_{[\tau^{(N+1)}=\tau_i]}+\frac{Ka+b}{N+b}\delta_{[\tau^{(N+1)}=\tau_{K+1}]}, \tag{3.12}$$

where $\tau^{(j)}$ refers to the table chosen by the jth customer, N is the total number of customers in the restaurant, K is the total number of occupied tables, indexed by $1 \le i \le K$; y_i is the number of customers seated at table i; δ_X is a δ-distribution which returns 1 if the proposition X is true and 0 otherwise, $0 \le a \le 1$ is the discount parameter of the model, and $b \ge -a$ is the concentration parameter of the model. To summarize Expression 3.12, customers sit at an already-occupied table with probability proportional to the number of individuals at that table (minus a constant a), or at a new table with probability proportional to $(aK+b)$.

Note that when a is equal to 1, every customer will sit at their own table. When a is equal to 0 the distribution reduces to the (single-parameter) Chinese restaurant process (Pitman, 2002). On average, a will be the limiting proportion of tables in the restaurant that have only a single customer, and the limiting probability of sitting at a new table, while the b parameter controls the rate of growth of new tables in relation to the total number of customers N (Teh, 2006).

Each table has a dish associated with it. Each dish v can be understood as a label on the table shared by all the customers at that table. When a customer sits at a new table $K+1$, a dish is sampled from another distribution μ—known as the base distribution of the Pitman-Yor process—and the sampled value v_{K+1} is shared by all future customers seated at that table.

Like the Dirichlet-multinomial distribution, the PYP defines a distribution over ways of partitioning N items (customers) into K partitions (tables), for all possible N and K. It differs from the Dirichlet-multinomial distribution in that it allows the number of partitions, K, to be unbounded. The probability of a particular partition of N customers over K tables is the product of the probabilities of the N choices made in seating those customers. It can easily be confirmed that the order in which elements are added to the partition components does not affect the probability of the final partition (i.e., the terms of the product can be rearranged in any order). Thus the distribution defined by a PYP is again exchangeable, like the other distributions described in this chapter, and the probability of a particular seating arrangement only depends on the counts of customers at various tables.

Given some vector of table counts $\vec{y}$ (i.e., partition component assignments), a closed-form expression for the probability of a partition can be given as follows. First, define the following generalization of the factorial function, which multiples m integers in increments of size s starting at x.

$$[x]_{m,s} = \begin{cases} 1 & m = 0 \\ x(x+s)\cdots(x+(m-1)s) & m > 0 \end{cases} \tag{3.13}$$

In particular, $[1]_{m,1} = m$. The probability of the partition given by the count vector $\vec{y}$ in a PYP is given by the following expression.

$$P(\vec{y}|a,b) = \frac{[b+a]_{K-1,a}}{[b+1]_{N-1,1}} \prod_{i=1}^{K} [1-a]_{y_i-1,1} \tag{3.14}$$

Like the Dirichlet-multinomial distribution, the Pitman-Yor process has a representation in terms of a hierarchical sampling scheme where we first draw a vector of probabilities and then sample from this vector of probabilities using a multinomial distribution. Abusing notation slightly, we will write $\mathrm{mem}\{G\} \sim \mathrm{PYP}(a,b,G)$ where $\mathrm{mem}\{G\}$ refers to the vector of probabilities. Unlike the vector of probabilities θ drawn from a Dirichlet distribution (i.e.,

$\theta \sim \textsc{dirichlet}(\vec{\pi})$), mem$\{G\}$ is a countably infinite probability vector and therefore must be enumerated lazily. We will not review the construction of this vector here, but see Sethuraman (1994) for the slightly simpler case when the probability vector is drawn from a Dirichlet process.

3.1.7 Adaptor Grammars

In the previous chapter, we showed how the adaptor grammars model of Johnson et al. (2007a) could be understood as a PYP-memoized version of the PCFG `unfold` procedure. The variant of the adaptor grammars model used in this book (i.e., Maximum a posteriori Adaptor Grammars, MAGs) extends the definition in the last chapter by adding Dirichlet priors to the rule weights in the CFG base system.

Formally, an adaptor grammar is a triple $\mathcal{A} = \langle \mathcal{G}, \{\vec{\pi}^{\mathtt{A}}\}_{\mathtt{A} \in V_{\mathcal{G}}}, \{\langle a^{\mathtt{A}}, b^{\mathtt{A}} \rangle\}_{\mathtt{A} \in V_{\mathcal{G}}} \rangle$ where $\mathcal{G}$ is a context-free grammar, $\{\vec{\pi}^{\mathtt{A}}\}_{\mathtt{A} \in V_{\mathcal{G}}}$ is a set of pseudo-count vectors for Dirichlet-multinomial distributions for each nonterminal $\mathtt{A}$, and $\{\langle a^{\mathtt{A}}, b^{\mathtt{A}} \rangle\}_{\mathtt{A} \in V_{\mathcal{G}}}$ is a set of hyperparameters for the Pitman-Yor processes associated with each nonterminal $\mathtt{A}$.

Adopting the hierarchical variants of the Dirichlet-multinomial and Pitman-Yor distributions, this model can be expressed by the following recursive stochastic equations.

$$G^{\mathtt{a}}_{\mathrm{AG}}(d) = \begin{cases} \displaystyle\sum_{r \in R_{\mathcal{G}} \mathtt{a} \rightarrow \mathtt{root}(\hat{d}_i), \cdots, \mathtt{root}(\hat{d}_k)} \theta_r \prod_{i=1}^{k} \mathrm{mem}\{G^{\mathtt{root}(\hat{d}_i)}_{\mathrm{AG}}\}(\hat{d}_i) & \mathtt{root}(d) = \mathtt{a} \in V_{\mathcal{G}} \\ 1 & \mathtt{root}(d) = \mathtt{a} \in T_{\mathcal{G}} \end{cases}$$

$$\vec{\theta}^{\mathtt{A}} \sim \textsc{dirichlet}(\vec{\pi}^{\mathtt{A}})$$

$$\mathrm{mem}\{G^{\mathtt{A}}_{\mathrm{AG}}\} \sim \textsc{pyp}(a^{\mathtt{A}}, b^{\mathtt{A}}, G^{\mathtt{A}}_{\mathrm{AG}})$$

Here, mem$\{G^{\mathtt{A}}_{\mathrm{AG}}\}$ represents memoization of the CFG recursion defined by $G^{\mathtt{A}}_{\mathrm{AG}}$. The recursion $G^{\mathtt{A}}_{\mathrm{AG}}$ always returns a fully expanded tree, and therefore adaptor grammars can only store and reuse complete derivation tree fragments.

Because all of the distributions used to construct the adaptor grammars model are exchangeable, so too is the overall model. Furthermore, the probability of a particular set of adaptor grammar analyses can also be given in terms of the counts associated with the individual Dirichlet-multinomial and Pitman-Yor distributions. For a corpus of derivation trees $\mathbf{D}$, let $\mathbf{X} = \{\vec{x}^{\mathtt{A}}\}$ be the set of count vectors of CFG rules used to construct the trees stored in the

PYP restaurants. Let $\mathbf{Y} = \{\vec{y}^{\mathtt{A}}\}$ be the set of count vectors of reused derivations stored on each table in each of the adaptor grammar's Pitman-Yor restaurants. We use the shorthand $\mathbf{A} = \langle \mathbf{X}, \mathbf{Y} \rangle$ for these two kinds of counts, taken together. The probability of the corpus in terms of counts of rule and subtree uses under the adaptor grammars model is given by the function **ag** below.

$$P(\mathbf{A}|\mathcal{A}) = \mathbf{ag}(\mathbf{A}; \mathcal{A})$$
$$= \prod_{\mathtt{A} \in V} \left[\frac{\prod_{i=1}^{K^{\mathtt{A}}} \Gamma(\pi_i^{\mathtt{A}} + x_i^{\mathtt{A}})}{\Gamma(\sum_{i=1}^{K^{\mathtt{A}}} \pi_i^{\mathtt{A}} + x_i^{\mathtt{A}})} \frac{\Gamma(\sum_{i=1}^{K^{\mathtt{A}}} \pi_i^{\mathtt{A}})}{\prod_{i=1}^{K^{\mathtt{A}}} \Gamma(\pi_i^{\mathtt{A}})} \frac{[b^{\mathtt{A}} + a^{\mathtt{A}}]_{K^{\mathtt{A}}-1, a^{\mathtt{A}}}}{[b^{\mathtt{A}} + 1]_{N^{\mathtt{A}}-1,1}} \prod_{i=1}^{K^{\mathtt{A}}} [1 - a^{\mathtt{A}}]_{y_i^{\mathtt{A}}-1,1} \right]$$

3.1.8 Fragment Grammars

As described in Section 2.3.6, fragment grammars generalize adaptor grammars by allowing the storage of partial derivations at tables in the PYP-memoizers. This is achieved by implementing a lazy version of the `unfold` procedure.

The actual implementation of the `FG` lazy recurrence used in this book differs slightly from the Church definition given in Section 2.3.6. In that section, the decision to recurse or halt at each nonterminal on the right-hand side of a rule was made by flipping a fair coin—a binomial distribution with parameter 0.5. In the actual implementation of fragment grammars used in this book, this decision is made by flipping a biased coin—a binomial distribution with parameter ν not necessarily equal to 0.5: $\text{BINOMIAL}(\nu)$. Furthermore, a beta-distributed prior is placed on the weight ν so that it too can be inferred during training. There is one such beta-binomial distribution for each nonterminal on the right-hand side of each rule in the underlying CFG. In other words, the system is able to learn a separate probability of recursing or delaying for each nonterminal on the right-hand side of each rule in the underlying CFG. Note that the beta distribution is the analog of the Dirichlet distribution when there are only two outcomes. Like the Dirichlet distribution it is parameterized by a vector of pseudo-counts, $\vec{\psi}$, which can be understood as imaginary prior observations of each of the two possible outcomes.

A fragment grammar is a tuple:

$$\mathcal{F} = \langle \mathcal{G}, \{\vec{\pi}^{\mathtt{A}}\}_{\mathtt{A} \in V_{\mathcal{G}}}, \{\langle a^{\mathtt{A}}, b^{\mathtt{A}} \rangle\}_{\mathtt{A} \in V_{\mathcal{G}}}, \{\vec{\psi}_{\mathtt{B}}\}_{\mathtt{B} \in \mathtt{rhs}(r \in R_{\mathcal{G}})} \rangle,$$

where $\mathcal{G}$ is a CFG, $\{\vec{\pi}^{\mathtt{A}}\}_{\mathtt{A}\in V_{\mathcal{G}}}$ are the vectors of Dirichlet-multinomial pseudo-counts for each nonterminal, $\{\langle a^{\mathtt{A}}, b^{\mathtt{A}}\rangle\}_{\mathtt{A}\in\mathcal{G}}$ is the set of Pitman-Yor hyper-parameters for each nonterminal, and $\{\vec{\psi}_{\mathtt{B}}\}_{\mathtt{B}\in\mathtt{rhs}(r\in R_{\mathcal{G}})}$ is the set of pseudo-counts for the beta-binomial distributions associated with the nonterminals on the right-hand side of each production rule in $\mathcal{G}$. In terms of the hierarchical variants of the Dirichlet-multinomial, beta-binomial, and Pitman-Yor distributions, fragment grammars can be described by the following recursive stochastic equations.

$$G^{\mathtt{a}}_{\mathtt{FG}}(d) = \begin{cases} \displaystyle\sum_{s\in\mathtt{prefix}(d)} \mathrm{mem}\{L^{\mathtt{a}}\}(s)\prod_{i=1}^{n} G^{\mathtt{root}(s'_i)}_{\mathtt{FG}}(s'_i) & \mathtt{root}(d) = \mathtt{a}\in V_{\mathcal{G}} \\ 1 & \mathtt{root}(d) = \mathtt{a}\in T_{\mathcal{G}} \end{cases}$$

$$L^{\mathtt{A}}(d) = \sum_{\mathtt{A}\to\mathtt{root}(\hat{d}_i),\cdots,\mathtt{root}(\hat{d}_k)} \theta_r \prod_{i=1}^{k}\left[\nu_{\mathtt{root}(\hat{d}_i)} G^{\mathtt{root}(\hat{d}_i)}_{\mathtt{FG}}(\hat{d}_i) + (1-\nu_{\mathtt{root}(\hat{d}_i)})1\right]$$

$$\vec{\theta}^{\mathtt{A}} \sim \mathrm{DIRICHLET}(\vec{\pi}^{\mathtt{A}})$$

$$\vec{\nu}_{\mathtt{B}} \sim \mathrm{BETA}(\vec{\psi}_{\mathtt{B}})$$

$$\mathrm{mem}\{L^{\mathtt{A}}\} \sim \mathrm{PYP}(a^{\mathtt{A}}, b^{\mathtt{A}}, L^{\mathtt{A}})$$

In these equations, $L^{\mathtt{A}}(d)$ implements the stochastically lazy recurrence described in the previous chapter. With probability $1-\nu_{\mathtt{B}}$, the recursion at nonterminal B on the right-hand side of some rule halts, and a partial tree is returned and stored by the memoizer. With probability $\nu_{\mathtt{B}}$, the recursion is continued and the stored tree will be further expanded. The probability of halting the recursion, $\vec{\nu}_{\mathtt{B}}$, is drawn hierarchically from a beta distribution; in practice, we employ the Pólya-urn representation of the process of sampling $\nu_{\mathtt{B}} \sim \mathrm{BETA}(\vec{\psi}_{\mathtt{B}})$, and the halting decisions. The distribution $\mathrm{mem}\{L^{\mathtt{A}}\}$ is the stochastically memoized version of this lazy recurrence, which stores partial derivation tree fragments. The top-level recursion is represented by $G^{\mathtt{A}}_{\mathtt{FG}}$, which implements a sum over all of the stored tree fragments in the memoizer, and forces computation at nonterminal leaves.

Once again, because all of the distributions involved in its definition are exchangeable, so too is the overall FG model, and the probability of a particular set of FG derivations can be computed directly from counts. Let **X** and **Y** be sets of count vectors for underlying rules and lexical items, as before. Let **Z** be the set of count vectors counting the number of times that each nonterminal on the right-hand side of a CFG rule was expanded in producing a lexical item, that is, the count vectors associated with the beta-binomial distributions described above. We will use the shorthand $\mathbf{F} = \langle \mathbf{X}, \mathbf{Y}, \mathbf{Z} \rangle$ for the set of all counts representing a corpus of fragment grammar derivations. The probability of a set of fragment grammar analyses is given by the function **fg** below.

$$
\begin{aligned}
P(\mathbf{F}|\mathcal{F}) &= \mathbf{fg}(\mathbf{F};\mathcal{F}) \\
&= \prod_{\mathtt{A} \in V} \left[\frac{\prod_{i=1}^{K^{\mathtt{A}}} \Gamma(\pi_i^{\mathtt{A}} + x_i^{\mathtt{A}})}{\Gamma(\sum_{i=1}^{K^{\mathtt{A}}} \pi_i^{\mathtt{A}} + x_i^{\mathtt{A}})} \frac{\Gamma(\sum_{i=1}^{K^{\mathtt{A}}} \pi_i^{\mathtt{A}})}{\prod_{i=1}^{K^{\mathtt{A}}} \Gamma(\pi_i^{\mathtt{A}})} \frac{[b^{\mathtt{A}} + a^{\mathtt{A}}]_{K^{\mathtt{A}}-1, a^{\mathtt{A}}}}{[b^{\mathtt{A}} + 1]_{N^{\mathtt{A}}-1,1}} \prod_{i=1}^{K^{\mathtt{A}}} [1 - a^{\mathtt{A}}]_{y_i^{\mathtt{A}}-1,1} \right. \\
&\qquad \left. \times \prod_{r \in R_{\mathcal{G}}^{\mathtt{A}}} \left(\prod_{\mathtt{B} \in \mathtt{rhs}(r)} \frac{\Gamma(\psi_{\mathtt{B}} + z_{\mathtt{B}}) \Gamma(\psi'_{\mathtt{B}} + (x_r - z_{\mathtt{B}}))}{\Gamma(\psi_{\mathtt{B}} + \psi'_{\mathtt{B}} + x_r)} \frac{\Gamma(\psi_{\mathtt{B}} + \psi'_{\mathtt{B}})}{\Gamma(\psi_{\mathtt{B}})\Gamma(\psi'_{\mathtt{B}})} \right) \right]
\end{aligned}
$$

3.2 Inference for Fragment Grammars

Inference for the fragment grammars model seeks to answer the question: what is the distribution over sets of stored tree fragments that best explains some observed data? For the simulations in this book, we will assume that we can observe the correct morphological derivation trees (parses), **D**, for a corpus of words. Given a fragment grammar $\mathcal{F}$ and these parse trees, we seek the distribution over sets of fragment grammar analyses for the parses **F**, which specifies the different ways that the parses can be divided into tree fragments. This distribution is given by Bayes' rule.

$$P(\mathbf{F}|\mathbf{D},\mathcal{F}) \propto P(\mathbf{D}|\mathcal{F},\mathbf{F})P(\mathbf{F}|\mathcal{F}) \tag{3.15}$$

In a following section, we describe a Metropolis-Hastings-based inference algorithm for fragment grammars that samples from this distribution and which we will use to search for an approximation to the maximum a posteriori fragment grammar given our corpus. This algorithm is a straightforward adaptation of the algorithms for Dirichlet-multinomial PCFG and adaptor grammar inference described in Johnson et al. (2007a,b). The reader should consult these sources for further details.

3.2.1 A Metropolis-Hastings Sampler

Markov Chain Monte Carlo (MCMC) is an approximate inference technique that samples from a hypothesis space by defining a Markov chain of local steps on that space. For the simulations reported in this book, a hypothesis consists of a set of all possible fragment grammar analyses **F**, given some input corpus.

MCMC works by defining a ***transition kernel*** $\kappa(\mathbf{F} \rightarrow \mathbf{F}')$, which defines a distribution over transitions from one hypothesis to another by making small changes to the current hypothesis. Using the Metropolis-Hasting technique, we define a transition kernel whose long-run behavior (i.e., whose ***stationary distribution***) converges to the desired target posterior distribution $P(\mathbf{F}|\mathbf{D},\mathcal{F})$.

The transition kernel exploits the fact that fragment grammars are exchangeable and, therefore, any word in the training corpus can be treated as if it were the most recently observed datapoint. In order to define $\kappa(\mathbf{F} \rightarrow \mathbf{F}')$, we first choose a target training data item $d^{(i)}$, and remove the tree fragments and other structures associated with its current analysis $f^{(i)}$ from our current state. We write $\mathbf{F}_{-f^{(i)}}$ for the state with $f^{(i)}$ removed. We then sample a new analysis $f^{(i)'}$, conditioned on $\mathbf{F}_{-f^{(i)}}$, $\mathcal{F}$ (the hyperparameters), and the parse tree $d^{(i)}$. This proposed analysis is then accepted or rejected according to the Metropolis-Hastings criterion (see below). Finally we add either $f^{(i)}$ or $f^{(i)'}$ (depending on whether the proposal was accepted or rejected) to the state of the system and the procedure is repeated for the next item in the training corpus. Thus, each local step in the Markov chain is defined by resampling an analysis for a particular observed form, conditioned on the current analyses for the rest of the forms in the corpus.

Ideally, it would be possible to exactly sample a new proposal analysis $f^{(i)'}$ from the posterior distribution $P(f^{(i)'}|d^{(i)},\mathbf{F}_{-f^{(i)}},\mathcal{F})$. However, for reasons discussed below, it is impossible to do this efficiently. Instead, we sample new analyses from a PCFG which approximates the distribution defined by the current state of the fragment grammar $\mathcal{G}'(\mathbf{F}_{-f^{(i)}},\mathcal{F})$.

$$f^{(i)'}|d^{(i)},\mathbf{F}_{-f^{(i)}},\mathcal{F} \sim P(\cdot|d^{(i)},\mathbf{F}_{-f^{(i)}},\mathcal{F}) \tag{3.16}$$

$$\approx P(\cdot|d^{(i)},\mathcal{G}'(\mathbf{F}_{-f^{(i)}},\mathcal{F})) \tag{3.17}$$

$$= \mathbf{pcfg}(\cdot;\mathcal{G}'(\mathbf{F}_{-f^{(i)}},\mathcal{F})) \tag{3.18}$$

Because these proposals are sampled from a distribution, $P(f^{(i)'}|d^{(i)},\mathcal{G}'(\mathbf{F}_{-f^{(i)}},\mathcal{F}))$, which only approximates the target distribution $P(f^{(i)'}|d^{(i)},\mathbf{F}_{-f^{(i)}},\mathcal{F})$, we correct to the true distribution by accepting or rejecting samples using the Metropolis-Hastings criterion; proposed analyses are accepted with probability $\mathcal{P}(f^{(i)},f^{(i)\prime})$, defined in Equation 3.19, and rejected otherwise.

$$\mathcal{P}(f^{(i)},f^{(i)'}) = min\left\{1, \left[\frac{\mathbf{fg}(\mathbf{F}_{-f^{(i)},+f^{(i)'}};\mathcal{F})}{\mathbf{fg}(\mathbf{F};\mathcal{F})} \times \frac{\mathbf{pcfg}(f^{(i)};\mathcal{G}'(\mathbf{F}_{-f^{(i)}},\mathcal{F}))}{\mathbf{pcfg}(f^{(i)'};\mathcal{G}'(\mathbf{F}_{-f^{(i)}},\mathcal{F}))}\right]\right\} \tag{3.19}$$

3.2.2 The Approximating PCFG

PCFGs make strong conditional independence assumptions. One important consequence of these conditional independence assumptions is that there are efficient algorithms available for solving the PCFG parsing problem. These algorithms rely on the fact that the distribution over parses for a string generated by a PCFG decomposes into a product over independent sums of substring parses.

Unfortunately, this decomposition does not hold in the case of the Pólya-urn representation for Dirichlet-multinomial PCFGs, Pitman-Yor adaptor grammars, or Pitman-Yor fragment grammars. For each of these formalisms, the probability of using a rule or tree fragment is not independent of other rule or fragment uses. For example, if an adaptor grammar tree fragment is used several times in the same analysis, then the probability of that fragment must jointly take into account all the uses in the analysis, and the distribution over parses no longer decomposes into a product over independent sums of sub-analyses. The lack of conditional independence guarantees for these models means that it is difficult to develop algorithms that efficiently compute the distribution over parses for a given string, and therefore it is not possible to exactly and efficiently sample analyses for these formalisms.

To address these problems, our fragment grammars inference algorithm samples proposal analyses by making use of a PCFG that closely approximates the true fragment grammars distribution but still preserves conditional independence assumptions and, therefore, can be efficiently parsed and sampled. Following Johnson et al. (2007a), we call this the *approximating PCFG*: $\mathcal{G}'(\mathbf{F}_{-f^{(i)}},\mathcal{F})$. The intuition behind this approximation is that a tree fragment like `(N (ADJ V` *-able*`)` *-ity*`)` corresponds to a context-free rule of the form `N` $\longrightarrow$ `V` *-able -ity*. We construct a context-free rule for each stored fragment in each fragment grammar Pitman-Yor restaurant, and assign it a probability proportional to the probability of the table. We also add context-free rules from the Dirichlet-multinomial PCFG base system, weighted by the new table probability for the restaurant corresponding to the rule's left-hand side. Formally, for a fragment grammar $\mathcal{F} = \langle \mathcal{G}, \{\vec{\pi}^{\mathtt{A}}\}_{\mathtt{A}\in V_\mathcal{G}}, \{\langle a^{\mathtt{A}}, b^{\mathtt{A}}\rangle\}_{\mathtt{A}\in V_\mathcal{G}}, \{\vec{\psi}_{\mathtt{B}}\}_{\mathtt{B}\in \mathtt{rhs}(r\in R_\mathcal{G})}\rangle$, we introduce a PCFG rule $\mathtt{A} \longrightarrow \gamma$ into $\mathcal{G}'(\mathbf{F}_{-f^{(i)}}, \mathcal{F})$ for each possible root $\mathtt{A}$ and each sequence of terminals and/or nonterminals γ found at the leaves of a stored derivation fragment or on the right-hand side of an underlying rule from $\mathcal{G}$. The probability of this rule is given as follows.

$$\theta_{\rho^{\mathtt{A}}_{\gamma}} = \sum_{\mathtt{v} \in \mathrm{mem}\{L^{\mathtt{A}}\} | \mathtt{yield}(v)=\gamma} \frac{y^{\mathtt{A}}_{\mathtt{v}} - a^{\mathtt{A}}}{N^{\mathtt{A}} + b^{\mathtt{A}}} + \sum_{r \in R^{\mathtt{A}}_{\mathcal{G}} | \mathtt{rhs}(r)=\gamma} \left[\frac{K^{\mathtt{A}} a^{\mathtt{A}} + b^{\mathtt{A}}}{N^{\mathtt{A}} + b^{\mathtt{A}}} \times \frac{x^{\mathtt{A}}_{r} + \pi^{\mathtt{A}}_{r}}{K^{\mathtt{A}} + \sum \vec{\pi}^{\mathtt{A}}} \right]$$

The first term represents the probability of all stored fragments with root category $\mathtt{A}$ with γ as their sequence of leaves. The second term represents the probability of all underlying CFG rules with $\mathtt{A}$ on their left-hand side that share γ as their right-hand side. The second term has two components. The first component represents the probability of sitting at a new table in the PYP restaurant associated with $\mathtt{A}$. The second component represents the probability of the rule $r \in R_{\mathcal{G}}$ in the underlying Dirichlet-multinomial base system.

Given this approximating grammar, we can efficiently compute the distribution over derivations given an expression. We propose a new fragment grammar analysis for a morphological parse $d^{(i)}$ according to the distribution $P(f^{(i)'} | d^{(i)}, \mathcal{G}'(\mathbf{F}_{-f^{(i)}}, \mathcal{F})$, following the steps below.

1. Compute the inside table over parse trees given $\mathcal{G}'(\mathbf{F}_{-f^{(i)}}, \mathcal{F}))$. This table includes the inside probability of every possible constituent in every possible parse of $d^{(i)}$. The inside table is computed with a version of the Cocke Younger Kasami algorithm (see, Manning and Schütze, 1999).
2. Sample a parse tree of approximating grammar rules using the inside table. Starting at the goal item, sample a parse for $d^{(i)}$ by recursively sampling from the distribution over back-pointers that results from normalizing inside scores. This algorithm is described in more detail in Goodman (1998) and Johnson et al. (2007b).
3. Sample a proposal fragment grammar analysis $f^{(i)'}$, using the approximating grammar parse. The rules used in the approximating grammar collapse across fragment grammar tree fragments and underlying CFG rules. A fragment grammar derivation is recovered by sampling conditionally on the collapsed derivation, undoing the marginalization that was performed when the probability of $\rho^{\mathtt{A}}_{\gamma}$ was calculated. This sampling is performed bottom-up along the structure of the approximating grammar parse tree. A new table

is created each time an approximating derivation node corresponds to an underlying CFG rule in the fragment grammar.

Once a proposal analysis has been sampled in this manner, the Metropolis-Hastings score is computed and the proposal is accepted or rejected accordingly.

3.3 Inference for Other Models

For the simulations in this book, Maximum A Posteriori (MAP) grammars were used for the Dirichlet-multinomial (`DMPCFG`) and Maximum a posteriori Adaptor Grammar (`MAG`) models. The MAP grammar refers to the most probable grammar in the posterior distribution over grammars, given the data. All of the training corpora used in this book provide labeled morphological parse trees for all inputs, and inference for all models is conditioned on these trees. One consequence of conditioning on these trees is that it is possible to directly compute a PCFG approximation to the MAP grammars for the `DMPCFG` and `MAG` models. For both `MAG` and `DMPCFG`, the posterior probability of a grammar is maximized when the number of tree fragments or context-free rules is minimized, that is, when probability is concentrated on the smallest number of shared fragments/rules across derivations. Therefore, we constructed an approximating PCFG for each of these grammars by simply counting the number of times each tree fragment or context-free rule appeared in the training corpus and introducing a rule for each in the approximating grammar. The resulting MAP grammars were used to compute all results reported later in this book.

Although MAP grammars were computed directly for the `MAG` and `DMPCFG` models, they were only approximated for the fragment grammars (`FG`) model by using the Metropolis-Hastings sampler described in the preceding section to perform stochastic search. To the degree that MAP approximations are the appropriate object to compute inference results, this should provide an advantage to the `DMPCFG` and `MAG` models over the `FG` model.

Predictions for the two DOP formalisms (Data-Orented Parsing 1, `DOP1`, and Equal-Node Estimator DOP, `ENDOP`) used in this book were computed using the PCFG reductions described above.

3.4 Computation of Conditional Probabilities

In succeeding chapters, we will make use of a large number of different conditional probabilities and expectations computed against the (approximating) PCFG grammars for each model. In this section, we briefly describe how these values were computed.

To marginalize over various positions in a morphological parse we implemented a generalized Cocke Younger Kasami parser that allowed wildcard values as string terminals. Using this parser, we could approximately compute the marginal probability of events such as "the set of all forms ending in *-ity*" by specifying a wildcard input of the form * *-ity*. Using such marginals, it was possible to compute conditional probabilities such as the probability of a particular base (e.g., *agreeable*), given a suffix (e.g., *-ity*; i.e., *P*(*agree -able* | *-ity*)). This conditional probability can be approximated as the ratio of the marginal score of *agreeability* divided by the marginal score of the wildcard input * *-ity*. The parser was further extended to allow for the computation of arbitrary expectations against the distribution over parses. This enabled the straightforward computation of quantities such as the entropy of the set of derivations, or the expected number of constituents for a particular word.

3.5 Practicalities of Inference for Fragment Grammars

Inference in the fragment grammars model is difficult. The space of possible analyses for a corpus of forms is large, and CYK parsing against large grammars is slow.[6] The sampler can also easily become trapped in modes of the posterior distribution. For these reasons, this section presents a number of practical techniques that we adopted for improving fragment grammar inference.

3.5.1 Initialization

A set of fragment grammar analyses for each expression in an input corpus must be initialized before the Metropolis-Hastings algorithm can be run. In

6. The Cocke Younger Kasami algorithm is hard against a $O(n^3|G|^2)$ bound, where n is the length of the input and $|G|$ is the size of the grammar. Approximating grammars frequently have tens of thousands of rules.

order to maximize the possibility of finding a good approximation to the MAP fragment grammar, we used the ***batch initialization*** procedure described in Johnson and Goldwater (2009). Under batch initialization, every node in the input corpus is assigned in parallel to its own table in a PYP restaurant, and all decisions about the distribution of other random variables are randomly sampled from the prior. This has the effect of starting the sampler at a random, low-probability point in posterior space. From this starting point, the sampler hill-climbs by merging Pitman-Yor tables that are labeled with the same tree fragments.

3.5.2 Type/Token Binning

A technique for mitigating the degree to which the sampler runs become trapped in posterior modes is ***type/token binning***. This is best illustrated with an example. The derivational morphology simulations reported in Chapter 7 were run over approximately 1.7 million word tokens. Resampling each of these tokens amounts to a very small step in posterior space. However, these 1.7 million word tokens are distributed over only approximately $20,000$ word types. One way of taking larger steps in posterior space is to resample many tokens corresponding to the same type simultaneously. In order to make this possible, the Metropolis-Hastings sampler was enhanced with the ability to take inputs in bins each of which corresponded to N tokens of some particular word type. The tokens in these bins were then resampled simultaneously by the sampler. This technique is related to the type-based MCMC inference scheme described in Liang et al. (2010).

Binning input tokens leads to a trade-off. The more aggressively one bins tokens into types, the larger the steps in posterior space, and the more work done per unit-time during sampling. However, all the tokens in a bin are treated by the sampler in an identical manner. Therefore, the more aggressive the binning, the greater the number of posterior states rendered inaccessible to the sampler. In particular, states where individual tokens within a bin are split between different analyses cannot be reached. In the limit, where all tokens of each type are in a single bin, many analyses become unreachable. To take advantage of binning while not eliminating too many possible states from the posterior, different Markov chains were run at a variety of bin sizes. The posterior state over all bin sizes with the highest posterior score was then used as our MAP approximation.

3.5.3 Conditioning on Constituent Information

As we mentioned above, in order to reduce the time complexity of parsing, our sampler conditions on input (gold-standard) morphological parses, rather than simple string forms. To condition on input parses, our CYK parsing algorithm was extended to allow what we call *span-wise conditioning*.[7] The parse chart was seeded with the set of spans present in the gold-standard tree. The parser was then only permitted to hypothesize spans that matched the gold-standard spans. By reducing the number of edges that could be entered into the parse chart, span-wise conditioning significantly improved parsing speeds.

Note that this technique does not exactly condition on full gold-standard parse trees. In particular, a rule in the approximating PCFG for a fragment grammar collapses across the internal structure of any fragments to which it corresponds. The approximating PCFG will represent a fragment such as (N (Adj V *-able*) *-ity*) with a rule like N $\longrightarrow$ V *-able -ity*, suppressing the internal Adj node. Under span-wise conditioning, the parser requires that an edge added to the parse chart be congruent with the gold-standard parse, but it does not require that every constituent in the gold standard parse be represented by an edge in the parse chart. This means, in particular, that it was possible for our algorithm to sample fragment grammar analyses where internal nodes in a parse tree, such as Adj in our example, were assigned different labels than in the gold standard.

3.5.4 Auxiliary Variable Slice Sampling

A final inference optimization to our Metropolis-Hastings algorithm was the use of *auxiliary variable slice sampling*. The main time bottleneck for inference is parsing and the main cause of high parsing times is the large size of approximating PCFGs. To mitigate this problem, we employed the following technique (Van Gael et al., 2008).[8] First, before removing the counts and resampling analysis f during the outer loop of the Metropolis-Hastings algorithm, we found the approximating PCFG rule with the lowest probability p used in that analysis. We then sampled $p' \sim \text{UNIFORM}(0,p)$. Before re-parsing f, we removed all rules from the approximating PCFG with probability less

7. We thank Mark Johnson for suggesting this technique.

8. We would like to thank Trevor Cohn for suggesting this technique.

than p'. This ensured that the Metropolis-Hastings reverse proposal probability for f was still computable, while, in some cases, significantly reducing the number of approximating PCFG rules used during parsing.

3.5.5 Hyperparameters

Hyperparameter values for all simulations in this book were identical wherever applicable.

1. $\vec{\pi}$: The Dirichlet-multinomial pseudo-counts $\vec{\pi}$ for CFG rules were set to 1 for all rules for the Dirichlet-multinomial PCFG, maximum a posteriori adaptor grammar, and fragment grammar models.
2. a: The Pitman-Yor a parameter was set to 0.5 for all restaurants in both the maximum a posteriori adaptor grammar, and fragment grammar models.
3. b: The Pitman-Yor b parameter was set to 100 for all restaurants in both the maximum a posteriori adaptor grammar, and fragment grammar models.
4. $\vec{\psi}$: The beta-binomial pseudo-counts $\vec{\psi}$, which govern the probability that a nonterminal on the right-hand side of a base rule is expanded or delayed, was set to 50 for all nonterminals on the right-hand side of all rules for the fragment grammar model.

The values for the Pitman-Yor process a and b parameters are relatively high, allowing for easy storage of novel structures. They were chosen as the result of informal experimentation prior to running the actual simulations reported in this book, and their values were not optimized for these datasets.

II EMPIRICAL APPLICATIONS

4 The English Past Tense: Abstraction and Competition

4.1 Introduction

In this chapter and the next, we turn to one of the most intensively studied phenomena in psycholinguistics: the English verbal system—in particular the English past tense and past participle. Because of the broad range of empirical facts which are known about this system, it provides a rich set of phenomena which can be used to test the models studied in this book. The English past tense system is especially suited to evaluating issues of productivity and reuse because it exhibits a sharp dichotomy in levels of productivity. The majority of past-tense forms make use of a highly productive regular rule, but there are a significant number of irregular forms that fall into a number of unproductive and semi-productive classes. This kind of system is not uncommon across the world's languages and an adequate theory of productivity and reuse must be able to account for such a pattern. Probabilistic methods in linguistics have sometimes come under criticism as approaches to certain phenomena because of their (putative) tendency to predict variability or gradience for phenomena humans treat more categorically. Thus, it is an important desideratum that our model be able to predict more sharply categorical phenomena, like the English past, despite its probabilistic underpinnings.

The literature surrounding the English past is vast, and we focus here on those aspects relevant to the modeling assumptions and empirical phenomena we address in the next chapter. On the empirical side, our evaluations in the next chapter focus on the violable pattern of *defaultness* of the regular rule, and *blocking* of the regular rule by irregular forms, and how this pattern can be predicted from the frequency distribution of words and morphemes in the input data. The first part of this chapter reviews these phenomena.

On the theoretical side, our formalization of the past-tense problem makes several modeling assumptions distinguishing it from majority of work in the literature. In particular, following the general theme of this book, we focus only on the problem of productivity: how to account for the pattern of competition between various generalizations exhibited by the English past system. We make use of the ability of fragment grammars to represent abstractions such as "attach *-ed* to any verb," but we do not model the process of discovering the phonological (or semantic) structure of the generalizations that characterize different past forms. This stands in contrast to most of the computational work in the literature (e.g., Albright and Hayes, 2003; Daugherty and Hare, 1993; Daugherty and Seidenberg, 1992; Derwing and Skousen, 1994; Eddington, 2000; Hare et al., 1995; Hoeffner, 1992, 1996; Ling and Marinov,

1993; MacWhinney, 1978; MacWhinney and Leinbach, 1991; Molnar, 2001; Mooney and Califf, 1995; O'Reilly, 1996; Pinker, 1984; Plunkett and Juola, 1999; Plunkett and Marchman, 1991, 1993; Rumelhart and McClelland, 1986; Westermann and Goebel, 1995; Yip and Sussman, 1996, 1997).

With this in mind, the second part of this chapter examines how productivity is handled by different theoretical accounts of the English past, organized around two high-level points. First, we emphasize that any theory of the past tense must provide mechanisms capable of representing abstract generalizations, a property shared by all of the formal models we study in this book. Second, we argue that merely providing the possibility of abstract generalizations is insufficient, and that a theory of the past tense must provide an account allowing for exceptions to generalizations and providing an explanation of how competition between various generalizations is resolved. In this section, we also emphasize how most accounts of the past tense have made use of a variant of the *elsewhere condition*, which states that when multiple generalizations are available, the most specific one should be preferred. In the next chapter, we will show how the elsewhere condition arises from the basic assumptions of the fragment grammars model and general principles of probabilistic inference, and, in contrast to most previous accounts, does not need to be stated as an independent postulate.

In the conclusion to this chapter, we mention several issues that have received attention in the past tense literature but which we have not addressed in our review. These include the learning of the phonological (and semantic) structure of past tense generalizations, the issue of whether the system is best accounted for using a single- or dual-mechanism approach, and the relationship between theoretical models of the past tense and biological evidence about how the system is represented in the brain. In the next chapter, after presenting modeling results, we will discuss our work with respect to these issues.

4.2 The English Verbal Paradigm

English verbs are obligatorily marked with inflections that carry four kinds of syntactic and semantic information:

1. **Tense**: The tense of the verb; for example, present (*walk*) or past (*walked*).
2. **Person**: The person of the subject of the verb; for example, first-person (*am*) or third-person (*is*)

3. **Number**: The number of the subject of the verb; for example singular (*is*) or plural (*are*)
4. **Aspect**: The completedness of the verb; for example, progressive (*walking*) or imperfect (*walks*).

Like many inflectional systems, the English verbal system is highly *syncretic*: many combinations of the features above can share a single marking. For example, aside from a few exceptional verbs—such as forms of *be*—person is only marked on third-person present singular verbs (e.g., *walks*). Similarly, the markings for perfective and past forms are the same for most verbs (+/d/ in *walked/have walked*), although a small number of irregulars distinguish these forms (e.g., *ate/eaten*). Linguists typically organize the set of verb forms that share a stem into a table, called a *paradigm*, which expresses the way in which combinations of features are marked.[1] Paradigms for several English verbs are shown in Figure 4.1.

This table shows six possible forms for seven English verbs. The first column gives the abbreviation for the verb form used in the tagset of the Penn TreeBank corpus (Marcus et al., 1999).[2] We will use these abbreviations throughout this and later chapters. Each column is labeled with its *stem* (e.g., GO). Note that stems refer to the abstract entities underlying the different forms of a verb in a paradigm and do not necessarily have any phonological content. For example, there is no phonological relationship between the present (*go*) and past (*went*) forms of GO.[3] English verbs fall into a number of distinct

1. We are using the term paradigm purely descriptively. The necessity for paradigms and their psychological reality are controversial topics in linguistics and psycholinguistics (see e.g., Bobalijk, 2008; Carstairs-McCarthy, 2001; Moscoso del Prado Martín et al., 2004; Stump, 2001b). In *word-and-paradigm* theories (Hockett, 1954), they are considered first-class theoretical objects, while in other theories, they are considered epiphenomenal (e.g., Kiparsky, 2005; Noyer and Harley, 1999).

2. We divide verbal paradigms into six cells. A more usual organization of English verbs collapses simple presents (VBP) and base forms (VB) (see, e.g., Carstairs-McCarthy, 2001). However, these six verb types are annotated in our training data (i.e., the Penn TreeBank Marcus et al., 1999), and nothing crucial hinges on the number of paradigm cells.

3. The use of the term "stem" is typical of so-called *item-and-arrangement* or *morpheme-based* approaches to morphology (see, Hockett, 1954). These are approaches to morphology which focus on morphemes and take concatenation as the most basic structure building operation. In so-called *word-and-paradigm* or *lexeme-based* approaches to morphology, the stem would be called a *lexeme*—the abstract node that represents whatever is shared between forms in a paradigm (but does not necessarily have any phonological content). Word-and-paradigm approaches stress the importance of operations that transform phonological structure between items in a paradigm (in various ways that can be more sophisticated than concatenation). Note that the term lexeme, as used in the word-and-paradigm literature, is not to be confused with the same term as used in the lexical processing literature. In the latter (see, e.g., Caramazza, 1997), the term typically (and confusingly)

	GO	EAT	GIVE	SING	RING	WALK	USE
VBD (*past tense*)	*went*	*ate*	*gave*	*sang*	*rang*	*walked*	*used*
VBZ (*3rd singular present*)	*goes*	*eats*	*gives*	*sings*	*rings*	*walks*	*uses*
VBN (*perfective* or *passive*)	*gone*	*eaten*	*given*	*sung*	*rung*	*walked*	*used*
VBG (*progressive*)	*going*	*eating*	*giving*	*singing*	*ringing*	*walking*	*using*
VBP (*simple present*)	*go*	*eat*	*give*	*sing*	*ring*	*walk*	*use*
VB (*base form*)	*go*	*eat*	*give*	*sing*	*ring*	*walk*	*use*

Figure 4.1

English Verbal Paradigm: This table shows the six forms of the English verbal paradigm, with examples for a number of verbs. Note that the labels on the left (i.e., VBD, VBZ, VBN, VBG, VBP, and VB) are the tags for these verbs types used in the the Penn TreeBank (Marcus et al., 1999).

inflectional classes. An inflectional class is a set of verbs that are marked in the different cells of a paradigm in the same way. Two classes visible in Figure 4.1 are shown in Table 4.1.

	Class I		Class II	
VBD	+/d/ (+/t/)	(*walk* → *walked*)	/ɪ/ → /æ/	(*sing* → *sang*)
VBZ	+/z/ (+/s/)	(*walk* → *walks*)	+/z/	(*sing* → *sings*)
VBN	+/d/	(*walk* → *walked*)	/ɪ/ → /ʌ/	(*sing* → *sung*)
VBG	+/ɪŋ/	(*walk* → *walking*)	+/ɪŋ/	(*sing* → *singing*)
VBP	+∅	(*walk* → *walk*)	+∅	(*sing* → *sing*)
VB	+∅	(*walk* → *walk*)	+∅	(*sing* → *sing*)

Table 4.1

Inflectional Classes: Two inflectional classes of English verbs. Class II contains verbs such as *sing* and *ring* while Class I contains the regulars. Note that the 3rd present marker +/z/ and past marker +/d/ are also attested as two regular, phonologically-conditioned voiceless allophones +/s/ +/t/.

Each cell of the table in Figure 4.1 shows a particular rule that is applied to the stem in that class to compute the surface form that appears in the cell. For example, Class I is the class of regular verbs, which simply concatenate /d/ to form their past participles (VBN). In contrast, the past participles (VBN) of Class II verbs (e.g., *sing*, *ring*) are formed by taking the stem vowel /ɪ/ and transforming it into /ʌ/.

The English verbal paradigm is of interest for the present study because of the great variability in the levels of productivity of the rules in the various inflectional classes. The progressive marker *-ing* that forms verbs in the VBG cell is completely productive and regular; there are no verbs that mark this cell in a different way. A few cells in the paradigm, such as the third-person singular present (VBZ), have a tiny number of exceptional forms (e.g., *have*/*has*, *say*/*says*), but are otherwise regular. The most interesting cases, however, are the past (VBD) and past participle (VBN) cells, which display high variability in both regularity and productivity. We discuss these in the next section.

means something nearly the opposite of what it does in word-and-paradigm approaches: a node which represents only the phonological structure of words (i.e., a node shared between homophones). Our use of the term stem does not indicate a commitment to the item-and-arrangement view of morphology. On the contrary, the modeling framework here is consistent with a variety of formal approaches. The particular formalization of the English past tense that we adopt in this book has aspects of both item-and-arrangement and word-and-paradigm approaches.

4.3 Empirical Review of the English Past

The vast majority of English verbs, the regulars, mark their past tense and past participle via concatenation of the default suffix *-ed* (+/d/). However, about 180 verbs, the irregulars, form their past tense and past participles in other ways. The existence of an irregular form usually blocks the possibility of using the regular form. For example, because of the existence of *went*, *goed* is considered infelicitous or ungrammatical by English speakers. There are, however, a small number of stems that seem to have both irregular and regular past forms. In some cases, the two forms have different meanings (e.g., *fit* in *The suit fit well* and *The tailor fitted the suit*), but in a small number of other cases, speakers seem to accept two variants of a stem with identical meaning (e.g., *dreamed/dreamt*, *leaped/leapt*, *dived/dove*). These latter verbs are known as **doublets**.

The following sections review the empirical literature on the English past tense, highlighting differences between regular and irregular forms in five areas: phonological regularity, productivity, frequency, processing, and development.

4.3.1 Phonological Regularity

The majority of English verbs form their past tense and past participle using a completely regular phonological rule concatenating the phoneme /d/ to the end of the stem. This morpheme appears in three regular, phonologically conditioned allomorphs:/d/, /t/ (after voiceless consonants), and /əd/ (after coronal stops). Irregular verbs, by contrast, vary widely in the degree of relatedness between their stems and past forms. For a small number of verbs, the past and present forms are phonologically unrelated, that is, they are **suppletive** (*go/went*). However, most irregular verbs exhibit one of the three following types of **semi-regularity** (see, Marcus et al., 1992; Pinker, 1999; Pinker and Prince, 1988).

The first type of semi-regularity comes from the fact that irregulars can be organized into a number of classes of verbs with systematic relationships between their present and past forms. For example, there are a number of irregulars that form their past participles according to the following rule: /ɪ/ → /ʌ/ (e.g., *sing/sung*). Another class of irregulars, the weak verbs, are identical in stem and past forms (e.g., *cut*, *spread*). A third class of irregulars lax their stem vowels and append a /t/ to form the past (e.g., *lose/lost*, *sleep/slept*).

The second kind of semi-regularity comes from the fact that stems in each irregular inflectional class often share phonological structure with one another. For example, most of the verbs that alternate between /ɪ/ and /ʌ/ (e.g., *fling/flung*) or /ɪ/ and /æ/ (e.g., *sing/sang*) end in the velar nasal /ŋ/. Similarly, all members of the class of weak verbs end in /t/ or /d/. The structure shared between stems, in many cases, constitutes a **family resemblance category** (Pinker and Prince, 1996). In a family resemblance category, there are a number of conditions predicting membership in a class, and stems meeting more of the conditions are more likely to be in the class. However, no one condition or set of conditions is both necessary and sufficient for predicting membership. For example, although verbs alternating between present /ɪ/ and past /ʌ/ (e.g., *fling/flung*) tend to end in the velar nasal consonant /ŋ/, the class also contains forms ending in non-nasal velars (*dig/dug*) as well as forms ending in non-velar nasals (*win/won*).

The third kind of semi-regularity exhibited by irregulars is that they tend to share structure with other past forms, including regulars. For example, the class of irregulars that laxes the stem vowel and appends a /t/ (e.g., *sleep/slept*), the weak verbs, and all regulars share a final coronal stop (i.e., /t/ or /d/).

4.3.2 Productivity

For the purposes of this book, the most important difference between regular and irregular verbs is their great disparity in productivity. A large number of studies have shown that the regular rule generalizes to novel stems in both production and judgment tasks (Albright and Hayes, 2003; Ambridge, 2010; Berko, 1958; Bybee and Moder, 1983; Gordon and Miozzo, 2008; Prasada and Pinker, 1993; Ramscar, 2002).[4] The regular rule can even be applied to forms that are phonotactically ill-formed for English (e.g., to a stem ending in the unvoiced velar fricative /x/, as in *The composer out-Bached Bach*, pronounced with the German pronunciation of *ch*). In fact, because the regular past ending can be attached to any verb stem, regular past-tense forms are less phonotactically regular, on average, than irregular past-tense forms (Burzio, 2000; Pinker

4. Note that these studies—with the exception of Berko (1958)—are concerned with how differing levels of phonological or semantic similarity between novel forms and existing forms influences generalization. However, all of these studies provide evidence for robust generalization of the regular rule.

and Prince, 1988). The regular rule also applies to stems that are formed by other morphological processes. For example, it applies when a verb has been created from a nominal form, even if the nominal form itself is derived from an irregular verb—for example, *fly* ⟶ *fly ball* ⟶ *fly out* ⟶ *flied out*, not *flew out* (Huang and Pinker, 2010; Kim et al., 1994, 1991). The regular rule is sometimes also over-generalized by both adults and children, resulting in over-regularization errors such as *breaked* (Marcus et al., 1992; Stemberger, 1983). The regular rule applies to newly coined verbs, and therefore new past forms entering the language typically enter as regulars. For these reasons, the regular rule is often described as a default, applying when no other form is available. Marcus et al. (1995) give a list of twenty-one cases where default generalizations might be expected to apply and where the English regular past inflection is used by speakers.

Irregulars, in general, are far less productive. Nevertheless, rating and novel stem, wug-test studies have shown that some classes of irregular verbs can be generalized under limited circumstances (Albright and Hayes, 2003; Ambridge, 2010; Bybee and Moder, 1983; Bybee and Slobin, 1982; Prasada and Pinker, 1993). These studies found that irregular classes containing greater numbers of stems and exhibiting more of the semi-regularity discussed in the last section tend to generalize more easily. Critically, the applicability of an irregular inflectional process to novel stems is highly sensitive to the phonological structure of the stem, and, in all cases, the regular past form is acceptable.

Although the the applicability of the regular rule is much less sensitive to the phonological makeup of the stem (Prasada and Pinker, 1993), it does show some sensitivity. In judgment tasks, participants have been shown to prefer novel past-tense forms that apply the regular rule to stems which are more canonically or prototypically regular (Albright and Hayes, 2003; Ambridge, 2010). However, the magnitude of this effect is much smaller than in the case of irregulars and it shows a different developmental trajectory (Ambridge, 2010). Even when a stem is not prototypically regular, the regular rule can always be applied to it—even if the output is phonotactically ill-formed. Therefore, although the regular rule shows some sensitivity to stem phonology, it is important to emphasize that it also systematically applies in rare boundary cases (e.g., phonotactically illicit stems, words formed by onomatopoeia or category conversion, etc.). Therefore, a model of the past tense must be able to represent a fully abstract, general regular rule (i.e.,"attach /d/ to any verb") that will apply when all else fails. It is not crucial, however, that this rule be the only way—or even most frequent way—of producing regular forms. We return to this point below. [5]

Historically, a small number of formerly regular verbs have become irregular (e.g., *wear*, *spit*, and *dig*, Bybee, 1985). By contrast, regularization has affected thousands of verbs (see, e.g., Lieberman et al., 2007). Accidental ***overirregularization*** (the misapplication of an irregular inflection to a regular stem, e.g., *bite/bote*) in speech occurs as well, but it appears to be a highly marginal phenomenon (Bybee, 1985; Xu and Pinker, 1995). In summary, while irregular inflectional classes show some generalizability, their generalizability is far less than the regular rule, and is highly constrained by the structure of the stem.

4.3.3 Frequency

Like most structures in natural language, English verb forms follow a frequency distribution with a relatively small number of highly-frequent forms and a large number of low-frequency forms (Yang, 2009). Such distributions are sometimes called ***large-number of rare event distributions*** (Baayen, 2001; Khmaladze, 1987), and are often characterized mathematically by Zipf's law or a more general power law (Baayen, 2001; Goldwater et al., 2011; Yang, 2009; Zipf, 1935). The large number of low-frequency verb forms is a direct consequence of productivity. Highly productive inflections, such as the past (+/d/) suffix, frequently give rise to new words. Because of this, the distributions associated with these suffixes have large numbers of low-frequency words. By contrast, less productive, irregular inflectional processes such as /ɪ/ → /ʌ/ (e.g., *sing/sung*) are primarily found amongst highly frequent verbs, and the most frequent past forms in English are irregular.

5. Another debate has concerned whether generalization of regular and irregular past inflections depends on semantic similarity with existing words. Some authors have claimed that the morphological effects found in studies such as Kim et al. (1991) can be accounted for in terms of semantic relatedness (Daugherty et al., 1993; Ramscar, 2002). However, the evidence seems to suggest that if there is a semantic effect, it is additional to a morphological one (Bandi-Rao and Murphy, 2007; Gordon and Miozzo, 2008; Huang and Pinker, 2010). In any case, similar observations to those about the effect of phonological similarity on regular rule generalization apply here. The concern of authors such as Daugherty et al. (1993) and Ramscar (2002) is to argue against morphological representations. However, an account that relies on morphological representations does not need to insist that there be no effect of semantics. In fact, it may be stipulative to do so, especially since it is well-known that semantic features affect other aspects of verb generalization, such as the alternation classes into which verbs enter (Pinker, 1989). The crux of the issue is not that semantics never has an effect, but rather that there exist some cases where the generalization is based on morphological structure.

Frequency is also predictive of (over)regularization during language development and use. Lower-frequency irregulars are more often overregularized during development (Marcus et al., 1992). Higher-frequency regulars and irregulars are more resistant to speech errors in adults (Bybee and Slobin, 1982; Stemberger and MacWhinney, 1986). Furthermore, lower-frequency forms are more susceptible to regularization over language change (Bybee, 1985; Lieberman et al., 2007).

4.3.4 Processing

Regulars and irregulars also pattern differently in a number of experimental paradigms that tap into the cognitive processes involved in producing and comprehending verb forms. First, priming studies have shown stem priming effects are, in general, stronger for regulars than irregulars, although there are sometimes weak priming effects for irregulars (Fowler et al., 1985; Kempley and Morton, 1982; Marslen-Wilson and Tyler, 1997; Münte et al., 1999; Napps, 1989; Stanners et al., 1979a). This fact is generally interpreted to mean that, while stems and past forms for irregulars tend to have independent lexical representations, regular stems are more often composed productively with suffixes. Moreover, priming effects for partially regular verbs (e.g., *sleep/slept*, which includes an allomorph of the regular suffix in addition to a vowel change) are stronger than for fully irregular verbs (e.g., *take/took*; Kielar et al., 2008).

Production tasks have repeatedly demonstrated that the effects of frequency are different for regular and irregular past forms. In these experiments, subjects are presented with a stem and asked to say its past tense form as quickly as possible. High-frequency, irregular, past-tense forms are named faster than low-frequency, irregular, past-tense forms. For regulars, there is either no effect of frequency, or, sometimes, it has been claimed there is an *anti-frequency* effect where low-frequency regulars are produced faster than high-frequency regulars (Beck, 1997; Prasada et al., 1990; Seidenberg and Bruck, 1990; Shenkman, 1994).[6] The difference in frequency effects is taken as evidence that irregular forms are stored independently and can, therefore, accumulate strength in memory, while regular forms are composed.

6. Prasada et al. (1990) is described in Pinker (1991) and Pinker (1999), and Seidenberg and Bruck (1990) is described in Seidenberg and Hoeffner (1998) and Pinker (1999).

This pattern of frequency effects also arises when participants are asked to judge the naturalness of past tense forms. Ullman (1993, 1999) regressed subjects' judgments of naturalness for past-tense forms against their frequency, controlling for judgments of naturalness for the bare stem forms. He found that while irregulars showed a positive correlation with frequency, regulars did not. Instead, the naturalness ratings for the regular past-tense forms were strongly predicted by the naturalness ratings for the stem, suggesting regulars are constructed via a compositional rule.

Finally, there is also evidence that high-frequency, regular, past forms can be stored in memory as wholes. Alegre and Gordon (1999) and Gordon and Alegre (1999) have demonstrated frequency effects for high frequency regulars using a lexical decision paradigm. This has been widely interpreted as showing that even fully regular forms can be stored if they are sufficiently frequent. Alegre and Gordon argue that the threshold for storage is about 6 per million, however, this has been disputed by Baayen et al. (2007), who argue for much more pervasive storage (although they do not provide evidence for the English past tense). Frequency-effect-based evidence for storage of fully regular inflected forms has also been demonstrated for a wide variety of other inflectional systems, including English nominal number marking (Sereno and Jongman, 1997), Italian nominal number marking (Baayen et al., 1997a), and Dutch and Finnish nominal (case and number) marking (Baayen et al., 1997b; Bertram et al., 1999).

4.3.5 Developmental Trajectory

A final way in which regular and irregular forms differ is in their developmental trajectory. One longstanding observation is that children sometimes overregularize irregular forms, producing forms such as *goed* in place of *went* (e.g., Chamberlain, 1906; Ervin, 1964; Ervin and Miller, 1963). An issue that has been the focus of much attention is whether overregularization follows a *U-shaped* developmental trajectory. Under a U-shaped developmental trajectory the first overregularization is preceded by a period of *early correct performance* on irregulars, and followed by eventual recovery to correct usage. The interest in U-shaped development lies in the way that it seems to support a view whereby children's grammatical representations reorganize when the regular rule is discovered. We will discuss the historical debate surrounding this issue in more detail below; here, we focus on the empirical evidence.

Although there were earlier studies of child language examining questions of overregularization and U-shaped development (e.g., Cazden, 1968; Kuczaj, 1977), the most comprehensive study of the phenomenon, and the only one to provide strong quantitative evidence for U-shaped development in the past tense, is the corpus study of child speech presented by Marcus et al. (1992).

Marcus et al. (1992) showed that rates of overregularization vary greatly between individual children and individual verb forms. At least one child in the corpus (Abe) overregularized at high rates throughout the sample, whereas some other children overregularized at much lower rates (e.g., Adam). There is also a large amount of variation between verbs, with some verbs nearly always overregularizing, some verbs nearly never overregularizing, some verbs showing very clear U-shaped development patterns, and some verbs oscillating chaotically between correct performance and overregularization (Marcus et al., 1992; pp. 54–55).

Marcus et al. (1992) also discussed several global (i.e., verb-general) properties of overregularization. First, they pointed out that overall overregularization rates are fairly low. In their sample, the overall median rate across children was only 2.5%. This result has been challenged, for instance by Maratsos (2000) and Yang (2002), who claimed that rates of overregularization are higher, nevertheless even these authors conceded that rates are relatively low (around 10%). A reanalysis of the Marcus et al. (1992) data by Hoeffner (1996) also supported this general conclusion.

Second, Marcus et al. (1992) found that, in the age range covered by their samples, there is no general evidence for robust recovery from overregularization errors, although one child (Abe), who had the highest rates of overregularization, did show evidence of recovery. Although recovery must occur eventually, the process appears to be gradual.

A more controversial claim made by Marcus et al. (1992) is that there is a global period of early correct performance before the onset of overregularization. In the Marcus et al. (1992) sample, three of the children showed a statistically significant early correct period. An aggregate test over all children was also significant. The controversy surrounding global early correct performance stems from the inability of some theoretical models (e.g., Hoeffner, 1996; MacWhinney and Leinbach, 1991; Plunkett and Juola, 1999; Plunkett and Marchman, 1993, 1996) to exhibit this effect without recourse to questionable assumptions about the input (see below). Some authors have argued that the empirical data is better characterized by so-called *micro-U-shaped* learning—the U-shaped learning of some verbs, but not others (e.g., Hoeffner,

1996; Plunkett and Marchman, 1993). While Marcus et al. (1992) provided evidence for global early correct learning, their data also supports micro-U-shaped learning, demonstrating a great deal of variability between verbs and between children. The reanalysis presented in Hoeffner (1996) also demonstrated high variability in early correct performance between verbs. Nevertheless, all authors agree that there is a clear period of early correct performance at least for some verbs, and that there is a clear period of global early correct performance at least for some children. Therefore, any model of past-tense learning must predict some degree of U-shaped development.

Another controversial claim made by Marcus et al. (1992) is that the onset of overregularization coincides with the emergence of reliable, obligatory past-tense marking of regular forms. English-speaking children go through a period during which they sometimes leave finite verb forms unmarked for tense (e.g., *Mary walk home*, in past context). Marcus et al. (1992) argued that the onset of the first overregularization is correlated with the emergence of reliable obligatory tense marking on regular forms. They performed two analyses to support this claim. First, they showed that rates of reliable regular marking are significantly lower than 50% before the first overregularization, but significantly higher after the first overregularization. Second, they showed that rates of reliable obligatory marking are higher after the first overregularization than before. Hoeffner (1996) criticized these analyses on the grounds that, because reliable marking is increasing throughout the period, the choice of the first overregularization as a dividing point is somewhat arbitrary. There are a large number of dividing points that could lead to the same results. Furthermore, he showed that obligatory regular marking is better predicted by age than the assumption of the emergence of an all-or-nothing regular rule. However, Hoeffner's analysis supports the more general observation that overregularization increases during the period when reliable regular marking is increasing. We discuss the theoretical significance of this debate in more detail below.

Another important (uncontested) finding of Marcus et al. (1992) is that overregularization is not predicted by sudden discontinuous increases in regular forms in the input to the learner, nor is it predicted by sudden increases in the use of regular forms by children. There is, of course, a logically necessary increase in the proportion of vocabulary that is regular. Only about 180 verb forms are irregular; therefore, as the child learns more verbs, the proportion that are irregular must decrease over time. However, there are no sudden jumps in the number of regulars corresponding to the onset of regularization.

Finally, Marcus et al. (1992) discussed several factors that protect individual verbs from overregularization. First, they found that higher frequency irregulars exhibit less overregularization. Second, they found that verbs from robustly attested inflectional classes (i.e., irregular verbs that are phonologically similar to other high-frequency irregulars) are protected from overregularization. However, similarity between an irregular stem and regular forms does not cause greater rates of overregularization. These results are consistent with our remarks on the differential sensitivity of regular and irregular inflectional processes to stem phonology. The applicability of irregular inflections depends on the phonology of the stem to a much greater degree than the applicability of the regular inflection. Marcus et al. (1992) also found that the phonological transparency of the stem-past mapping (i.e., the amount of change or complexity inherent in the irregular inflectional class) does not predict the rate of overregularization. Finally, Xu and Pinker (1995) showed that developmental overirregularization—the misapplication of irregular inflection to regular stems (e.g., *bite/bote*)—is very rare.

4.4 Theories of the Past Tense

In this section we review previous theories of the past tense, with a special focus on computational models. In preparation for our presentation of modeling assumptions and simulation results in the next chapter, our review is organized around two high-level points.

First, much of the historical discussion of the past tense occurred in the context of the debate between connectionist and rule-based approches to morphology. Although many issues were raised in this discussion, we will focus on one: the necessity for mechanisms that support abstract generalization. In particular, theories of the past tense must provide the means to represent an abstract generalization like "attach +/d/ to any verb." We will show how traditional linguistic models, dual-mechanism accounts, more recent probabilistic rule-based models, rich analogical models, and some connectionist models allow for such generalizations. However, the classical feedforward connectionist networks that sparked the original past-tense debate do not (e.g., Rumelhart and McClelland, 1986). The failure of these models to adequately account for the empirical data, especially with regard to the regular rule, is a direct result of not providing the means for this abstraction.

The second part of our discussion will emphasize that merely providing the means to represent abstract generalizations is not enough. An empirically adequate theory must also provide a mechanism for resolving competition between generalizations at differing levels of abstraction. Furthermore, phenomena such as overregularization, overirregularization, and the differential generalizability of past-tense inflectional classes indicate that the mechanism must be quantitative. An important principle of competition resolution which arises in this discussion is the elsewhere condition. This principle states that when multiple generalizations are available, the most specific one should be preferred. We will show how many different approaches have advocated variants of this principle. In the next chapter, we will demonstrate how a probabilistic version of the principle follows as a natural consequence of the modeling assumptions advocated in this book.

4.4.1 The Necessity for Abstract Representations

In an influential paper, Rumelhart and McClelland (1986) presented a connectionist model of the acquisition of the English past tense, which, they claimed, was able to account for past tense acquisition in a way that obviated the need for traditional linguistic rules.

Although the Rumelhart and McClelland (1986) model and similar connectionist approaches have been critiqued on many grounds (see, e.g., Lachter and Bever, 1988; Marcus, 2001; Marcus et al., 1992; Pinker, 1999; Pinker and Prince, 1988; Sproat, 1992), we will focus on one specific aspect of these critiques: an adequate theory the past tense must be capable of representing abstract generalizations. This fact is one of the motivating factors behind the first of the psychological assumptions behind our model: the adoption of a discrete, hierarchical, and recursive underlying system of computation, which is capable of representing abstract generalizations. All of the models we examine in later chapters share this property.

4.4.1.1 Rumelhart and McClelland (1986) and Related Connectionist Theories

To understand the difficulties faced by Rumelhart and McClelland (1986) and related theories, we must clarify the aspects of linguistic rules that are absent in these models. Traditional linguistic rules can be understood as functions that accept arguments from a certain domain and map them to outputs according to a specified algorithm.[7] A fundamental feature of the types of rules used in linguistic theories is that their input domains and mapping algorithms are are often specified in terms of abstract features or categories. For example, traditional treatments of English verbal inflection have frequently made use of a very general rule that accepts any verb (stem) and concatenates the regular ending +/d/ to the stem.

Using abstract categories in rules (such as the categories VERB or STEM) is critical because it allows a rule to treat an entire class of inputs uniformly, ignoring any structure (e.g., the phonology or semantics of the stem) irrelevant for the operation it performs. Importantly, the kinds of abstract categories or features used in traditional rules are not usually directly observable from any particular input. A stem is a stem by merit of its relation to the rest of the system of English grammar; this fact cannot be deduced with certainty merely by examining the phonology of a single verb form.

The Rumelhart and McClelland model also provided a mapping from inputs to outputs. However, it defined this mapping over a very specific class of representations that did not make use of abstract categories such as VERB or STEM. The Rumelhart and McClelland model is a simple *feedforward network*—also called sometimes called a *multilayer perceptron*.[8] Such networks operate by strengthening connections between a distributed representation of input forms and a distributed representation of output forms. In the case of the Rumelhart and McClelland model, the input and output representations were sets of trigrams of phonological features. Thus any generalization possible in their model could be described in terms of associations between directly observable, phonological features of the input and output. Because this representational scheme did not utilize abstract categories to distinguish classes of inputs or outputs (e.g., VERB or NOUN) or subparts of words (e.g., STEM or SUFFIX), but was able to learn some aspects of English past marking (see below), Rumelhart and McClelland claimed that their modeling results showed that there was no need for such categories to account for linguistic generalizations.

However, there are a number of problems with their argument. For their model, every correspondence between a stem and its past form correlates some phonological subsequences in the input with some phonological subsequences

7. A particularly clear and general definition of (probabilistic) rules is provided by Skousen (1992).

8. See Marcus (2001) for a discussion of this class of network architectures from a psychological perspective. For a mathematical overview, see Bishop (1995).

in the output. Because irregular past tense mappings such as /ɪ/ → /ʌ/ (e.g., *fling/flung*) often reside in (relatively) coherently defined parts of phonological space, such as before /ŋ/ (i.e., *-ng*), the model often successfully acquired these irregular inflection patterns. However, the regular rule is a default—it does not apply only in some specific part of phonological space, but rather can apply to any verb form that is not irregular (and in the case of overregularization to some irregular forms). However, each particular instance of a regular stem-past mapping, like *stop/stopped*, provides evidence of some phonological context in which the regular rule can apply (e.g., in the case of *stop/stopped*, after /p/). Since the Rumelhart and McClelland model tracked all such correlations, it was only able to acquire a fully general regular rule when there was sufficient variability in the phonological structure of regulars to "wash out" spurious generalization (e.g., +/d/ after /p/). This fact led to a number of systematic problems with the model's ability to generalize.

First, because there are accidental phonological correspondences between stem and past forms that occur with some frequency in the input corpus, the model sometimes hypothesized bizarre mappings called **blends** that do not resemble any structure that a native speaker would recognize (Pinker and Prince, 1988; give the example: *mail/membled*). Second, since the model computed past-tense mappings only on the basis of the phonology of the stem, it often had trouble generalizing the regular rule to stems that are phonologically very dissimilar to existing regulars (Prasada and Pinker, 1993; give the example: *ploamph*). As we discussed earlier, in such cases native speakers typically employ the regular rule. Third, because the model employed only phonological structure, it could not distinguish between phonologically identical words of differing morphological origin. For example, it would uniformly produce *flew* as the past tense of *fly* regardless of whether it was used with its standard meaning or in the phrase *fly out*.

Subsequent connectionist approaches that made representational assumptions similar to Rumelhart and McClelland's model (e.g., Daugherty and Hare, 1993; Daugherty and Seidenberg, 1992; Hoeffner, 1996; MacWhinney and Leinbach, 1991; Plunkett and Juola, 1999; Plunkett and Marchman, 1991, 1993), also suffered from the same problems with the regular rule (see, Marcus, 2001; Pinker, 1999; for reviews). Although these models made various enhancements to the original Rumelhart and McClelland work—such as adding additional layers to the network, or providing semantic or orthographic features on inputs and outputs, they failed to provide mechanisms by which abstract categories such as VERB or STEM could be represented, and therefore they suffered similar problems to those just described.

Before moving on to a discussion of how developmental concerns informed the past tense debate, we wish to emphasize two points. First, the weaknesses with these feedforward networks follow from their representational assumptions, and do not necessarily invalidate the other insights of this modeling idiom. In particular, early connectionist models of morphology emphasized the importance of giving a precise account of competition between generalizations and of the importance of providing a model of acquisition that could handle phenomena such as U-shaped development. We take these issues up below.

Second, we wish to emphasize that to claim that abstract categories are necessary is not to claim that they cannot be partially predicted from phonological, semantic, or distributional information. Rather, phonological, semantic, and distributional structure are surely necessary cues to category membership. However, without abstract categories that exist independently of these cues, the way in which humans generalize in boundary cases, where such cues are less available, cannot be predicted.

4.4.1.1.1 Feedforward Networks and Overregularization

Earlier in the chapter, we mentioned several controversies regarding the interpretation of the developmental data presented in Marcus et al. (1992). In particular, advocates of connectionist networks have called into question the evidence for a global period of early correct learning, and the correlation between the emergence of reliable regular marking and overregularization. In this section, in order to clarify the issues underlying these debates, we focus in detail on the way in which feedforward networks account for overregularization.

Because feedforward networks require high variability in the structure of regular stems in order to wash out spurious generalizations, the original Rumelhart and McClelland model was able to account for overregularization only by introducing a discontinuity in the training regime. After initial training on a set of mostly irregular forms, the model was flooded with a large number of regular forms. The sudden increase in the variability associated with the regular generalization led to overregularization errors. This move was criticized as empirically implausible (see, e.g., Marcus, 1995; Marcus et al., 1992). Children do not receive discontinuous input in this way, but, rather, are exposed to a dataset of gradually increasing size with an overall cumulative increase in the number of regular verbs.

There have been many subsequent attempts to achieve U-shaped learning in feedforward networks without an input discontinuity. MacWhinney and Leinbach (1991) showed that a feedforward network can exhibit overregularization and recovery without an input discontinuity, but were unable to demonstrate early correct performance. Plunkett and Marchman (1991) achieved similar results, showing, in addition, that their model's rates of overregularization were highly dependent on the number of distinct verb types in irregular inflectional classes (e.g., the class: /ɪ/ →/ʌ/ before /ŋ/). They were able to achieve early correct learning for some verbs, but no global effect (i.e., micro-U-shaped learning). Hoeffner (1992) showed that U-shaped learning is possible without abrupt input discontinuities in an attractor-network-based connectionist model. However, the overall rates of overregularization and the rate of recovery in this model are inconsistent with the findings in Marcus et al. (1992).

Plunkett and Marchman (1993) were able to achieve overall early correct learning, but this appears to be attributable to a more subtle training discontinuity. They first trained the network to perfect performance on a subset of twenty (mostly irregular) verbs before expanding the input set (Marcus, 1995; Plunkett and Marchman, 1996). Furthermore, their model exhibits significant rates of overirregularization. Based on a systematic analysis of the Plunkett and Marchman models, Hoeffner (1996) argued that the Plunkett and Marchman (1996) results do not generalize to more realistic input settings.

Hoeffner (1996) further explored a large number of factors that may affect U-shaped development in feedforward networks. The overall outcome was that none of the manipulations provided robust U-shaped learning except for manipulations that involved qualitative changes in semantic and phonological representations during learning. In particular, he found that phonological manipulations, which suppressed some of the phonological structure of regulars, and semantic manipulations, which obscured the identity of the form being processed, could lead to correct U-shaped trajectories. Such manipulations of phonological and semantic structure allowed the model to ignore aspects of the inputs that were irrelevant to their past form in a way similar to the use of an abstract default rule. A final feedforward model, Plunkett and Juola (1999), demonstrated correct U-shaped learning, but only by employing a training schedule where the size of the training set expanded exponentially rather than linearly—an obvious absurdity for real language learning.

4.4.1.2 Related Approaches Allowing for Abstraction

Although feedforward connectionist networks in the tradition of Rumelhart and McClelland (1986) cannot represent abstractions such as `VERB`, `NOUN`, `STEM`, or `AFFIX`, this is not a necessary property of the connectionist framework. In the first part of this section, we briefly review some connectionist networks that extend the feedforward architecture with mechanisms able to make abstract, categorical distinctions such as these. Connectionist approaches are often grouped together with analogical models (see, e.g., the introduction of Skousen et al., 2002), and, in the second part of this section, we review the application of analogical approaches to the English past tense. For both sets of models, the ability to represent abstract classes of inputs means that they can avoid some of the problems of simple feedforward networks.

4.4.1.2.1 Connectionist Approaches with Abstract Generalization

A number of connectionist models have introduced mechanisms going beyond the simple architectures described above. As a result, these models have been able to avoid some of the generalization problems (e.g., blends, failure to generalize to phonologically odd stems, etc.) characteristic of earlier approaches (see, Marcus, 2001; Pinker, 1999; for more detailed reviews of many of models mentioned below).

Several connectionist models have addressed the problem of generalization of the regular rule by adding components to the network that explicitly represent morphological or lexical information beyond the level of phonology (Hare et al., 1995; Westermann and Goebel, 1995). In such models, the abstract identity of stems and/or affixes can be used to override low-level phonological associations in cases where a default rule should apply. Another approach to addressing the problems with feedforward connectionist networks has been the use of *classifier networks* (Forrester and Plunkett, 1994; Hare et al., 1995; Nakisa and Hahn, 1996; Plunkett and Nakisa, 1997; Westermann, 1998). Most feedforward networks implement a function mapping from a phonological representation of input stems to a phonological representation of past-tense forms. Classifier networks, instead, start out with a set of past-tense inflectional classes (e.g., /ɪ/ → /ʌ/ or +/d/) and learn to select the correct class for each input stem. These networks do not learn the segmentations of affixes from stems, nor do they learn the specifics of the phonological transformations that map from stem to output form. Instead, they simply select which features of a stem are diagnostic of each inflectional class. By separating the problem of identifying an inflectional class from the problem of generating output forms, such models can explicitly learn which input features to ignore, implementing a form of abstraction. These approaches are similar in spirit to the formalization of the past tense that we adopt in the next chapter; however, since they do not explicitly represent or learn mappings from stem to past form, these models are not direct heirs of the Rumelhart and McClelland's proposal.

Westermann (1998) applied a *constructivist* classifier network to the problem of the English past tense (see, also, Westermann, 1998). This network learned its structure (i.e., the number and connectivity of nodes) in addition to association strengths during training. When applied to the English past tense, the model was able grow a network topology in which regulars and irregulars were handled by different components of the system. The regular component was then free to generalize in an abstract way to arbitrary stems without interference from the phonological properties of irregulars. Another related system, that of O'Reilly (1996), employed two different learning mechanisms in the same network: error-driven (back-propagation) learning of the kind used by Rumelhart and McClelland and association-based (Hebbian) learning. In this work, the network was provided with an additional +PAST feature, marked on past-tense forms. The associationist learning mechanism was able to learn to associate this feature directly with the common ending of regulars (i.e., +/d/), effectively ignoring the phonological structure of regular stems and simulating an abstract rule. The error-driven learning component, which was similar to that of Rumelhart and McClelland (1986), was able to account for irregulars.

In all of the preceding examples, the addition of a mechanism that could be used to make a categorical distinction between regulars and irregulars allowed the models to produce more accurate generalization of the regular rule. Connectionist approaches that provide such mechanisms are also, in some cases, better able to account for the developmental trajectory of overregularization. For example, constructivist neural networks can achieve U-shaped performance because they gradually learn to divide the input space between specific irregular processes and a general rule, exhibiting reorganization during training (Westermann, 1998). Likewise, the model of O'Reilly (1996) was able to simulate U-shaped learning because regulars, which were initially handled by the error-based learning mechanism, gradually came to be associated with the +PAST feature in the associative learning system.

4.4.1.2.2 Analogical Models

In addition to rule-based and connectionist approaches, the English past tense has also been studied using a number of analogical models. These models specify algorithms for classifying input forms based on their similarity with large sets of stored exemplars. For example an input stem might be compared with a large number of stored stems, each of which is labeled with an inflectional rule (e.g., +/d/, or /ɪ/ → /ʌ/). When a stem is sufficiently similar to some particular stored stem (or set of stems), the rule associated with that stored stem will be used to form the past of the input stem.

Skousen (1992) developed an exemplar-based approach, known as *analogical modeling of language*. Roughly, analogical modeling of language operates by finding a set of stored exemplar forms (called a context) whose features match the input and all predict the same output rule. For example, if presented with an input stem *fring*, a context might be a set of stored stems that all end in *-ng* and all form their past tense with the rule: /ɪ/ → /ʌ/. The set of all such contexts is computed, and the largest context that does not contain a smaller context with a different rule is chosen as the appropriate generalization for the input (see, also, Skousen et al., 2002). Derwing and Skousen (1994) applied analogical modeling of language to the English past tense, but found that their model was unable to reliably assign the correct past-tense inflections to stems. Furthermore, their model produced overregularization rates far in excess of those seen empirically. However, another study applying analogical modeling of language to the past tense, Eddington (2000), was more successful (likely due to using more realistic phonological representations, see below). It produced accurate classification of input stems and captured the fact that the generalizability of irregular inflections depends upon the phonological structure of the stem to a greater extent than the generalizability of the regular rule (Prasada and Pinker, 1993).

Eddington (2000) explored another analogical model: *memory-based learning*. Memory-based learning is a form of *nearest-neighbor* classification augmented with mechanisms that discount unimportant dimensions of similarity (see, Daelemans and van den Bosch, 2005; for details). This model performed similarly (although slightly worse) to analogical modeling of language on the data in Eddington's study.

Albright and Hayes (2003) applied Nosofsky's exemplar-based *generalized context model* to the problem of the English past tense (Nosofsky, 1990). The generalized context model computes feature-based similarity between an input and stored exemplars, weighting each exemplar by a measure of its representativeness. Albright and Hayes (2003) found that, for adult, novel verb generalization data, this model did not account as well as an alternative probabilistic, rule-based model (see below for a discussion of the rule-based model). Albright and Hayes (2003) argued that the generalized context model and related approaches fare poorly on the English past tense because they cannot ignore spurious generalizations that result from incidental similarities between inputs and stored exemplars—the problem addressed by the addition of mechanisms of abstract generalization discussed above.

4.4.1.3 Summary

In this section, we have reviewed an important lesson that emerged from the debate over connectionist models of the English past. An adequate theory of this domain must provide mechanisms capable of representing abstract generalizations over categories such as `VERB` or `STEM`. Without such abstract categories, models can capture neither the way in which adults generalize in boundary cases, nor the trajectory of overregularization during development. We further showed how incorporating the possibility of such abstractions allowed some connectionist and analogical architectures to be more successful in modeling some aspects of the past tense. Following our basic psychological assumptions outlined in Chapter 1, the models we explore in this book all have the capability of representing such abstract generalizations.

We also emphasized two additional points. First, to claim that abstract categories exist is not to claim that their membership cannot be partially predicted from phonological, semantic, or distributional cues. On the contrary, such cues are crucial for learning. Second, the weaknesses of feedforward connectionist models are a consequence of their lack of abstract representations. However, historically such models emphasized the importance of giving a precise account of competition between generalizations and of the importance of providing a model of acquisition that could handle phenomena such as U-shaped development. We turn to these issues in the next section.

4.4.2 Competition between Generalizations

The preceding sections discussed how many connectionist and analogical models fail to account for the English past because they lack abstract generalizations in their hypothesis spaces. In this section, we will identify the fundamental challenge which face systems that do provide the capacity for such abstract generalizations: the necessity for a theory of competition between multiple hypotheses. The possibility of abstract generalizations at multiple levels of specificity—from rules like +/d/ that apply to any verb stem, to phonologically-conditioned stem changes like /ɪ/ → /ʌ/ before /ŋ/ (*fling/flung*)—leads to a large number of hypotheses consistent with any dataset and multiple ambiguities for many particular forms; for example, should *fring* form its past with +/d/, *fringed*; /ɪ/ → /æ/, *frang*; or /ɪ/ → /ʌ/, *frung*. An adequate theory must provide mechanisms for deciding which of the many, potentially incompatible, generalizations is the right one for a given verb. As we will discuss below, phenomena such as overregularization, speech errors, and the existence of doublets also mean that mechanisms of competition resolution must also be able to account for exceptional cases where more than one—possibly incorrect—generalization is sometimes applied at different rates.

Following the general theme of this book, our discussion will focus only on the problem of productivity, in particular, how different theories account for the pattern of competition between various generalizations exhibited by the English past system. There is an additional, very important learning problem faced by the child acquiring English, namely how to discover the phonological structure of the generalizations characterizing each inflectional class. In particular, the child must learn the phonological relationship between a stem and its past form, and potentially the relationship between stems in each inflectional class and the relationship between past forms across classes.[9] This problem has been the focus of most of the computational modeling work in the area (e.g., Albright and Hayes, 2003; Daugherty and Hare, 1993; Daugherty and Seidenberg, 1992; Derwing and Skousen, 1994; Eddington, 2000; Hare et al., 1995; Hoeffner, 1992, 1996; Ling and Marinov, 1993; MacWhinney, 1978; MacWhinney and Leinbach, 1991; Molnar, 2001; Mooney and Califf, 1995; O'Reilly, 1996; Pinker, 1984; Plunkett and Juola, 1999; Plunkett and Marchman, 1991, 1993; Rumelhart and McClelland, 1986; Westermann and Goebel, 1995; Yip and Sussman, 1996, 1997). Although the problems of acquiring phonological rules and determining the productivity of those rules interact, they are logically distinct (see, also, Yang, 2005, 2010, 2002). In the current

9. These are the three kinds of semi-regularity exhibited by the English past (see our empirical review above).

section we will restrict attention to those theories, and those aspects of theories, that directly address the problem of productivity and competition between generalizations.

Two important points will emerge from our discussion. First, most theories that can account for the English past have made use of a probabilistic version of the elsewhere condition. We will show in the next chapter how the elsewhere condition arises as a natural consequence of the modeling assumptions made in this book. Second, we will argue that theories providing quantitative accounts of rule competition, like the fragment grammars model, are consistent with a wide-variety of different mixtures of generalizations at different stages of learning (including the adult state). The particular systems learned by such models will depend largely on the makeup of the input data.

4.4.2.1 Classical Linguistic Theories

Classical linguistic theories of the past tense made use of rules, annotated with input conditions, for both regular and irregular inflectional classes (e.g., Chomsky and Halle, 1968; Halle and Marantz, 1993; Kiparsky, 1982b; Mohanan, 1986). For example, such theories posited rules for the *sing/sang* class of irregular verbs that mapped the final vowel nucleus /ɪ/ to /æ/ subject to the additional input constraints that (i) the stem was verbal, (ii) the stem ended with a velar nasal (i.e., /ŋ/), and (iii) the stem was one of a set of listed stems for which this rule was applicable (e.g., *sing*, *ring*, etc.). By contrast, the regular generalization was modeled by a rule that simply appended +/d/ to any stem, subject only to the weaker condition that the stem was verbal. When an input form matched the input conditions on several different rules (for instance, when it matched the two rules just described), various mechanisms were proposed to resolve rule competition.

One strategy for handling competition between rules in classical linguistic theories was to posit that the lexicon is organized into ordered*strata* or *levels*. Morphological processes that operate in earlier levels take precedence over processes assigned to a later level (Halle and Mohanan, 1985; Kiparsky, 1982b,c; Mohanan, 1986). In the case of the past tense, the regular rule was assigned to a later stratum than the irregular rules, and thus only applied if no irregular rules had applied earlier (see, also, Chapter 6 for a detailed discussion of stratal theories).[10] In order to handle rule competition that arose within a level, stratal approaches often needed to make additional assumptions—such as a variant of the elsewhere condition (e.g., Kiparsky, 1982b; see below) or an *avoid-synonymy principle* (e.g., Kiparsky, 1982c).

However, the most frequently invoked and most important mechanism for resolving competition between rules in linguistic theories is the elsewhere condition, also called Pāṇini's principle, pre-emption, the subset principle,[11] or (in certain cases) blocking (e.g., Anderson, 1969; Andrews, 1982; Aronoff, 1976; Halle, 1997; Kiparsky, 1973; Pinker, 1984; Stump, 2001b). The elsewhere condition states that a rule with more specific input conditions takes precedence over a rule with more general conditions (see, e.g., Kiparsky, 1982b; Stump, 2001b; for use in ranking morphological rules). For example, in theories that allow rules to be annotated with lists of stems to which they can apply (e.g., Chomsky and Halle, 1968; Noyer and Harley, 1999), irregular rules such as /ɪ/ → /ʌ/ are associated with specific sets of stems (e.g., *fling/flung*, *dig/dug*), whereas the regular rule is not annotated for specific stems, and can apply to any verb. Because the input conditions on the regular rule (i.e., any verb), are less specific than the input conditions on the irregular rule (e.g., specific stems), the elsewhere condition prefers the irregular rule when it can apply, and the regular rule elsewhere.

The elsewhere condition represents an important generalization about linguistic structure. More specific rules or items tend to be preferred to less specific rules or items. However, classical formulations of this principle suffered from two important problems. First, and most importantly, it was not defeasible, and thus could not account for cases of multiple generalization. Second, as a necessary correlate of the first problem, it did not provide an account that explained the rates at which different generalizations apply for adults or during development. We discuss each issue in turn.

Exception-free versions of principles like the elsewhere condition cannot explain cases where more than one rule seems to apply to a given stem. For example, for some adult speakers, *dreamed* and *dreamt* are both acceptable

10. Note that this further assumes that multiple marking of the same inflectional process is disallowed.

11. This term is used in Distributed Morphology (Halle, 1997; Halle and Marantz, 1993; Noyer and Harley, 1999), and is not to be confused with the use of the same term with different meaning in computational learning theory (Angluin, 1980; Berwick, 1985; Manzini and Wexler, 1987; Wexler and Manzini, 1987).

past tenses of DREAM. Similarly, during development, individual children produce both overregularized and correct irregular past forms of irregular stems (Marcus et al., 1992). Moreover, in controlled, wug-type experiments, both adults and children will sometimes generalize irregular rules to novel stems (Albright and Hayes, 2003; Ambridge, 2010; Bybee and Moder, 1983; Bybee and Slobin, 1982; Prasada and Pinker, 1993). All of these phenomena should be impossible under an exception-free variant of the elsewhere condition.

A necessary consequence of not allowing for multiple generalization in cases like those discussed above, is that such theories also cannot predict the rates at which different generalization will be applied. For a given speaker, there is often preference for one form of a doublet or the other. In novel verb rating experiments, speakers' judgments of acceptability often reside on a cline (e.g., for the past tense of *fring*, *fringed* is better than *frang*, which may be better than *frung*). Since they do not allow variation between generalizations in the first place, classical theories often have no direct way of incorporating information that seems to affect differential generalization, such as the frequency of forms, or similarity to other forms. Furthermore, they have no way to explain how these kinds of information influence the trajectory of development, and difficulty explaining how these kinds of information can inform models of language change.

These problems with classical accounts of competition resolution are widely recognized and a number of approaches have been proposed to address these issues. We turn to these defeasible, quantitative theories of competition next.

4.4.2.2 The Blocking-and-Retrieval-Failure Hypothesis of Marcus et al. (1992)

Recognizing the limitations of classical linguistic theories, many researches have adopted dual-mechanism approaches in which classical rules are paired with additional mechanisms that can handle quantitative aspects of competition between generalizations. The most well-known instantiation of the dual-mechanism approach is the blocking-and-retrieval failure hypothesis of Marcus et al. (1992). The blocking-and-retrieval failure hypothesis posits an architecture consisting of two components. Regulars are handled via a rule component, similar to classical linguistic theories, and irregulars are stored in an associative memory system similar to a connectionist architecture. Competition between the regular rule and irregular past tense forms is handled by the blocking principle, which states that when both the regular rule and a stored irregular form are available for a particular stem, the stored form should be preferred. Note that the blocking principle is a specific variant of the elsewhere condition. Since a stored irregular form is more specific than the regular rule, it will be preferred when both are available. Quantitative variation between different generalizations, such as overregularization during language acquisition, is accounted for by assuming that the associative memory system is probabilistic, and stored forms have a probability of being correctly retrieved related to their frequency, similarity to other stored forms, and other factors. When the memory trace for a stored irregular is insufficiently strong, retrieval can fail, and, as a result, blocking will not apply, causing an overregularization error.

The blocking-and-retrieval failure hypothesis is designed to simultaneously account for the productivity of the abstract regular rule and the variability in rates of overregularization for different irregular forms. Under this account, because adults have been exposed to large amounts of linguistic evidence, they posses robust memory traces for irregular verbs, and therefore nearly always produce the appropriate past forms. Occasional overirregularization (e.g., *bite*/*bote*) in this theory proceeds via generalization processes made available by associative memory, explaining its low frequency in learners, and even lower frequency in adult speech. U-shaped development is explained by an interaction between associative memory and the acquisition of the regular rule. At the onset of learning, the regular rule has not yet been acquired. Because irregular verbs constitute the highest-frequency verbs in English, learners will often initially store them whole, and they will be used correctly. At the point in time when the regular rule is acquired, it will begin to compete with these stored irregulars. However, rates of overregularization will in general be low because blocking will only fail probabilistically, and failure will be less common for more frequent verbs. Furthermore, because irregulars are stored in a structured, associative memory, they will be protected from overregularization by similarity to other irregulars. As the child hears more examples of irregular forms, memory traces will strengthen, and recovery will gradually occur.

4.4.2.2.1 The Blocking-and-Retrieval Failure Hypothesis and Rule Learning

Because the morphological rules of a particular language cannot be innate, to provide a complete account of past tense acquisition, a dual-mechanism theory like blocking-and-retrieval failure must be paired with a theory of how productive rules, such as the regular past-tense rule, are learned. In this section, we consider a detailed example of how this might be accomplished, with the goal of examining the kinds of rule systems and developmental trajectories that are compatible with learning theories like the blocking-and-retrieval failure hypothesis.

There is some controversy in the literature over the developmental trajectory of the regular past-tense generalization. It is often assumed by critics of rule-based theories that the acquisition of a productive rule must result in a sudden increase in forms generated by that rule (see, e.g., Braine, 1976; Brown, 1973). Thus, some critics of rule-based systems have sought to demonstrate that empirical phenomena associated with the emergence of the regular generalization (e.g., reliable tense marking and overregularization) have a gradual developmental trajectory (e.g., Hoeffner, 1996). We will argue, however, that when integrated with a plausible account of rule learning, the blocking-and-retrieval failure hypothesis is consistent with a wide variety of developmental trajectories, including gradual changes in rates of tense marking and rule use, and therefore the idea that rule-based theories must predict sudden discontinuities is a misconception. Moreover, we will argue that the precise trajectory of rule learning will be highly contingent on the structure of the input data.

There have been a number of theories of inflectional rule learning that could plausibly serve as the rule-learning component for the blocking-and-retrieval failure model (e.g., MacWhinney, 1978; Pinker, 1984). Here we consider the algorithm presented in Pinker (1984). This algorithm assumes a paradigm-based representation of inflectional structure where paradigms are tables with dimensions labeled by semantic, phonological, or formal attributes, each with a number of specific levels. For example, a dimension might be labeled by the semantic attribute `TENSE` with levels `PAST`, `PRESENT`, and `PROGRESSIVE`.[12] The theory makes use of both word-specific paradigms, where cells are filled with individual word forms, and word-general paradigms, where cells are filled with inflectional rules that apply to classes of words. For example, the model might represent a word-specific paradigm for forms of the word `GO`, which would contain *went* in the paradigm cell indicated by `TENSE:PAST` and `NUMBER:SINGULAR`. The model might represent a word-general paradigm that could apply to any verb stem, with entries for the regular

12. Formal attributes, that is, attributes that do not represent phonological or semantic properties, are included to allow the algorithm to learn the structure of languages with arbitrary subclasses of inflections (e.g., nominal declensions, verbal conjugations, genders, etc.) and are marked on the lexical entries for individual words.

progressive (i.e., *-ing*) and past (i.e., *-ed*) suffixes. Paradigms have associated strengths related to factors such as their frequency in the input data.

Words are initially stored whole in the lexicon and associated with an appropriate cell in a word-specific paradigm. For example, if a child hears *walked* in a past-tense context, then this word will be stored, and entered into a word-specific paradigm in a cell with the appropriate levels marked on each dimension (i.e., with `TENSE` marked as `PAST`). If certain dimensions or levels are underspecified, for example, if the context did not make `NUMBER` clear, then the relevant dimensions and levels are sampled randomly from a universal set according to prior probabilities that encode the prominence of each dimension and level across inflections in the world's languages. As children hear forms used in more contexts, this sampling process will gradually expand the number of dimensions associated with each word-specific paradigm. Independent mechanisms segment word forms and extract the common phonological material across word-specific paradigms in order to construct word-general paradigms (i.e., rules) from the shared material.

This process of building word-specific and word-general paradigms will necessarily give rise to many, potentially inconsistent ways of inflecting particular words. For example, the algorithm might reach a state where it has stored the past tense form *flung* for the verb `FLING` in a word-specific paradigm, while simultaneously generalizing a word-general rule /ɪ/ → /ʌ/ that produces the same past form, and a word-general rule +/d/ that produces the alternate past form *flinged*. Competition between these alternate ways of forming the past of `FLING` must be resolved in some way.

Pinker's algorithm has two mechanisms for resolving competition. First, multiple words are prevented from being stored in the same cell of a word-specific paradigm, thus preventing, for instance, both *flinged* and *flung* from being stored as the past form for `FLING` in a word-specific paradigm (but see below). This is known as the ***unique entry principle***. Second, a criterion similar to the elsewhere condition—marked by a ***pre-emptability*** diacritic—is maintained between forms generated from word-general paradigms (i.e., computed by rule) and words retrieved from memory. When a retrieved form is available, it pre-empts the rule-generated form, implementing blocking. Thus, if *flung* has already been stored in the word-specific paradigm for `FLING`, then any word-general rules, such as the irregular rule /ɪ/ → /ʌ/, or the regular rule +/d/, will be blocked from applying to this stem. Note that both principles are needed; the unique entry principle handles competition between multiple stored forms, while the pre-emptability diacritic handles competition between stored forms and rules.

In Pinker's theory, both the unique entry principle and the pre-emptability diacritic are probabilistic and defeasible. If the memory traces of two forms are strong enough (i.e., there is sufficient evidence for both forms), they may both be stored in a single word-specific paradigm cell. Likewise, if the strength of a word-specific paradigm is insufficiently strong, it may not preempt the use of a word-general paradigm. The defeasibility of these two principles of competition resolution solves the empirical problems associated with classical theories, such as doublets (e.g., *dreamed/dreamt*), overregularization (e.g., *goed*), and overirregularization (e.g., *bite/bote*).

Unsurprisingly, many of the basic assumptions of Pinker's theory are similar to those of the blocking-and-retrieval failure model. Under both models words are initially stored whole in memory, retrieval from memory is probabilistic, and the existence of a stored form can block the use of a regular rule to express a particular meaning. The Pinker (1984) algorithm, however, provides additional mechanisms that allow rules to be abstracted from input forms in a way that remains consistent with early correct performance. The algorithm starts by storing all forms and form-specific paradigms, and only later abstracts general rules.

When paired with the blocking-and-retrieval failure hypothesis, the Pinker (1984) algorithm is consistent with a wide variety of different mixtures of generalizations over development. In the model, both individual forms and paradigms are weighted, the principles handling competition are probabilistic and defeasible, and the process of abstracting rules is iterative. These facts mean, for example, that it is possible for this algorithm to reach a state where a version of the regular rule has been learned, but where this version is not yet fully general (i.e., it may include an unnecessary phonological, semantic, or formal condition), and therefore does not apply across the board to all past-tense forms. In fact, there is nothing to stop the model from learning several versions of the regular rule with disjoint or partially overlapping domains of applicability. The model can also learn rules corresponding to irregular inflectional classes (e.g., /ɪ/ → /ʌ/), and which, therefore, can compete with stored irregulars in associative memory.

Because the unique entry principle and pre-emptability diacritic are probabilistically defeasible, the model allows for a variety of interactions between the rules and stored forms. The model can produce overregularization if a memory trace for a form in a word-specific paradigm is insufficiently strong to pre-empt a word-general rule. Since the weights associated with stored forms and the pre-emptibility diacritic depend on frequency (amongst other factors), the rates with which the model predicts overregularization will depend on the input data. Most importantly, the model does not predict a necessary, sudden increase in the use of the regular rule at the moment when this rule is first abstracted. While it can produce this behavior in principle, a sudden increase is not a logical necessity.

4.4.2.3 The Probabilistic Rule-Based Theory of Yang (2002)

Yang (2002) described a model of the English past extending the traditional, linguistic rule-based approach with probabilistic rule-application. In Yang's system, all inflectional classes, including irregular classes, are handled by rules (as, for example, in Chomsky and Halle, 1968), but these rules are enhanced with two mechanisms not present in traditional accounts. First, the matching of stems to rules is probabilistic. When a stem is consistent with the phonological input conditions on a rule (e.g., *fring* is consistent with the condition "ends with ŋ" on a rule like /ɪ/⟶/æ/), the rule is matched with some probability p. When matching fails, the default, regular rule is applied. Second, even after matching a stem to a rule, the application of the rule is noisy, and with some probability it fails, again leading to the use of the default. Yang (2002) argued that the model can account for the observed effects of frequency and similarity on the generalization of regular and irregular classes. Furthermore, as Yang (2002) pointed out, these assumptions give rise to a probabilistic variant of the blocking principle.

Yang's framework is similar to the blocking-and-retrieval failure hypothesis and the models in this book in that it focuses on learning which inflectional classes can apply to a stem, rather than learning the phonological structure of different inflectional classes. However, Yang did not specify how the learner decides which inflectional class should be the default, a problem at the heart of the present work. Because this aspect of his model is underdetermined, it is not yet possible to generate quantitative predictions about competition or learning trajectory using his framework.

Yang (2002) did not define a specific way in which his learner estimates the probability that a stem matches a rule or the probability that a rule fails to apply once selected. However, it is clear from his discussion that frequency plays a role. Given this assumption, when paired with an appropriate rule-learning mechanisms, Yang's model is also likely to be consistent with a wide variety of differing mixtures of probabilistic rules and learning trajectories given differing distributions of input forms.

4.4.2.4 The Dual-Mechanism Model of Taatgen and Anderson (2002)

Another computational model of productivity in the past tense was explored by Taatgen and Anderson (2002). This model, formulated in the ACT-R framework (Anderson and Lebiere, 1998), makes use of a distinction between declarative and procedural memory. In ACT-R, declarative memory consists of a database of stored facts (also called chunks), while procedural memory consists of a set of production rules that operate on those facts. The stored facts have an `ATTRIBUTE`=*value* format, so they can store information such as `PAST[SING]`=*sang*. Production rules are sometimes referred to as condition-action pairs, and consist of `IF ... THEN` statements, which can encode rules like `IF` *the goal is to generate the past tense of category* `VERB`, `THEN` *add* /d/.

Like Marcus et al. (1992), Yang (2002), and the models studied in this book, Taatgen and Anderson did not model the process of discovering the phonological form of past tense rules. Instead, to formalize the past tense, Taatgen and Anderson provided their model with three initial production rules corresponding to three strategies for generating past tense forms. The first strategy directly retrieves facts that encode the past tense of stems from memory, and corresponds to verbatim memorization of past forms. For example, this rule might retrieve a fact like `PAST[SING]`=*sang*. The second strategy randomly chooses another past form stored in memory, and uses the inflection associated with that form. The third production rule implements a "do-nothing" strategy, which, rather than inflecting an input stem, simply returns the stem without inflection.

Similar to the fragment grammars model, the variant of ACT-R used by Taatgen and Anderson (2002) provides mechanisms that compile frequently co-occurring sequences of computations into single production rules (see, Taatgen and Lee, 2003; for more details). Using these mechanisms, a version of the regular past-tense rule can be discovered by combining steps that randomly sample the inflection on another form and apply it to a new stem. Thus, in Taatgen and Anderson's ACT-R model of the past tense, the system has three options in deciding how to inflect a stem: (i) it can do nothing, (ii) it can retrieve the corresponding past form stored for a stem directly from memory, or (iii) it can choose an inflection randomly from the inventory of stored forms. This last possibility is enhanced by the compilation mechanism, which allows the process of sampling a random inflection and applying it to a new stem to become single production rule in particularly valuable cases. This allows the system to learn a general rule in the case of the regular inflection.

As with other models of past inflection, there will often be several different ways of inflecting the same stem. The model resolves this competition by choosing a production rule based on that rule's expected outcome. In ACT-R, the expected outcome of a production rule is based on four factors: (i) the prediction accuracy of the rule on previously generated past tense forms, (ii) the utility of the output of the rule (constant for all forms in Taatgen and Anderson's study), (iii) a fixed cost (measured in time) associated with each production rule, and (iv) random noise. The rule with the highest expected outcome is chosen to inflect an input form. The high frequency of irregulars means that direct retrieval is favored for these forms because of its high prediction accuracy (Factor i). For regulars, the model eventually learns that the compiled, abstract, regular rule has the highest prediction accuracy. Because rules with more specific conditions will tend to have higher prediction accuracy, Taatgen and Anderson's model implicitly enforces a probabilistic variant of the elsewhere condition.

Taatgen and Anderson's model was able to learn to the correct pattern of blocking and defaultness in the English past system. It associated irregular stems with their correct past-tense inflectional allomorph and generalized the regular rule to novel stems. It was also able to correctly capture U-shaped development. In the model, the do-nothing rule was assigned lower a priori cost than analogy or retrieval. As a result, like children, early in learning, the model produced uninflected verb forms at high rates. Although it had low cost, the do-nothing rule also had poor accuracy in predicting the past-tense forms of stems. By contrast the other rules were able to predict the past tense of stems with higher accuracy, despite their cost. Therefore, as more data was observed, they began to have higher expected outcomes and be preferred to the do-nothing rule. This occurred first for high-frequency irregular forms, leading to a period of early correct performance. Eventually the model discovered the regular rule, and began correctly inflecting low-frequency and novel regular stems. At the same time, overregularization began to occur. Consistent with the empirical data, recovery was gradual, and relative rates of overregularization were realistic.

Because the Taatgen and Anderson model resolves competition using expected outcome, which depends to a great extent on a rule's prediction accuracy in the input data, it is consistent with a variety of differing mixtures of probabilistic rules and learning trajectories—like the blocking-and-retrieval failure hypothesis of Marcus et al. (1992), the model of Yang (2002), and fragment grammars.

Taatgen and Anderson's model is similar in several other respects to the application of fragment grammars to the English past that we report in the next chapter. It used similar representations, a related mechanism of rule-compilation, and treated the problem of learning the past tense as an inference in an optimization-based framework. As a result, it produced a similar pattern of defaultness for the regular inflection and blocking by irregulars at the end of learning. However, unlike the simulations reported in the next chapter, Taatgen and Anderson's model predicted a global U-shaped developmental curve. This is the direct result of their provision of a do-nothing strategy, which can produce unmarked verb forms early in learning. We discuss this difference and its significance after presenting our simulation results in the next chapter.

4.4.2.5 Other Models of Rule Induction

Before moving on, we briefly mention several other computational models of the past tense. Unlike those discussed so far, these models are not primarily aimed at the problem of productivity and competition between generalizations, but rather are meant to address the issue of learning the phonological structure of different inflectional classes. Nevertheless, several of them implement mechanisms to resolve competition between irregular and regular generalizations, and, when such a mechanism is present, it consists of a variant of the elsewhere condition. We focus on this aspect of the these models.

Albright and Hayes (2003) described a probabilistic, rule-based model based on the minimal generalization learning algorithm (see, Albright and Hayes, 2002). The algorithm constructs phonological rules by aligning training examples and abstracting common phonological structure. For example, the algorithm might learn a rule mapping /ɪ/ to /æ/ in the context of /ŋ/ by aligning the past-tense pairs *sing/sang* and *ring/rang*. The algorithm makes minimal generalizations in the sense that it hypothesizes the rule preserving the maximal amount of phonological material shared by aligned forms. In the case of *sing/sang* and *ring/rang*, it will preserve the shared context, /ŋ/, rather than hypothesizing the more general rule that transforms /ɪ/ to /æ/ in any context.

Albright and Hayes handled rule competition by assigning scores to rules and choosing the rule with the highest score in each case. Rule scores were calculated using a statistic comparing the number of items in the dataset to which a rule can apply (i.e., its abstractness or level of generality) to the number of items to which it does apply (i.e., its reliability). This quantity was further modulated by a confidence value, which reduces the strength of rules abstracted from fewer datapoints.

One fundamental difference between the model of Albright and Hayes (2003) and the models discussed in preceding sections is that it does not provide an explicit mechanism whereby irregular rules can be directly associated with stems, although they do argue that such a mechanism must exist. Therefore, their approach to rule competition is best understood as a theory of rule generalization to novel stems; it does not make predictions with regard to rules used to generate existing past forms. Because of this, it seems likely that their model will overgeneralize irregular classes to existing stems that match the input conditions on an irregular rule, but are regular (e.g., hypothesizing *ping*⟶*pang* instead of *pinged*), while underestimating the probability of the regular rule for novel stems (e.g., hypothesizing *fring*⟶*frang* instead of *fringed*).[13] Albright and Hayes (2003) evaluated their model on experimental data consisting only of novel stems, thus this prediction is untested as yet.

Yip and Sussman (1996, 1997) provided a model of phonological rule learning in the constraint propagation framework. In this model, inputs verb stems and output past forms are encoded as vectors of features, and constraints are learned between these vectors of features. For example, a constraint might be learned that requires that when an input stem ends in -/ɪŋ/, the output past form must end in -/æŋ/. Their system contains two kinds of constraints: rote-classifiers, which handle relations between specific input and output forms, and rule-classifiers, which generalize across rote-classifiers and implement general linguistic rules. Competition between constraints is handled by weighting each constraint in proportion to the number of input feature values matched on a given input vector. Thus, when two constraints apply, the constraint with a more specific input condition will be preferred—implementing a version of the elsewhere condition.

13. Note that a correlation with human data on only novel stems, like that reported in Albright and Hayes (2003), will not reveal this issue since the baseline productivity of all inflections—when evaluated on novel stems—is absorbed by the intercept of the correlation. Such a correlation will only reveal whether the relative fit of different processes on novel stems is accurate.

Yip and Sussman provided an algorithm that iteratively constructs increasingly general constraints from a corpus of examples and showed that their algorithm is able to efficiently build a system of rules capable of correctly inflecting stems for past tense using only a small number of training examples. They also noted that learning became faster as more constraints were generalized and that their system was not sensitive to the token frequency of forms in the input. However, they did not provide evaluation on specific experimental or developmental datasets and the insensitivity of their model to token frequency makes it is unlikely to be able to account for many of the phenomena of rule competition we have reviewed.[14]

Ling and Marinov (1993) presented a model of phonological rule induction that learns a mapping between input and output vectors of phonological features similar to those used by Yip and Sussman (1996, 1997). Unlike the model of Yip and Sussman, the mapping between input and output features were structured as a decision tree; input features are considered in a specific order, with later decisions depending on earlier ones. Their algorithm learned the ordering over features based on an information gain criterion: input features that provide greater amounts of information about outputs are used before those that provide less information. Because input features associated with irregulars (e.g., features encoding the irregular stem sequence -ɪŋ) are more predictive of specific output features (e.g., features encoding the irregular past sequence -æŋ) they were ordered first in the decision tree. This leads their algorithm to enforce a version of the elsewhere condition whereby more specific, predictive features are checked first. Ling and Marinov (1993) showed that their system was able to produce past-tense forms for existing stems with high accuracy. However, it was not able to produce U-shaped learning without additional manipulation of a noise parameter.

A final phonological rule learning algorithm for the past tense, developed in an inductive logic programming (ILP) framework, was presented by Mooney and Califf (1995). The algorithm iteratively generalizes logical predicates corresponding to the regular and irregular inflectional classes. The complexity of the ILP framework puts the details of this algorithm beyond the scope of the

14. Molnar (2001) presented a related algorithm for learning phonological rules consisting of two components, called generalize and sift. Generalize iteratively constructs increasingly abstract rules from the input corpus by finding and factoring shared phonological substructure between forms. The sift component associates a probability with each rule, proportional to the frequency with which it is used to generate forms in the input. However, Molnar (2001) did not provide any evaluation on experimental or developmental data.

present discussion. We note, however, that in addition to standard ILP generalization mechanisms, Mooney and Califf (1995) found that they needed to enforce a constraint whereby more specific predicates were ordered before more general ones. This, again, constitutes a variant of the elsewhere condition.

4.4.3 Summary

In this section, we have reviewed how abstract, rule-based theories have handled the problem of competition between generalizations with varying levels of productivity. Rule-based approaches make available a large variety of generalizations consistent with the data, and therefore must provide mechanisms for deciding which rules apply in which cases. Phenomena such as the existence of doublets, overregularization/overirregularization by both children and adults, and the generalization of both regular and irregular inflections in controlled experiments mean that such mechanisms must be defeasible and must be able to take into account input-data factors such as frequency and similarity. In nearly all accounts, competition has been handled by a probabilistic version of the elsewhere condition—whether explicitly encoded or arising as a consequence of other modeling assumptions. We have also argued that the resulting class of models can, in principle, be consistent with a large variety of mixtures of rules at different stages of learning. The details of how easily and in what contexts specific rules can generalize will depend on features of the input data.

4.5 Conclusion

In this chapter, we have reviewed the empirical and theoretical literatures on the English past-tense system from the point of view of productivity and reuse. The system is characterized by a violable pattern of defaultness and blocking. The regular +/d/ rule acts as a default, applying when no other form is available, while the availability of irregular forms blocks its application. The English past is an interesting test case for a model of productivity and reuse precisely because of the sharp dichotomy it exhibits between the productivity of the regular rule and the non-productivity of generalizations that characterize irregular forms.

The general applicability of the regular rule means that any theory that can account for the past tense must be able to represent abstract generalizations that allow the regular suffix to be attached to any verb stem, regardless of its phonological (or semantic) structure. However, the empirical literature also shows that blocking is a probabilistic phenomenon and that irregular generalizations such as that characterizing the *ring/rang* class of verbs (i.e., /ɪ/⟶/æ/) can sometimes apply to novel stems. This, together with phenomena such as the existence of doublets, overregularization and overirregularization errors by adult speakers, and U-shaped learning in children indicate that any adequate theory of the past tense must be able to provide a violable, quantitative account of blocking and defaultness. We have emphasized how one common solution to this problem is a probabilistic version of the elsewhere condition—whether explicit or an implicit consequence of other mechanisms. We have also argued that theories that provide such a mechanism can be consistent with a wide-variety of different mixtures of (semi-)productive generalizations and developmental trajectories. The exact predictions they make will depended on the input data.

As we mentioned in the introduction to the chapter, the literature on the past tense is vast, and we have restricted our review to those topics of direct relevance to the modeling results presented in the next chapter. Important topics we have not covered in detail include the learning of the phonological structure of rules, whether the system is best accounted for using a single- or dual-mechanism approach, and the relationship between theoretical models of the past tense and biological evidence. After presenting our results in the next chapter, we will discuss how this work relates to this wider literature.

5 The English Past Tense: Simulations

5.1 Introduction

The previous chapter argued that an adequate theoretical account of the English past tense must provide both the capability of representing abstract generalizations and a quantitative account of how such generalizations compete. The models studied in this book build on tools from the theory of computation and classical cognitive science, and therefore all provide mechanisms for abstract generalization. This chapter examines the question of whether these models can provide an account of competition between generalizations in the English past tense system.

Throughout this chapter we compare the performance of the five models introduced in Chapters 1 and 2: (i) Our inference-based model, Fragment Grammars (FG), (ii) the full-parsing model, implemented using Dirichlet-Multinomial PCFGs (DMPCFG), (iii) the full-listing model, implemented using Maximum a posteriori Adaptor Grammars (MAG), and two versions of the exemplar-based approach, (iv) Data-Oriented Parsing I (DOP1), and (v) Data-Oriented Parsing: Equal-Node estimator (ENDOP).

To provide an adequate account of competition between inflectional classes for the English past tense, a model must explain several phenomena. First, a model must treat the regular rule as a default, applying in cases where no other inflectional process is available. Second, it must exhibit blocking of the regular rule by the irregular forms. Third, it must be able to capture cases where blocking fails, for example, cases of overregularization during language acquisition. Fourth, it must explain how the distribution of forms in the input gives rise to this pattern of probabilistic defaultness and blocking.

In the following sections, we will first give a general overview of the modeling assumptions used for the simulations of the English past-tense system. We will then evaluate the five models on a corpus of adult speech. These evaluations will show that, of the five models, only fragment grammars capture the correct pattern of defaultness and blocking in the English past-tense system. We will also show that only fragment grammars make use of the frequency distribution of forms in the input in a way that is consistent with empirical evidence from processing experiments. After examining these adult speech simulations, we will present a developmental study of the five models. These simulations demonstrate that only fragment grammars predict overregularization phenomena consistent with findings in the literature.

5.2 The Simulations

In this section, we describe the linguistic representations and input data used by all of the models studied in this chapter.

5.2.1 Input Representations

Each of the five models explored in this book start from an underlying Context-Free Grammar (CFG) defining the space of all possible tree structures that can be generated by each system. In this section, we describe the CFG used for the past-tense simulation results reported below. The modeling decisions behind the CFG and how they correspond to assumptions about learning, can be summarized by the following points.

- Segmentation of forms into stems and inflections is provided a priori for all verbs.
- An inventory of all possible inflections in the English verbal paradigm, such as +/d/ or /ɪ/ → /æ/, is provided as part of the starting context-free grammar for the models.
- The correspondence between verb forms or tenses (e.g., past tense [VBD], or progressive [VBG]), and each inflectional process is provided by the grammar. For example, the underlying grammar specifies that +/d/ can be used to mark either the past tense (VBD) or the past participle (VBN) of a verb.
- The training data specify for each verb form its stem (e.g., EAT, GO, WALK, etc.), its inflectional process (e.g., +/d/, /ɪ/ → /æ/, etc.) and its tense from the set: past tense [VBD], bare form [VB], progressive form [VBG], past participle [VBN], simple present [VBP], or third singular form [VBZ].
- The models are not provided with information about how productive each process is or which stems should be paired with which inflectional markers (beyond co-occurrence in the training data).
- No phonological or semantic structure is represented. In particular, the models cannot represent selectional restrictions on rules. For example, each model is free to hypothesize that /ɪ/ → /æ/ can apply to the stem WALK, despite the fact that the application of such a rule to that stem is phonologically impossible (since it does not contain the vowel /ɪ/).

V	⟶	Stem	Inflection
Stem	⟶	EAT	
Stem	⟶	GO	
Stem	⟶	WALK	
	...		
Inflection	⟶	+∅	VB
Inflection	⟶	+∅	VBP
Inflection	⟶	+/ɪŋ/	VBG
Inflection	⟶	+/z/	VBZ
Inflection	⟶	/ei/ → /ɛ/ ∧ +/z/	VBZ
Inflection	⟶	WENT-SUPPLETION	VBD
Inflection	⟶	/ɪ/ → /æ/	VBD
	...		

Figure 5.1

Context-Free Grammar for Past Tense Simulations: A CFG for the English past-tense simulations.

These assumptions lead to an underlying CFG like the one in Figure 5.1. A full list of the inflectional classes used in the simulations is given in Appendix A. The rules for this CFG were constructed by hand, starting from the lists of irregular inflectional classes given in Pinker and Prince (1988). Note that the rules in Figure 5.1 for each inflection are of the form "`Inflection` ⟶ INFLECTIONAL-PROCESS TENSE," where the INFLECTIONAL-PROCESS specifies the phonological changes to the stem for the form (e.g., /ɪ/ → /æ/) and the TENSE specifies which cell of the English verbal paradigm the form comes from (e.g., progressive form [`VBG`]).[1]

Conforming to this grammar, each input training item contains terminal symbols identifying both the inflectional class associated with the form and the tense of the verb from the set: past tense (`VBD`), past participle (`VBN`), bare form (`VB`), progressive form (`VBG`), past participle (`VBN`), or third singular form (`VBZ`). Figure 5.2 shows several example input trees.

1. More precisely, the list of inflectional classes given in Appendix A divides the INFLECTIONAL-PROCESS into two parts. The first part specifies the transformation applied to the stem (if any). The second part specifies the suffix attached to the stem (if any). However, all stem transformations and suffixes are bound into single rules—the system cannot learn them independently.

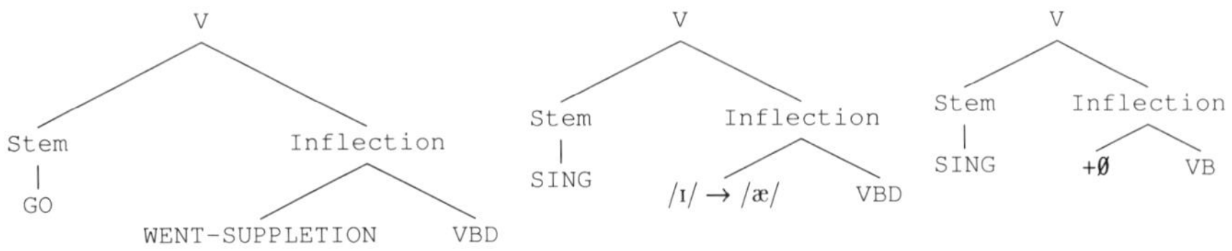

Figure 5.2

Example Past-Tense Trees: This figure shows some examples of the kinds of trees used as inputs to the simulations in this chapter.

From a psychological perspective, the assumptions just outlined mean that the models are provided with both more information and less information than is available to the child learner. On one hand, unlike in real speech, all input forms are segmented into stems and inflections and labeled with their tense. This representation also enforces the constraint that all English verbs must be marked for tense. Unlike a child learner, the models can neither represent verbs as unanalyzed wholes, nor choose to not inflect a stem for tense. On the other hand, this representation fails to provide an important kind of information available to the child learner: it allows any `Inflection` to appear with any `Stem`, including phonologically impossible (i.e., unpronounceable) cases. In short, the child learner faces several learning problems not faced by our models, such as identification of tense, segmentation of verb forms, and discovery of phonological rules. However, the child learner also has several kinds of additional phonological and semantic information available to assist in these tasks.[2]

Taken together, these assumptions mean that the starting state for the models allows many more forms than are actually possible in practice. Since we do not model phonology or semantics, the only kind of information available to constrain this productivity over the course of learning is the distribution of stem/inflection co-occurrences in the input.

These modeling choices were made in order to allow us to isolate and study the relationship between the distribution of morphological units in the input data and productivity, free from other confounding factors. For example, if

2. We believe that the lack of constraining phonological and semantic structure is likely to be primarily adversarial to the problem of correctly inferring the pattern of productivity and reuse in the past tense system. Because these additional sources of evidence are correlated with productivity (e.g., because less phonologically transparent rules tend to be less productive, and vice versa), including this information should only improve simulation performance. See Section 1.2.3.1.

one of the models infers that an irregular rule, such as /ɪ/ → /æ/, cannot be generalized to novel stems, we can be confident this is because of the pattern of use of this rule with various stems in the input data, not because the rule itself was not discovered, or forms that use it (e.g., *sing/sang*) were not correctly segmented.

5.2.2 Training

The following sections describe the results of training the models with inputs derived from two corpora: the SWITCHBOARD corpus (Godfrey et al., 1992; Marcus et al., 1999) and the CHILDES database (MacWhinney, 2000). We describe data preparation common to both sets of simulations in this section.

Both corpora provide part-of-speech tagged utterances. From these utterances, all verbs were extracted along with their part of speech (i.e., `VB`, `VBP`, `VBG`, `VBD`, `VBN`, `VBZ`). Each verb was stemmed based on its form and part of speech. A first pass of stemming was performed using the WordNet stemmer (Fellbaum, 1998) provided as part of the *Natural Language Toolkit* (Bird et al., 2009), and then mistakes and missing stems were corrected by hand. The stemmed verbs were then paired with the appropriate inflectional marker (e.g., `SING:VBD` was paired with the rule /ɪ/ → /æ/).

Verbs from the training samples, together with their frequencies, were input as trees of the following form:

`(START (V (Stem` *stem*`) (Inflection` *phonology tense*`)).`

Simulations were run as described in Chapter 3. Also see Chapter 3 for a detailed mathematical description of the parameters and parameter values for the five models.

5.3 Adult Data: SWITCHBOARD Simulations

To provide an adequate account of competition between the inflectional classes of the English past tense, a model must explain several phenomena. First, a model must explain why the regular rule operates as a default. Second, it must explain *blocking*—the phenomenon whereby the availability of an irregular makes a regular form infelicitous. Third, it must provide an account of competition between generalizations. There are exceptions to both defaultness and blocking in adult speech. For example, there are cases where both a regular and an irregular form exist side-by-side (i.e., doublets such as *dreamed/dreamt*). Adults also sometimes produce speech errors, such as overregularizations (e.g., *go/goed*) and, more rarely, overirregularizations (e.g., *bite/bote*; Bybee, 1985; Stemberger, 1983; Xu and Pinker, 1995). Similarly, while novel verb rating and elicitation studies consistently demonstrate robust generalization of the regular rule to novel verbs, they also find that irregular classes can sometimes be generalized (Albright and Hayes, 2003; Ambridge, 2010; Bybee and Moder, 1983; Bybee and Slobin, 1982; Prasada and Pinker, 1993). Importantly, the model should allow for such variability while simultaneously capturing the fact that cases of overregularization and overirregularization are marginal. An account of competition should also explain the relationship between defaultness, blocking, and the frequency of linguistic structures in the input. In particular, it should be able to explain the differential frequency effects for regulars and irregulars found in the literature on past tense processing.

We will show in this section that the `FG` (inference-based) model is the only model, of the five considered in the book, which can explain both defaultness and blocking and can do so in a way that explains frequency effects discussed in the experimental literature.

5.3.1 Data Preparation and Simulations

The SWITCHBOARD corpus (Godfrey et al., 1992) consists of a large number of telephone conversations between strangers. The Penn Treebank provides a version of this corpus annotated for part-of-speech and other information (Marcus et al., 1999). To create the SWITCHBOARD corpus, pairs of strangers were chosen randomly, connected, and given a topic of conversation to discuss. Thus, this corpus provides a large sample of naturalistic adult speech. All verbs, except forms of *be*, *have* and *do*, were extracted from the corpus, stemmed, and paired with their past-tense inflectional class as described above.[3] Figure 5.3 summarizes the frequency of forms our training set.

110 simulations using the `FG` (inference-based) model were run on the training set, each with 10000, 15000, 20000, and 25000 inputs, type/token-binned

3. *Be*, *have*, and *do* are often used as both auxiliary and main verbs, and thus are extremely frequent. We excluded them for the practical reason that, because of their high frequency, doing so significantly reduced the size of the training corpus, reducing simulation run time. However, there is no theoretical reason why excluding these verbs should have any effect on the simulation results, and a large pilot study, in which they were included, exhibited exactly the same pattern of results that we report below.

	Tokens		
	Regular	Irregular	Total
VBD (past tense)	4480	6096	10576
VBZ (3rd singular)	4382	153	4535
VBN (past participle)	4044	2858	6902
VBG (progressive)	10890	0	10890
VBP (present)	30661	0	30661
VB (bare stem)	26528	0	26528
Total	80985	9107	90092

	Types		
	Regular	Irregular	Total
VBD (past tense)	601	107	708
VBZ (3rd singular)	514	1	515
VBN (past participle)	853	105	958
VBG (progressive)	907	0	907
VBP (present)	747	0	747
VB (bare stem)	1297	0	1297
Total	4919	213	5132

Figure 5.3

Frequencies of Verbs: This table shows the token (left) and type (right) frequencies of verbs from the SWITCHBOARD corpus used in the simulations reported here. There were 2117 stem types in total. Note that the one irregular VBZ form is *says*.

as described in Section 3.5.2. These ran for between 20 and 700 sweeps through the whole corpus, depending on time available on the MIT CSAIL cluster. We approximated the maximum a posteriori fragment grammar by using the highest probability approximating PCFG discovered by the search procedure described in Section 3.2. The approximating PCFGs for the other models were computed directly from the input corpus as described in Section 3.3.

5.3.2 Simulation Results

In this section, we will first discuss the overall performance of the models, and then turn to a more detailed analysis of their behavior on regular and irregular past tense forms. Following this, we will consider how the `FG` (inference-based) model can account for frequency effect data from the literature. We will discuss these results, and then conclude by examining two empirical phenomena, anti-frequency effects and storage of regular doublets, which are not directly reflected in the SWITCHBOARD simulations, but which can also potentially be explained by the `FG` (inference-based) model.

5.3.2.1 Overall Performance

The appropriate quantity for measuring the ability of each model to correctly inflect verb stems is the conditional probability of an inflectional marker given the stem. This quantity was computed for past tense (`VBD`) and past participle (`VBN`) forms. Because results were similar for both sets of forms, all numbers and plots in the rest of the chapter collapse across the two sets. We excluded forms for which there was no clear correct past marker, such as doublets, and forms for which there were errors in the training data that were not discovered prior to running the simulations (e.g., alternate British and American spellings of words such as *focussed/focused*, incorrect past forms, etc.). Three kinds of `VBD` and `VBN` test items were considered:

1. **Attested items:** The set of regular and irregular forms in the training data.
2. **Novel items with known stems:** A set of forms that did not appear within the training data, but for which English past forms exist. This set was constructed by locating stems in the input training corpus that appeared in other tenses (i.e., `VBZ`, `VB`, `VBP`, or `VBG`) but did not appear as either a past tense (`VBD`) or a past participle (`VBN`). The correct `VBD` or `VBN` form was identified by hand. There were approximately 1100 such forms.

3. **Wug tests:** We tested generalization to a completely novel "wug" stem that did not appear in the training set. The regular past tense inflection (i.e., +/d/) was considered correct for this stem.

Table 5.1 shows the probability that each model correctly inflected stems from each of the categories above, broken down into regular and irregular subsets (where appropriate). These numbers are approximations to the maximum a posteriori conditional probabilities of inflection given stem as computed against the (approximating) PCFGs for each model. Note that these numbers weight more frequent forms more highly than less frequent forms, and therefore place greater importance on correctly inflecting more common stems.

The first row shows performance over all categories of tests. This overall test set is dominated by forms that appeared in training and, therefore the MAG (full-listing) model performed best in this row. This is unsurprising since MAG (full-listing) memorizes all forms in the input. The next three rows show performance on forms that appeared in the training set, excluding novel items. Again, the MAG (full-listing) model performed best due to memorizing all training items.

In the next two rows, the pattern changes. These two rows show the performance of each model on novel VBD (past tense) and VBN (past participle) forms with stems that appeared in other tenses in the training data. Note that the majority of such forms are regular. In these categories, the best performance is exhibited by the DOP1 (exemplar-based) model. However, FG (inference-based) comes second in each case. This pattern is repeated for the next line: completely novel wug stems. These results show that DOP1 (exemplar-based) has learned a robustly generalizable +/d/ regular rule. However, DOP1 (exemplar-based) performs very poorly on irregulars from the training set, correctly inflecting at a rate of only 0.057%. DOP1 (exemplar-based) overgeneralizes the regular rule to these irregular items.

Examining the pattern of results more generally, FG (inference-based) demonstrates the best performance with regard to the pattern of defaultness and blocking characterizing the English past-tense system. It is able to correctly inflect irregular forms—that is, block the regular rule—while correctly treating the regular as a default and generalizing it to novel test cases. Although DOP1 (exemplar-based) also generalizes the regular rule robustly, it does so at the expense of correct performance on irregulars. Of the remaining models, DMPCFG (full-parsing) and ENDOP (exemplar-based) both perform relatively

Test Set	FG	DMPCFG	MAG	DOP1	ENDOP
Overall	0.953	0.227	**0.993**	0.875	0.380
Attested Overall	0.955	0.209	**0.995**	0.874	0.375
Attested Irregular	0.935	0.042	**0.995**	0.057	0.175
Attested Regular	0.981	0.488	**0.995**	0.998	0.662
Novel (Known Stems) Overall	0.906	0.454	0.804	**0.937**	0.488
Novel (Known Stems) Regular	0.930	0.472	0.824	**0.983**	0.510
Regular Wug	0.888	0.486	0.828	**0.989**	0.536

Table 5.1

Performance of Models on Past Tense Dataset: This table shows the probability that forms from different test sets will be correctly inflected by each model. The best scoring model is highlighted in bold in each row.

poorly on all tests. MAG (full-listing) shows the same overall pattern of performance as FG (inference-based) but the regular rule in this system is not as strongly a default.

One potential concern about the performance of the FG (inference-based) model is that while it correctly generalizes the regular rule to novel stems, it still assigns substantial probability to other, irregular inflections for these stems. For example, in the novel wug-test cases above, the regular +/d/ ending has a conditional probability of around 0.89. This, however, means that approximately 11% of the time the model still assigns one of the irregular endings to this wug stem. Although we do not know of any studies directly addressing rates of overirregularization in adults, it is a clearly a rare phenomenon, and very likely to have a rate much lower than 11%.

However, this may not be as large of a concern as it first appears. This conditional rate of overirregularization is divided between more than 30 irregular inflectional processes. For a given novel stem, most of these processes will be phonologically impossible. For instance, the process /ɪ/ → /æ/ cannot apply to any stem which does not contain the stem vowel /ɪ/. Moreover, most irregular past tense processes impose additional selectional restrictions on the stems to which they can apply. For example, the process mentioned above is sometimes formulated as applying only to monosyllabic stems ending in the velar nasal /ŋ/. However, because we do not model either phonological selectional restrictions or phonological rule discovery, we cannot capture such facts. Nevertheless, if the FG (inference-based) model were combined with a suitable model of phonology, most instances of overirregularization contributing to the rate of 11% would be ruled out immediately. For cases of novel stems which do conform to input conditions on irregular rules (e.g., *spling*), the rate of overirregularization would still be significantly lower than 11% since only a subset of the 30 irregular inflectional process could apply to even these stems. Nevertheless, we leave to future research the question of whether or not, after these facts are taken into account, the residual rates of overirregularization produced by the model are empirically viable.

The five models studied here differ in the kinds of structures that they store, and this difference is what leads to the patterns of performance just described. In the next section, we examine these patterns of storage and reuse in more detail.

5.3.2.1.1 Expected Number of Constituents

Each of the five models makes different choices about which computations to store, and what structures to compute on the fly. To understand the pattern of results above, we must investigate these choices in more detail. One useful way of doing this is to examine the expected number of constituents for each form under each model. Given a particular form, each model defines a distribution over all of the derivations that could give rise to that form. By counting the number of fragments used in each derivation, weighting this number by the probability of the derivation, and, finally, summing over all derivations, the expected number of constituents can be computed for the form. If a form is retrieved whole with high probability, then this number will be low. Alternatively, if a form has a high probability of being built out of small pieces, then the number will be higher. If, instead, a model assigns equal probability to a number of derivations, some of which contain many distinct fragments, and some of which contain few, the expected number of constituents will be intermediate.

Each form in the training data has a tree structure like that shown below.

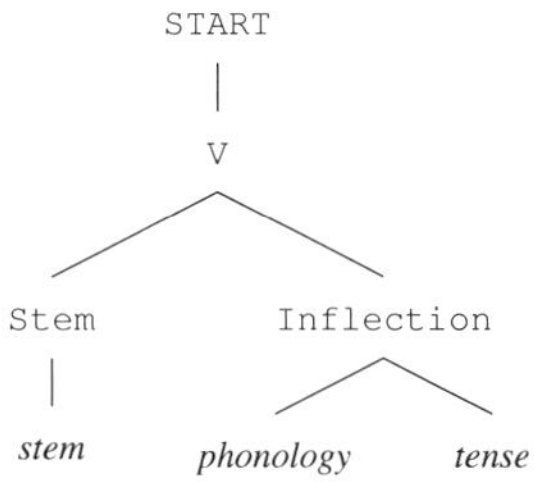

The maximum number of independently computed constituents for a past-tense tree is four since the models can represent a variable at each of the non-leaf nodes of the tree above (i.e., at the `START`, `V`, `Stem`, and `Inflection` nodes). Because a fragment must be chosen at the `START` node, the minimum number of constituents is one.

Figure 5.4 plots the expected number of constituents predicted by the `FG` (inference-based) model for the forms in the test sets. The *x*-axis is the frequency (log scale) with which the form appeared in the training set. Note that novel forms appear at frequency 0 on the left-hand side of the plot. Irregular forms are plotted as open blue squares while regulars are plotted as filled red circles. Wug-test items are considered regular. The size of the point for each form is scaled according to the probability that the form was correctly inflected by the model. Smaller points represent forms less likely to be inflected correctly. Larger points represent forms more likely to be correctly inflected.

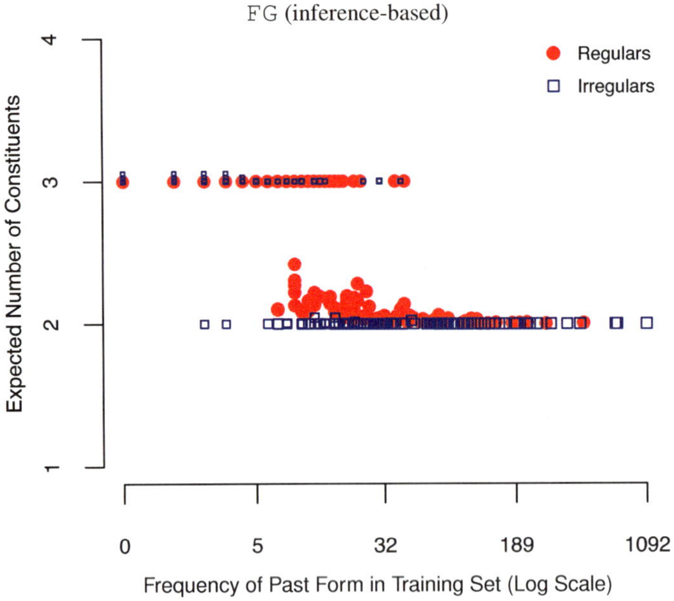

Figure 5.4

Expected Number of Constituents for `FG` (inference-based): The expected number of constituents for each test form. Irregulars are shown as open blue squares, and regulars are shown as filled red circles. Point sizes are scaled such that forms that were more likely to be correctly inflected are larger.

First, consider the irregulars (open blue squares) in Figure 5.4. Correctly inflected irregulars, represented by larger blue squares, have two expected constituents, corresponding to computation at the START and V nodes of the past tense tree, but storage of stems together with inflections. That is, for correctly inflected irregular forms, the FG (inference-based) model has stored tree fragments like the following:

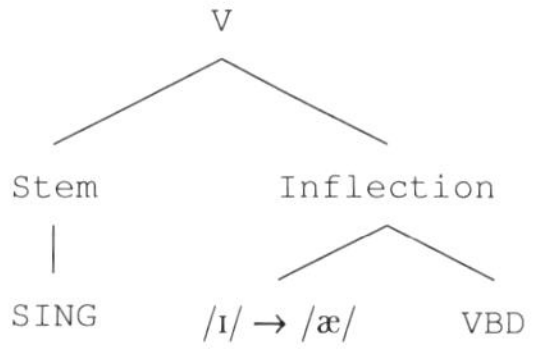

Note that the FG (inference-based) model is treating the V node as a variable because this part of the tree is shared amongst all forms—regular and irregular, but its internal makeup varies between these two sets.

There are also a number of incorrectly inflected irregulars (smaller blue squares) towards the lower end of the frequency spectrum, which appear with three expected constituents. These forms are overregularized by the model, due to their low token frequency in the training set. The regular rule is represented by a fragment like the one below.

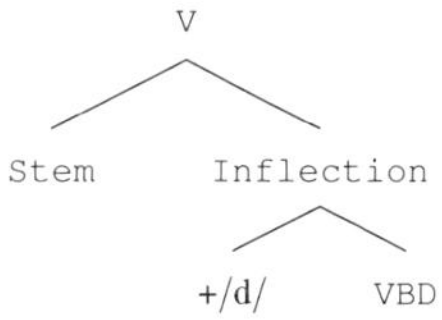

As a result, these overregularized forms have three expected constituents corresponding to computation at the START, V, and Stem nodes of the past tense tree.

Turning to the regulars (filled red circles), Figure 5.4 shows that the majority of regulars, including novel (frequency 0) forms, also have three expected constituents. As was the case with the overregularized irregulars, this reflects the use of the regular rule, with choices at the START, V, and Stem nodes in each derivation. However, there are also a considerable number of regulars—increasing with the frequency of the form—which have less than three expected constituents. These high-frequency regulars are sometimes stored as wholes like irregulars. We will return to the storage of high-frequency regulars in more detail later in the chapter.

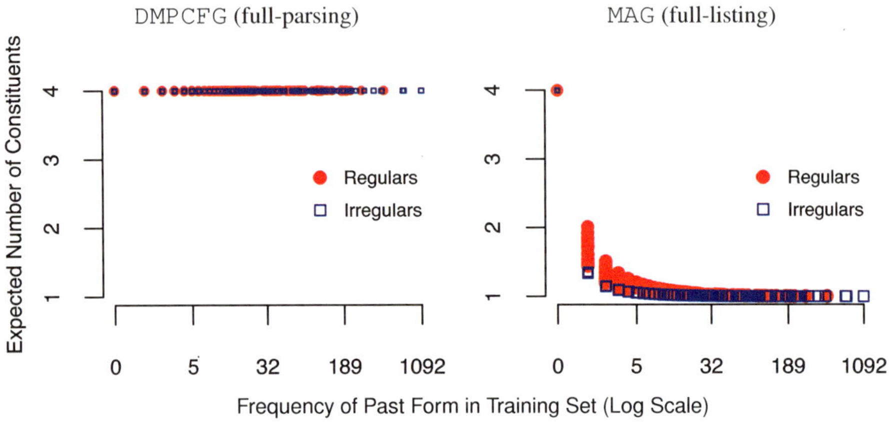

Figure 5.5

Expected Number of Constituents for DMPCFG (full-parsing) and MAG (full-listing): The expected number of constituents for each past tense form. Irregulars are shown as open blue squares, and regulars are shown as filled red circles. Point sizes are scaled such that forms which were more likely to be correctly inflected are larger.

Figure 5.5 shows the expected number of constituents for the DMPCFG (full-parsing, left) and MAG (full-listing, right) models. Recall that in our evaluation of model performances (Table 5.1), the DMPCFG (full-parsing) model performed poorly on all tests. This follows from the fact that this model parses all forms in full, and therefore cannot learn contingencies between stems and inflections. This is shown on the left-hand side of Figure 5.5—all forms necessarily have exactly four expected constituents under this model.

Because MAG (full-listing) memorizes each form in the dataset, it performed well on attested items in our evaluation (Table 5.1). The right-hand side of Figure 5.5 demonstrates that most forms in the MAG (full-listing) model are treated as wholes, and have an expected number of constituents near one, especially for more frequent forms. Note that unlike FG (inference-based), MAG (full-listing) cannot treat the V node of each tree as a variable, and so most forms have an expected number of constituents close to one rather than two.

In addition to storing the entire derivation for each input form, MAG (full-listing) also recursively stores all (complete) sub-derivations of each form, and can use such stored subforms to account for later derivations. In the case of infrequent forms, the probability associated with the complete stored tree is relatively low, and therefore analyses treating the item as new and making use of stored sub-derivations can contribute significant probability mass to the total score of a form. The left-hand side of the MAG (full-listing) plot demonstrates this effect. Forms that appeared infrequently in the training set have an expected number of constituents greater than one, although this number never reaches noticeably beyond two. By contrast, forms that did not appear in the training set at all (i.e., frequency 0 forms) are at the upper limit of four expected constituents, because MAG (full-listing) cannot reuse any stored subtrees for these forms and must build them completely from scratch using its underlying PCFG.

As mentioned above, MAG (full-listing) also performed well on (regular) novel items. This is a result of the way in which MAG estimates the weights of rules in its underlying PCFG. Each underlying rule of the MAG (full-listing) model has a probability (approximately) proportional to the type frequency (i.e., the number of distinct verb forms) of the items in which it appears. Because the regular rule is the most type-frequent inflectional marker in the English past-tense system, MAG (full-listing) can generalize this rule correctly with high probability.

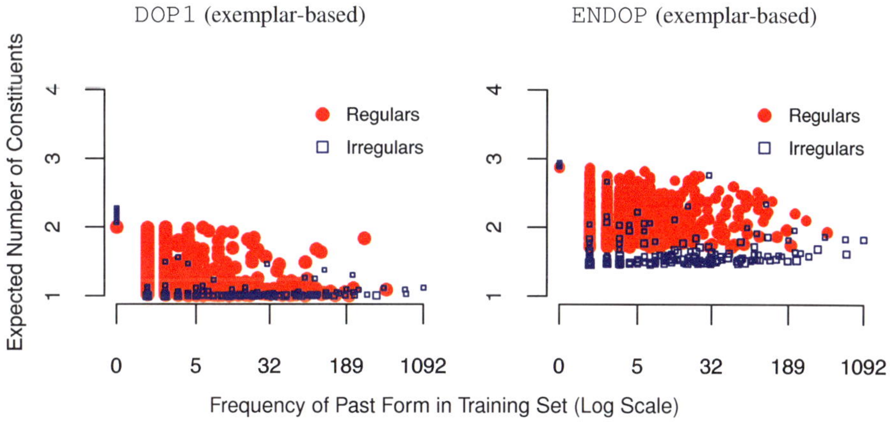

Figure 5.6

Expected Number of Constituents for DOP1 (exemplar-based) and ENDOP (exemplar-based): The expected number of constituents for each past tense form. Irregulars are shown as open blue squares, and regulars are shown as filled red circles. Point sizes are scaled such that forms that were more likely to be correctly inflected are larger.

The performance of the two exemplar-based models (DOP1 and ENDOP) is more complex. Figure 5.6 shows the expected numbers of constituents for each. Both models show significantly more variation than the previous three models. This is a result of the fact that exemplar models do not commit to a single analysis of each form, but rather, spread their probability mass over all generalizations consistent with the input. In both models, there is a tendency for irregulars to consist of fewer constituents, and regulars to consist of a greater number of constituents. However, this dichotomy is not nearly as sharp as with the FG (inference-based) model, which accounts for the inability of these models to simultaneously capture defaultness and blocking (as seen in Table 5.1).

The DOP1 (exemplar-based) estimator favors larger subtrees and therefore, on average, it produces lower expected numbers of constituents than ENDOP (exemplar-based). This fact explains the robust performance of DOP1 (exemplar-based) on the regular rule. Because it favors larger subtrees, it is able to place more probability mass on the frequent subtree corresponding to the regular rule. By contrast, ENDOP (exemplar-based) places more mass on depth-one trees corresponding to rules in the underlying CFG, and therefore overestimates the probability that stems will combine productively with irregular inflectional classes.

5.3.2.2 Whole-Form Frequencies and Stem Frequencies

In Section 4.3.3 and 4.3.4, we discussed the differential frequency effects for regulars and irregulars that arise in production and naturalness judgment tasks. In production tasks, high-frequency irregulars are named faster than low-frequency irregulars, while there is no such effect for regulars (Beck, 1997; Prasada et al., 1990; Seidenberg and Bruck, 1990; Shenkman, 1994). In a set of naturalness judgment tasks, Ullman (1993, 1999) found that, when stem judgments are partialed out, participants' judgments of irregulars are predicted by their whole-form frequencies, while their judgments of regulars are not. These differences in frequency effects are taken as evidence that irregular forms are stored independently and can therefore accumulate strength in memory, while regular forms are composed.

In this section, we examine the relationship between stem and whole-form frequencies and the probabilities assigned to individual forms by each model. The specific quantity on which we focus is the marginal probability of a form. The marginal probability of a form is the total probability assigned to that form by a model—the sum of the probabilities of all possible derivations of the form under the model.

Table 5.2 shows the results of a partial correlation analysis similar to that presented in Ullman (1999). The numbers in the table represent the correlations between marginal probability for each form and its frequency in the training set, with the stem frequency partialed out of both values. The experiments in studies like Ullman (1999) were performed on adults, who know the correct past tense form of verbs. To compare those partial correlation analyses to similar analyses on the outputs of our models, we need to restrict attention to forms where each model "knows" the correct past tense. Therefore, the analyses in Table 5.2 are restricted to past forms with at least 85% probability of being correctly inflected by the corresponding model.[4] Note that in some cases there were no forms meeting this criterion, and therefore that the corresponding table entries are marked "NA."

Forms	FG	DMPCFG	MAG	DOP1	ENDOP
Irregulars	0.999	NA	0.988	NA	NA
Regulars	0.344	NA	0.831	0.650	0.551

Table 5.2
Partialed Correlation of Marginal Probability and Whole-form Frequency Controlling for Stem Frequency

Of the five models, only FG (inference-based) and MAG (full-listing) correctly inflected a sufficient number of both regulars and irregulars to perform the analysis for both sets of forms. For MAG (full-listing), the probability of both regulars and irregulars show a strong correlation with whole-form frequency when stem frequency is controlled. By contrast, for FG (inference-based), there is a large difference between the correlations for these two subsets of forms. Irregulars show an almost perfect correlation with their stem frequencies, while irregulars show a much lower correlation. Thus the FG (inference-based) result is more consistent with the results in Ullman (1993) and Ullman (1999).

To understand the pattern of results in Table 5.2, consider the way that the FG (inference-based) model accounts for regulars and irregulars. Most regulars are composed; the probability of each derivation is the product of the probabilities of its parts—in particular, it is the product of the probability of the stem and the regular past-tense ending. These probabilities, in turn, are proportional to the frequencies of the stem and regular past ending in the training data.[5] Across

4. Out of 3961 potential existing (attested and novel with known-stem) English past forms from in the preceding section, the following numbers of forms were excluded for failing to meet this criterion: FG: 143, DMPCFG: 3961, MAG: 1348, DOP1: 229, ENDOP: 3640.

5. This is statement is true, on average, across high-probability posterior states.

composed regulars, the past-tense ending is shared, and therefore has constant probability. Thus, any variance in the probabilities of composed regulars will be driven primarily by stem frequency, which is partialed out in these analyses. By contrast, irregulars are stored as wholes by the FG (inference-based) model. Their probabilities are (roughly) proportional to their whole-form frequencies in the input.

Unlike the FG (inference-based) model, the MAG (full-listing) model stores all forms as wholes, and thus shows a strong correlation between marginal probability and whole-form frequency for both regulars and irregulars. The results in Table 5.2 are plotted for the FG (inference-based) and MAG (full-listing) models in Figure 5.7.

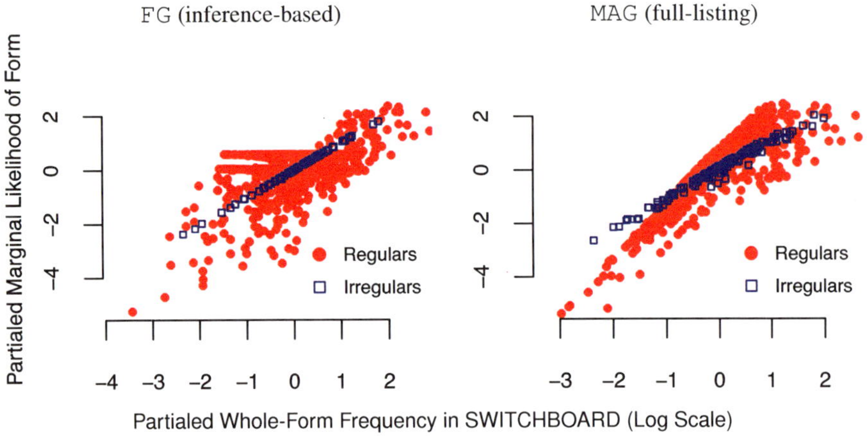

Figure 5.7

Effect of Full-Form Frequency on Marginal Probability for FG: The x-axis shows the log full-form frequency, with stem frequency partialed out. The y-axis represents the log marginal probability of the form, with stem frequency partialed out. Irregulars are shown as open blue squares, and regulars are shown as filled red circles.

The y-axis in Figure 5.7 plots the residuals of the marginal probabilities of each form after partialing out stem frequency. The x-axis plots the residual whole-form frequencies after partialing out stem frequency. Irregulars are plotted with open blue squares and regulars are plotted with filled red circles. The differences between FG (inference-based) and MAG (full-listing) emerge clearly in these plots. For FG (inference-based), irregulars fall along an almost perfect line with positive slope, while regulars seem to fall into two large groups. Most low- and medium- frequency regulars show little (partialed) correlation with whole-form frequency. A smaller number of regulars in the FG (inference-based) plot seem to show a somewhat stronger positive relationship. By contrast MAG (full-listing), shows a strong positive relationship between (partialed) marginal probability and (partialed) whole-form frequency for all forms.

We showed in our analysis of expected number of constituents, that the FG (inference-based) model stores some high-frequency, regular forms as wholes (see Figure 5.4). Experimental evidence for the storage of high-frequency, regular forms has been reported by Alegre and Gordon (1999) and Gordon and Alegre (1999), who showed, using a lexical decision paradigm, that these forms exhibit a frequency effect not exhibited by lower frequency items. Under the FG (inference-based) model, forms that are stored whole should exhibit a whole-form frequency effect. This effect is confirmed by dividing the regulars into stored and composed subsets based on expected number of constituents and correlating each with partialed frequency. Stored regulars correlate more highly with partialed frequency ($r = .743$) than composed regulars ($r = 0.255$).

To summarize, of the five models, only FG (inference-based) produces relationships between regulars and irregulars and between high- and low-frequency regulars matching the experimental literature.

5.3.2.3 Discussion

In the preceding sections, we showed that the FG (inference-based) model was the only model able to provide an empirically adequate account of the past tense when trained on a corpus of adult speech, exhibiting the correct pattern of blocking and defaultness, and the same relationship between stem and whole-form frequencies that arises in the experimental literature. In this section, we discuss the origin of this behavior.

Why does the FG (inference-based) model achieve the correct pattern of defaultness and blocking, and why is this pattern linked to the frequency of whole forms and stems in a way that matches the experimental literature? These patterns arise because of three properties of the model. First, as discussed in Chapter 1, the FG (inference-based) model balances a trade-off between productivity and reuse with respect to the input data. Each tree fragment stored by the model can be understood as a prediction about the relative frequency of that fragment in the input data and the probability of future forms containing that fragment. Second, in high probability posterior states (i.e., grammars), the probability of a stored tree fragment will be (approximately) proportional to the frequency of that tree fragment in the input data. Third, all else being equal, forms accounted for by composing a greater number of stored fragments will tend to be assigned lower probability than forms accounted for by composing a smaller number of stored fragments. This follows straightforwardly from the fact that the probability of a derivation is the product of the probability of its component tree fragments. Since a product of probabilities declines geometrically in the number of terms in the product, forms built with a larger number of tree fragments will tend to have lower probability, and therefore lower predicted (relative) frequency in the input data (see Section 1.3).

In the simulations reported above, the `FG` (inference-based) model learned to account for most regular forms by composing a stem and a fragment corresponding to the regular suffix. The probabilities of these composed forms was, therefore, given by a product of the probability of the stem (proportional to its frequency across tenses) and the probability of the regular suffix (proportional to its frequency across composed regulars). Since these quantities are products, they are lower on average than they would be if these forms had been stored as wholes. However, for most regulars, this is the solution that best predicts the frequency distribution in the input. Most individual regulars are relatively low frequency, but the regular ending itself has high probability of being used to generate new verbs. By treating most regulars compositionally, the model can assign high probability to the regular rule itself, allowing it to function as a default. By contrast, most irregulars in the input occur with a frequency that is greater than would be expected if they were composed out of their stems and inflectional markers. Thus, the model learned to store these forms as wholes. Blocking arises because these stored irregulars have higher probability than the corresponding compositional derivation that combines their stem with the regular rule. We discuss this phenomenon in more detail in Section 5.5.3.

The differential treatment of regulars and irregulars also explains the relationship between their marginal probabilities and their stem/whole-form frequencies. Regulars are composed; their probability is given as a product of their stem probability, which is proportional to the frequency of the stem in the input, and the probability of the regular ending, which is proportional to its frequency across composed regulars. Since the latter is a constant factor across composed forms, variance in the probability of regular forms is accounted for primarily by stem frequency. When this is partialed out, little variance remains.

By contrast, irregulars are stored whole and, therefore, their probability is proportional to their whole-form frequency. Their probability is still positively correlated with their whole-form frequency even when their stem frequency is partialed out.

As we discussed in the preceding sections, the FG (inference-based) model also stored a number of high-frequency regulars. This can also be understood in terms of the three properties above. Each regular can be derived compositionally. However, for some of these forms, their observed frequency still exceeds the predicted frequency under the compositional route. To account for this discrepancy, the model stores additional copies of these forms as wholes in memory. Importantly, the model only needs to account for the difference between the observed frequency in the input and the predicted frequency under the compositional route. This means that, in general, a regular form will have to be more frequent in order to be stored as a whole than a corresponding irregular. Furthermore, because the model only needs to account for the difference between observed and predicted frequencies, some individual instances of each high-frequency regular will be assigned whole-form derivations, and some instances of each will be composed, leading to competition between the two possibile ways of deriving these verbs (see Section 5.3.3.2, below).

The conditions under which FG (inference-based) stores high-frequency regulars explain an additional empirical effect reported by Alegre and Gordon (1999). Alegre and Gordon showed that storage of high-frequency regulars is modulated by the overall stem frequency of the form. Regular forms transparently share their stems with other tenses in the English verbal paradigm (i.e., with tenses such as the 3rd singular [VBZ] and progressive [VBG]), and there is a large amount of variability between verbs in the total frequency of the stem across all tenses. Alegre and Gordon (1999) performed an analysis of their reaction time data where they compared subsets of high-frequency regulars with high and low cross-tense stem counts. They found that high-frequency regulars with high cross-tense stem counts showed less of a frequency effect than high-frequency regulars with low cross-tense stem counts. In other words, when a stem was shared robustly across many forms of a verb, individual forms of the verb were less likely to be stored. This can be explained by again considering observed and predicted frequencies. As the frequency of the stem increases so does the predicted frequency of the verb form under the compositional route. This, in turn, reduces the discrepancy between observed and predicted frequencies that the model must account for via whole-form storage. Therefore, even if we hold the whole-form frequency of a particular past tense item constant, an increase in the frequency of the stem in other tenses should reduce the rate of whole-form storage, which is exactly the reported effect.

5.3.3 Other Phenomena

In this section, we consider two additional phenomena from the past-tense literature: the storage of regular doublets and anti-frequency effects. We present these analyses in a separate section, because, while they relate the behavior of the FG (inference-based) model to phenomena discussed in the literature, they do not directly involve evaluation of simulations on naturalistic data like those above.

5.3.3.1 Doublets

A closely-related phenomenon to the storage of high-frequency regulars is the phenomenon of storage of doublets—verbs which appear in the past tense in both irregular and regular forms. For example, for many native speakers of English, both *dreamed* and *dreamt* are grammatical as past tenses of the stem DREAM.

Ullman (1993) provided evidence from a naturalness rating task that both the regular and irregular past-tense forms of doublets correlate in naturalness with their whole-form frequencies (when stem judgments are partialed out), suggesting that both forms are stored in memory. While it is unsurprising that irregular doublet forms (e.g., *dreamt*) are stored in memory, the storage of regular doublets (e.g., *dreamed*) may be unexpected. In general, these forms are not extremely frequent, which rules out an explanation that they are stored for the same reasons as other high-frequency regular forms (Pinker, 1999; Ullman, 1993)

An alternative account was offered by Pinker (1999), who argued that storage of regular doublets is the result of the need to overcome blocking, which would otherwise result from the stored irregular form. Intuitively, since *dreamt* must be stored, *dreamed* should be blocked. However, *dreamed* nevertheless appears in the input. The only mechanism available to the learner to account for its appearance is to store *dreamed* in memory alongside *dreamt*.

Doublets are relatively rare, and the sample from the SWITCHBOARD corpus only contained a single example (i.e., *proved/proven*), which was insufficiently frequent for the model to correctly learn both the regular and irregular forms. Therefore, to examine how the FG (inference-based) model handles doublets, we conducted simulations on a series of artificial datasets generated from a vocabulary of 1000 artificial stems. All but one of these training stems appeared with a single inflectional marker (corresponding to the regular rule, i.e., +/d/). One training stem appeared with both the regular inflectional marker, and an irregular inflectional marker (e.g., /ɪ/⟶/æ/), representing an (idealized) doublet. Stems were assigned ranks, and random training data sets (of size 2001) were constructed by sampling from a power-law distribution over ranks. The percentage of the doublet forms that appeared as regulars was varied between 0% and 100% across training sets. For example, in some training sets, the doublet appeared 50% of the time as a regular and 50% of the time as an irregular, in others 60%/40%, 90%/10%, etc. Simulations were run on twenty randomly generated input datasets at each doublet percentage. The sampler was run for 200 sweeps in each simulation, and the results were averaged across the twenty simulations at each doublet percentage.

Figure 5.8 shows the breakdown of storage and composition for input forms in these simulations. The x-axis represents the proportion of doublet instances that appeared as regulars. At the far left-hand side of the plot, all forms of the doublet stem were irregular, and therefore therefore the form was simply a high-frequency irregular form. At the far right-hand side of the plot, the doublet stem did not appear as an irregular at all, and therefore was simply a high-frequency regular form. In between, the doublet stem appeared in intermediate mixtures of regular and irregular instances.

For training sets where the doublet stem appeared in both irregular and regular forms there are three quantities to consider: (i) the rate at which irregular forms of the doublet were stored as wholes, shown with open blue squares, (ii) the rate at which regular forms of the doublet were built compositionally using the regular rule, show with filled red circles, and (iii) the rate at which regular forms of the doublet were stored as wholes, shown with open red circles. There is an additional logically possible case: the rate at which irregular forms of the doublet were built compositionally using an irregular rule. However, because all irregulars were always stored by the model, there were no such forms in these simulations.

As shown in Figure 5.8, the FG (inference-based) model stored all irregulars forms of the doublet, and thus the open blue squares form a perfect line with

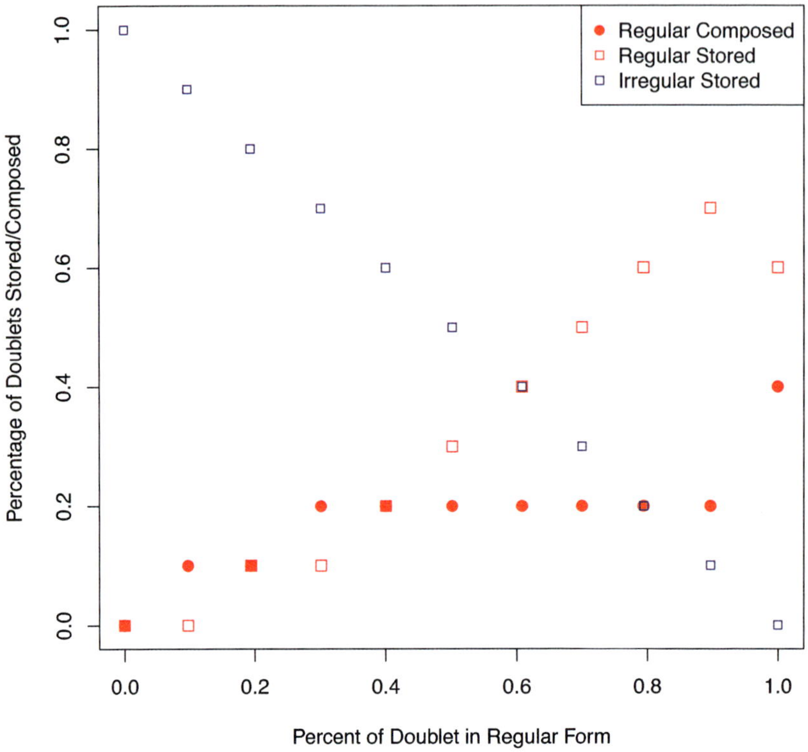

Figure 5.8

Doublets: The *y*-axis of this figure shows the percentage of doublet forms that were treated as (i) composed regulars (filled red circles), (ii) stored regulars (open red squares), and (iii) stored irregulars (open blue squares). The *x*-axis is the percent of doublet forms that were regular in the training input.

negative slope. When the percentage of regular doublets in the input is at 10% or less, all regular forms of the doublet are composed. As the percentage of regular forms of the doublet increases, there is a small increase in the rate of composition for these forms. Interestingly, however, the percentage of the regular doublet forms that are composed plateaus at 20%. Beyond this proportion, all additional regular forms are stored whole (with the exception of the case where the "doublet" is 100% regular).

This behavior can be understood in terms of the comparison of observed and predicted relative frequencies that we discussed in Section 5.3.2.3. The predicted relative frequency of a composed, regular doublet form is proportional to the number of composed stem instances multiplied by the frequency of the regular rule. In these simulations, this predicted relative frequency is approximately 20%. The model must account for any additional, unpredicted instances of the regular form by storing them whole. This can be seen clearly in Figure 5.8. However, when the doublet is 100% regular, the entire training set is regular; the regular rule is used at a higher frequency and can predict a higher percentage of the input forms, reducing the need for whole-form storage.

As we discussed above, the `FG` (inference-based) model stores some high-frequency regular forms, for the same reason as it stores some regular instances of the doublet—to account for the discrepancy between observed and predicted frequencies. However, there is an important difference between the case of doublets and the case of other high-frequency regulars. Because the irregular forms of a doublet are always stored, their stems are not shared with other tenses of the verb. This reduces the probability of composing these stems with the regular rule and, thus, increases the discrepancy between observed and predicted frequencies, increasing the need for whole-form storage. Thus, competition with irregular doublet forms causes `FG` (inference-based) to store regular forms of the doublet at a higher rate than other regular forms. This can be seen as a formal validation of the account of doublet storage offered by (Pinker, 1999). Regular forms of the doublet are stored in order to avoid being blocked by irregular forms of the doublet.

5.3.3.2 Anti-Frequency Effects

In production experiments where participants are asked to produce a past tense form from its stem, regular forms are sometimes produced faster than irregulars (see, Beck, 1997; Prasada et al., 1990; Seidenberg and Bruck, 1990;

Shenkman, 1994).[6] Furthermore, regular forms sometimes exhibit an anti-frequency effect in which high-frequency regulars are produced more slowly than lower-frequency regulars (see, Beck, 1997; Prasada et al., 1990). Pinker (1999; p. 303) discusses an account of anti-frequency effects based on competition between stored and composed regular forms (see, also, Balota et al., 2000). Because high-frequency regular forms are sometimes stored whole in memory, and stored forms block the output of the regular rule, competition is expected between stored and composed high-frequency regulars. This competition, in turn, slows the processing of these forms.

As we discussed above, high-frequency regulars are sometimes stored as wholes by the `FG` (inference-based) model. This leads to a representational ambiguity for these forms, and, therefore, a kind of competition that is not present for irregulars or low-frequency regulars. In this section we examine this competition in more detail. To do this, we compute the *entropy*—a measure of the average amount of uncertainty in the distribution over derivations for each form. The entropy of the distribution over derivations will be low if the model chooses to represent a form in a single way (i.e., composed or stored whole) with probability close to one. The entropy will be highest when the model puts equal probability on all derivations of a form, that is, when it is maximally ambiguous about storage versus computation. Figure 5.9 plots the entropy for items from the SWITCHBOARD training set as a function of their whole-form frequency.

In Figure 5.9, both irregulars and low-frequency regulars have low entropy. The model is very confident that they should be retrieved whole from memory, and composed, respectively. However, as the frequency of regulars increases, so does their entropy. Entropy is highest for those regulars in the middle–high frequency range. For these forms the whole-form retrieval route is available, but is not yet strong enough to overwhelm competition with the compositional route.

These results are broadly consistent with the theory proposed by Pinker (1999). Some higher-frequency regulars exhibit competition between storage and retrieval routes. However, this analysis makes a prediction that, to our

6. While these elicited naming experiments show an advantage for regulars, this is not generally the case for all latency-based methods. For instance, in the primed lexical decision experiments reported in Stanners et al. (1979a), regular past-tense primes are associated with longer latencies than irregulars. Thus the speed advantage for regulars may be task-specific.

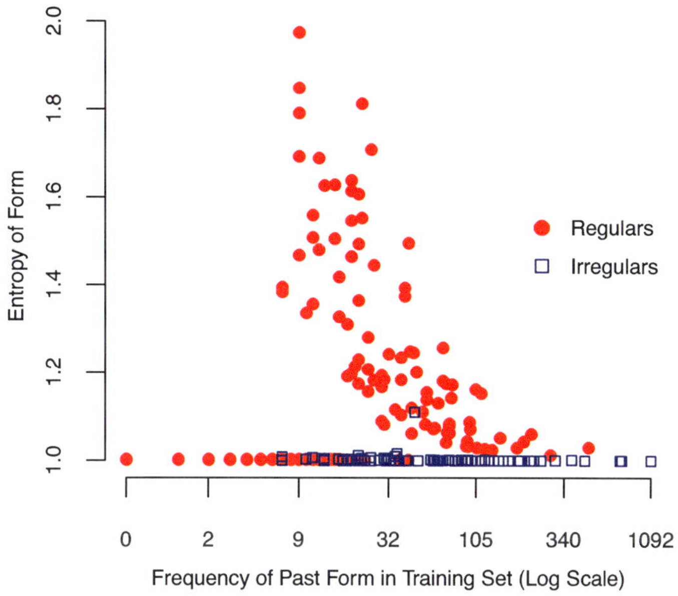

Figure 5.9

Entropy of Regular Forms by Frequency in FG (inference-based) Model: This figure shows the entropy (uncertainty) over derivations for regular and irregulars forms as a function of their whole-form frequency (log scale). Irregulars are shown as open blue squares, and regulars are shown as filled red circles.

knowledge, has never been proposed before. In Figure 5.9, the entropy is highest for regulars in the middle–high frequency range, but is lower for both low-frequency regulars and the highest-frequency forms. This predicts that anti-frequency effects should appear only in this mid-frequency range. Previous studies (e.g., Beck, 1997) have reported mean production latencies for two groups of verb forms: low-frequency regulars (whose frequency falls in the low-frequency range in Figure 5.9), and high-frequency regulars (whose frequencies fall in both the middle–high and high frequency ranges in figure 5.9), and so cannot be used to evaluate this prediction. Testing this prediction is an interesting avenue for future work.

5.4 Developmental Data: CHILDES Simulations

In this section, we study the developmental predictions of the five models. To do this, we will examine the behavior of each model as it is trained on a sequence of naturalistic samples of English verbs, of increasing size. Each step in this sequence can be understood as an idealization of the increasing amount of data that children are exposed to as they grow older. The simulations, therefore, examine how predictions about productivity and reuse change for each model as it is exposed to more information about the English verbal system.

The most important empirical phenomenon in past tense development is the overregularization of irregular verbs. In Section 4.3.5, we discussed five important empirical properties of overregularization. First, all English-speaking children overregularize at some point during development. Second, the overall rate of overregularization is relatively low. Third, irregulars are protected from overregularization by their frequency and their similarity to other irregulars. Fourth, at least for some children and some verbs, there is a period of early correct performance preceding the onset of overregularization. Fifth, the onset of overregularization is correlated with an increase in the rate of correct, obligatory marking of regular forms.

As we will discuss below, some of these phenomena are beyond the scope of the present simulations because they involve unmodeled kinds of structure (e.g., phonology) or developmental change in the space of possible representations (e.g., the discovery of the regular rule), rather than changes in productivity and reuse. However, we will show that for those phenomena concerning competition between productive and unproductive past-tense forms only the `FG` (inference-based) model is able to produce an account consistent with the empirical facts.

5.4.1 Data Preparation and Simulations

The following simulations were performed on data extracted from the CHILDES corpus (MacWhinney, 2000). Data from the following corpora were used: Bloom (1970, 1973), Brown (1973), Clark (1978), Demetras (1989a,b), Gleason (1980), Higginson (1985), Kuczaj (1977), Nelson (1989), Demuth et al. (2006), Sachs (1983), Suppes (1974), MacWhinney (2000), and Weist et al. (2009). Files that were annotated for part-of-speech information (i.e., were marked with MOR tags, MacWhinney, 2008) and that were marked for target-child age were chosen from these corpora and processed as described below.

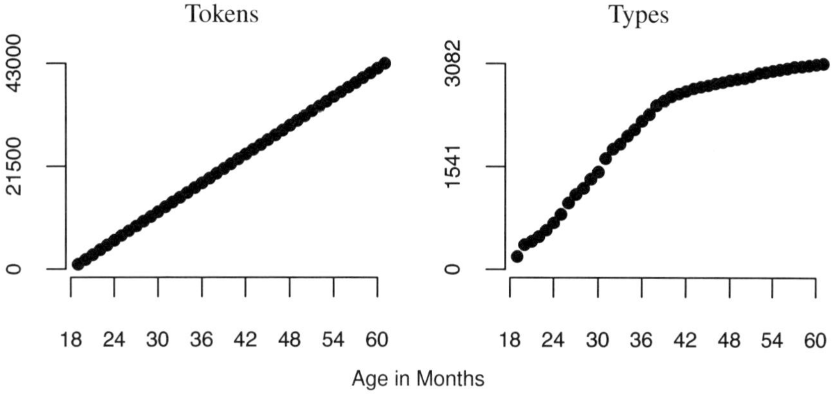

Figure 5.10

Total Cumulative Regular and Irregular Types and Tokens: This figure shows the cumulative total number of past-tense tokens (left) and types (right) for each training increment used in the CHILDES simulations.

To simulate the input available to a child, all relevant adult speech was used.[7] All verbs excluding forms of *be*, *have* and *do*, were extracted from the corpus. MOR tags for verbal categories were mapped into their corresponding Penn TreeBank verbal categories (i.e., `VBD`, `VBN`, `VB`, `VBP`, `VBZ`, `VBG`; Marcus et al., 1999), taking into account MOR categories (i.e., `v`, `aux` and `part`) and distinguishing fusional markings and suffixation. The verbs were lemmatized and paired with the appropriate past-tense inflection as described in Section 5.3.1.

Verb tokens were divided into sets according to the age of the target child in the file in which they were found, and a corresponding empirical relative frequency distribution was constructed for each child age in months from 1;6 to 5;0. The result was an empirical estimate of the relative probabilities of each verb form in the speech of adults speaking to (or around) children of that age. These relative frequency distributions were then used to construct a series of forty-three training sets, each with approximately 1000 tokens, by multiplying the relative frequency of each verb by the total number of tokens. The input

7. In particular, any utterances tagged with the following CHILDES roles were excluded: `Target_Child`, `Child`, `Playmate`, `Non_Human`, `Environment`, and `Camera_Operator`. All other utterances were included.

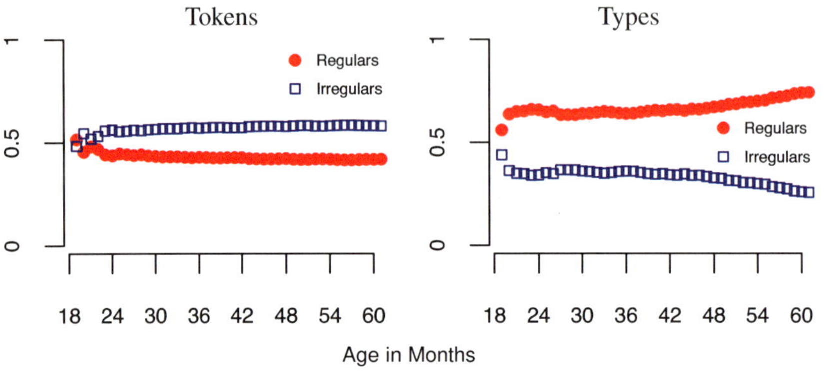

Figure 5.11

Percent Cumulative Regular and Irregular Tokens and Types: This figure shows the cumulative proportion of irregular (open blue squares) and regular (filled red circles) tokens (left) and types (right) for each training increment used in the CHILDES simulations.

to each model at each time step consisted of the cumulative verb tokens in all of the training sets up to, and including, the target age. Figure 5.10 shows the cumulative types and tokens in the input datasets for each age in months.

Figure 5.11 shows the breakdown in cumulative percentages of tokens and types for regulars (filled red circles) and irregulars (open blue squares). What can be seen is that, aside from a small amount of variability in the initial few samples—likely due to the small sample sizes associated with these time steps—there are no sudden discontinuities in the proportion of regulars in either types or tokens. Rather, the input data is characterized by a gradual overall increase in the proportion of regular types, and an overall greater proportion of irregular tokens at most ages—consistent with the known empirical facts about the input received by children learning English (Marcus et al., 1992).

For these analyses, six fragment grammar simulations were run, each with 10000, 15000, 20000, and 25000 inputs, for each training set (i.e., for each age) as described in Sections 3.2 and 3.5. These ran for between twenty and 700 sweeps through the whole corpus, depending on time available on the MIT CSAIL cluster. The approximating PCFG for the highest scoring fragment grammar for each training set was used to compute the results below. Approximating PCFGs for the other models were computed directly from the input corpora as described in Section 3.3.

5.4.2 Results

In this section, we first consider the performance of each model in terms of the trajectory of overregularization. We will show that only the FG (inference-based) model produces a trajectory is consistent with empirical findings. We then show that the FG (inference-based) model also predicts the importance of frequency in protecting irregular verbs from overregularization. Next, we consider two other important empirical phenomena: early correct performance and the correlation between reliable obligatory regular marking and overregularization. We will argue that, in both cases, these phenomena involve representational capacities or changes that cannot be captured by the present models, although they are not inconsistent with the results presented here. Finally, we discuss these findings.

5.4.2.1 Trajectory of Overregularization Rates

Figure 5.12 plots the rates of overregularization produced by each model on each training set. The upper left-hand plot in this figure shows one minus the overregularization rate for the children Adam, Sarah, Eve, and Abe as reported in the appendices of Marcus et al. (1992). This plot was produced by taking the total number of overregularizations across these four children at each age in months and dividing by the total number of past-tense forms produced at that age. Because sample sizes at different ages across the four children from Marcus et al. (1992) varied greatly, the exact overregularization rates at each age should not be interpreted as specific empirical targets for the models. This plot is included just to give a rough sense of the trajectory of overregularization during acquisition. The overregularization rates for each model are displayed in blue in the remaining plots in the figure. Each plot also includes the Marcus et al. (1992) data plotted with dashed lines for comparison.

As we discussed in Section 4.3.5, two empirical facts about overregularization for which there is a broad consensus in the literature are (i) all children exhibit some overregularization for some verbs, and (ii) the overall rate is relatively low (estimated by various authors from around 2.5% to around 10%; Hoeffner, 1996; Maratsos, 2000; Marcus et al., 1992; Yang, 2002). Examination of the plots in Figure 5.12, shows that only the FG (inference-based) model is in accord with these two empirical findings. The DMPCFG (full-parsing), DOP1 (exemplar-based), and ENDOP (exemplar-based) models produce overregularization rates far in excess of empirical observations. By contrast, MAG (full-listing) produces virtually no overregularization at any point during learning, simply memorizing all irregulars perfectly.

These results show that under an inference-based approach to productivity and reuse, there is sufficient evidence in the data for the correct inflection of irregular forms the majority of the time. However, there is also sufficient uncertainty about the appropriate inflection for there to be occasional overregularization throughout learning.

Our SWITCHBOARD simulations earlier in the chapter showed that with sufficient evidence the `FG` (inference-based) model can eventually correctly learn the pattern of productivity (generalization of the default +/d/ past marker) and reuse (blocking of the default by irregulars) in the English verbal system. A trend in this direction is apparent in Figure 5.12: rates of overregularization for `FG` (inference-based) decline as the size of the training set increases. However, this decline is gradual and chaotic, suggesting that in these datasets there is still significant uncertainty about the non-applicability of the regular rule to some irregular stems.

Marcus et al. (1992) found considerable variability in overregularization rates between verbs in their samples, and this result has been supported by other studies (e.g., Hoeffner, 1996). The `FG` (inference-based) model produces similar behavior, showing a wide variety of different trajectories for different verbs. Figure 5.13 shows trajectories for a number of examples. These are informally sorted into four classes: (i) verbs that show overregularization followed by recovery (row I), (ii) verbs that show little overregularization (row II), (iii) verbs that show high rates of overregularization throughout (row III), (iv) and verbs that behave chaotically (row IV). The verbs in the figure were classified into these four classes by hand inspection of plots for all irregulars in the CHILDES training set. The four classes are the same as used by Marcus et al. (1992) when presenting examples of Abe's overregularization trajectory on individual verbs. The verbs in the first column—*ate*, *said*, *drew*, and *won*—are the same examples as given in Marcus et al. (1992) for the class, suggesting that, at least for these verbs, Abe's overregularization trajectory may follow from the pattern of productivity and reuse implicit in the frequency distribution of these forms in English.

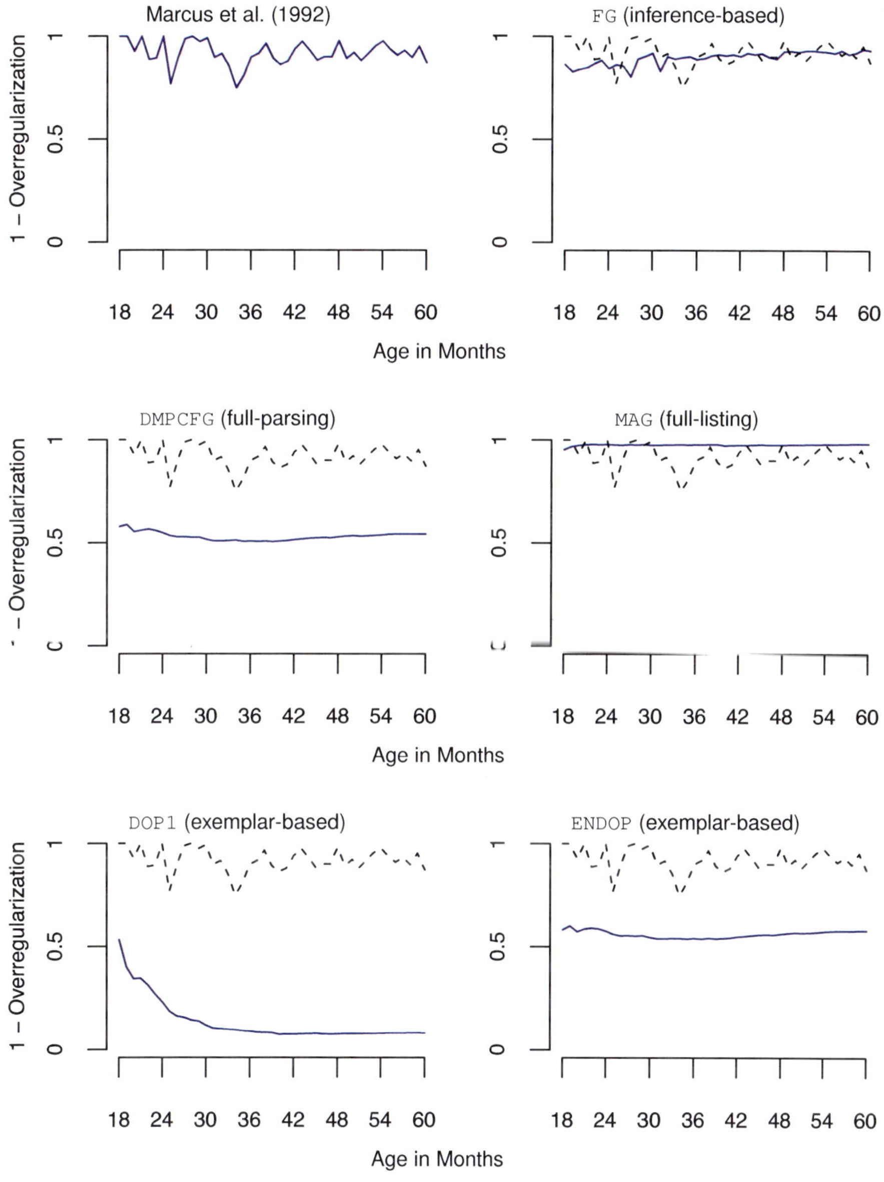

Figure 5.12

One minus Overregularization Rate: The upper left plot shows one minus the overregularization rate for irregulars (in blue) by the children Abe, Sarah, Adam, and Eve as reported in the appendices of Marcus et al. (1992). The remainder of the plots show the same quantity (in blue) for each of the five models. For comparison, these plots also include the Marcus et al. (1992) data plotted with dashed lines.

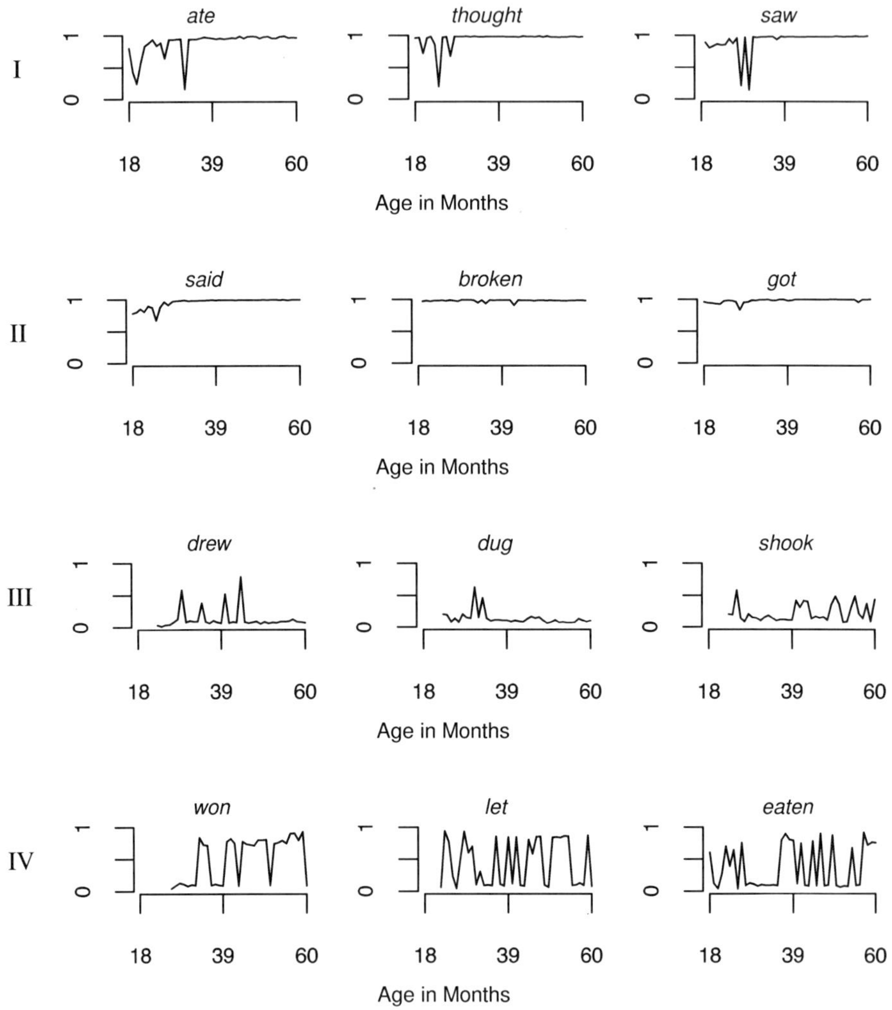

Figure 5.13

Examples of One Minus the Overregularization Rate for Individual Verbs: This figure shows examples of the trajectory of overregularization errors for individual verbs. These are informally sorted into four classes: verbs that show overregularization followed by recovery (row I), verbs which show low rates of overregularization throughout (row II), verbs which show high rates of overregularization throughout (row III), and verbs which behave chaotically (row IV). The verbs in the first column—*ate*, *said*, *drew*, and *won*—are the same examples as given in Marcus et al. (1992) for the corresponding classes I–IV.

5.4.2.2 Protection from Overregularization

Marcus et al. (1992) showed that individual verbs are protected from overregularization by two factors. First, higher-frequency irregulars are overregularized at a lower rate than lower-frequency irregulars. Because the FG (inference-based) model stores irregular forms based on the high frequency of co-occurrence of their stems and inflections, high-frequency irregulars should be less likely to overregularize. This is confirmed by a Pearson correlation of the log rate of overregularization with the log frequency of each past-tense form in the input ($r = -0.90, p < .001$).

Second, Marcus et al. (1992) reported that irregulars are protected from overregularization by similarity to other irregulars. Specifically, irregulars that come from inflectional classes which are attested more frequently in the input are overregularized at lower rates. Because the models studied in this chapter do not represent the phonological selectional restrictions associated with irregular inflection, or the internal phonological structure of stems and inflections, it is not possible to test the model for this empirical phenomenon, and we leave it to future extensions of the framework to explore this prediction.

5.4.2.3 Early Correct Performance and the Discovery of the Regular Rule

In the last chapter, we described two controversial empirical phenomena described by Marcus et al. (1992): the existence of a global (i.e., verb-general) period of early correct performance, and a correlation between the onset of overregularization and an increase in the rate of reliable marking of regular forms.

Examination of the results for the FG (inference-based) model in Figure 5.12 reveals no evidence for a clear global period of early correct performance, nor for a clear qualitative jump in overregularization rates at any point during development. However, these results are not surprising given the simplifications made for the simulations in this chapter. Most models of past-tense acquisition capture early correct learning and overregularization onset using a combination of two mechanisms. First, under many accounts (e.g., MacWhinney and Leinbach, 1991; Pinker and Prince, 1988; Yang, 2002), early correct performance is driven by the memorization of irregular past-tense forms as morphologically simple, unanalyzed wholes. Because all theories of language learning must provide mechanisms for the acquisition of morphologically simple (i.e., monomorphemic) words, once such a mechanism is provided, there is nothing to prevent it from accidentally treating some morphologically complex forms as monomorphemic. This will be especially true for high-frequency forms like irregulars. Early correct learning is often attributed to this memorization process.

Second, under both rule-based and connectionist accounts, the onset of overregularization is predicted to correlate with the discovery of the regular generalization, whether this generalization is represented as a rule or as a pattern of connection weights (and whether or not its emergence requires manipulation of training schedules—see Chapter 4). At the point where the generalization becomes available, it will begin to apply erroneously to some irregular verbs, leading to overregularization. At the same point in time, an increase in correct marking for regular forms is predicted. While these forms are often left unmarked for tense in the speech of young children, at the point where the regular rule emerges, we expect such non-marking errors to become rarer. This is the phenomenon reported in Marcus et al. (1992).

However, the simulations presented in this chapter made a number of assumptions that make it impossible for them to capture these two phenomena. First, all inputs to the models were segmented a priori into stem and inflection components, and verb representations necessarily comprised both components. This means that none of the models had the option to either treat irregulars as monomorphemic and store them as unanalyzed wholes, or to leave any forms unmarked for tense. Second, in the inputs to the simulations, all (and only) the inflections of the English verbal system were present from the beginning of learning. What is learned by the models is not the set of available generalizations, but the productivity of existing generalizations. Therefore, any increase in overregularization or reliable tense marking that depends on the sudden discovery of the regular generalization could not be captured by the models here.

However, it is possible that the `FG` (inference-based) model may exhibit weaker versions of early correct learning and the emergence of the regular rule that do not require the kinds of mechanisms just described. One of the key ideas underlying the model is that the mere existence of an abstract generalization does not mean that the generalization must be used productively. The model must learn which generalizations are productive from patterns in the input data. In the present simulations, at early stages of learning, many abstract generalizations were available (corresponding to the various past tense inflectional classes), and the model was uncertain about the productivity of the regular rule. At the first training step, its performance on novel-stem, wug tests was just below 60% correct. At the same time, there were many high-frequency irregular verbs inflected correctly (defined as probability 0.95 or higher correct inflection). As the amount of training data increased, the FG (inference-based) model's confidence in the productivity of the regular rule also increased. During the same set of training steps, some of the initially high-performing irregulars showed an increase in their overregularization rates.

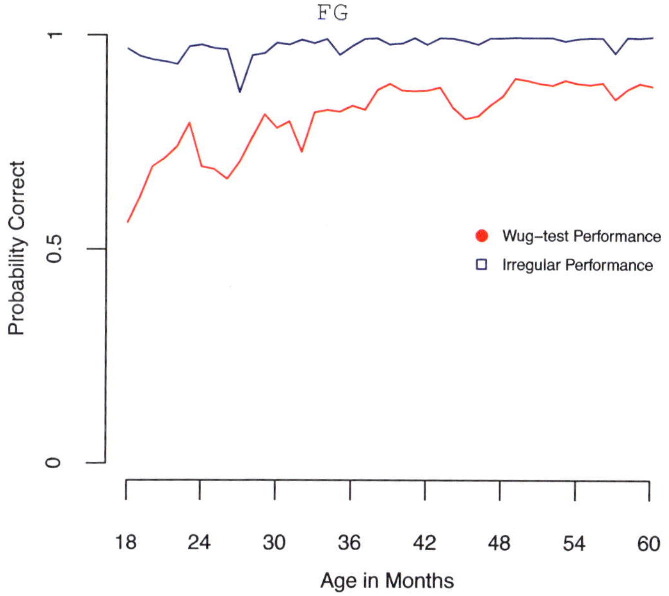

Figure 5.14

Wug-test Performance and One minus Overregularization Rate and Irregulars with Early High Performance (FG): This figure shows one minus the overregularization rate (blue) for irregulars that were correctly inflected at a rate of 95% or higher on their first appearance in the training set. Performance of the model in correctly inflecting a novel wug stem is shown red.

This is illustrated in Figure 5.14, which shows one minus the overregularization rate for irregular verbs that were correctly inflected in the sample in which they first appeared at a rate of 95% or better (in blue). Also shown is the rate of correct performance on novel-stem, wug-test items (in red). What can be seen in this figure is a small increase in overregularization as the productivity of the regular rule increases (as evidenced by correct wug-test performance).

The example trajectories for individual irregular verbs given in 5.13 also suggest that the increase in productivity of the regular rule shown in Figure 5.14 may be the cause of the increase in overregularization rates for some initially correctly marked verbs. For example, in Figure 5.13 the verb *saw* appears to increase in overregularization around the same training steps where the model is showing improved wug-test performance in Figure 5.14.

However, this relationship must remain somewhat speculative. The statistical reliability of this interaction between the productivity of the regular rule and overregularization is difficult to investigate. It is necessary to first establish that the rates of overregularization for each verb, at each training step, are statistically significant, and not the result of noise during inference or idiosyncrasies of the input data. While it is possible, in principle, to examine the statistical reliability of models like FG (inference-based), in practice, this requires a prohibitive amount of computation. We therefore leave the possibility of such an interaction as a suggestion, and its detailed investigation as a problem for future work.

5.4.2.4 Discussion

In this section, we showed that of the five models examined in the book only the FG (inference-based) model provides an account of the productivity of past tense inflections which is consistent with observed developmental trajectories. Specifically, in simulations designed to idealize the increasing input data available to the child learner, only the FG (inference-based) model produced a low but significant rate of overregularization. Furthermore, we showed that for the inference-based model—like children—overregularization is less likely for more frequent verbs.

In Chapter 1 we argued for a view of productivity in which this property is inferred from the patterns of novelty and repetition in the input data. Under such a view, the availability of an abstract generalization does not oblige a system to treat the generalization as productive. In the present simulations, even irregular inflectional classes were represented by fully abstract rules in the underlying computational system. Despite this fact, the FG (inference-based) model only produced moderate rates of overregularization, even on the earliest and smallest training sets. Because the space of underlying rules in the model does not develop or change during learning, this low rate of overregularization can only be a consequence of the frequency distribution in the input data. As we have discussed, extensions to the model may allow it to more accurately capture phenomena like early correct performance. However, the fact that the FG (inference-based) model produces the qualitatively right pattern of overregularization, even within the limits of its representational assumptions, is strong evidence that there is rich information about productivity available to learners, and therefore that an inferential approach to productivity and reuse is plausible on developmental grounds.

5.5 General Discussion

In the preceding sections, we have presented the results of two sets of simulations on the English past tense. These results show that, of the five models considered, only fragment grammars, the inference-based model, can correctly capture the relevant patterns of productivity and reuse, accounting for the sharp dichotomy between the highly generalizable regular rule and the irregular inflectional classes. Furthermore, of the models considered, only fragment grammars produces a plausible account of the developmental trajectory of overregularization during development, up to the limitations imposed by the simplifying assumptions we have made in our representations.

In this section, we step back and consider the place of the fragment grammars model more generally with respect to the past-tense literature. First, we discuss its relationship to several of the other models we examined in Chapter 4. Second, we consider the place of fragment grammars in the debate between dual- and single-mechanism accounts of the English past. Finally, we conclude with a discussion of the probabilistic elsewhere condition, which arises as a consequence of the modeling assumptions made in this book.

5.5.1 Relation to Other Models in the Literature

In the previous chapter, we discussed two properties that must be possessed by any model of the English past. First, any model must be able to represent generalizations involving abstract categories such as `STEM` or `VERB`. Because we assume a common, underlying, computational system that can build hierarchical structures over abstract categories, all of the models in this chapter posses this property. Second, we argued that any model of the English past must provide an account of competition between generalizations that explains the conditions under which one generalization is preferred to another, while allowing for exceptions and quantitative variation. We have shown that, of the models considered here, only fragment grammars, which infers the pattern of defaultness and blocking in the English past from the input data, can provide such an account.

We have reviewed three models from the literature which also possess these properties: the blocking-and-retrieval failure hypothesis of Marcus et al.

(1992), Charles Yang's probabilistic rule-based model (Yang, 2002), and Taatgen and Anderson's ACT-R model (Taatgen and Anderson, 2002). Like the present work, all three of these theories allow for abstract generalization and provide a quantitative account of competition. Moreover, these models also focus on how the pattern of competition can be inferred from the input data, but do not explicitly model the problem of discovering the phonological, morphological, or semantic form of competing rules. In this section, we compare fragment grammars to these theories.

Unlike Yang (2002) and Taatgen and Anderson (2002), the blocking-and-retrieval failure hypothesis of Marcus et al. (1992) is not a formalized mathematical model. Nevertheless, it is similar to fragment grammars in making explicit a notion of competition between stored representations in memory and productive rules. However, the exact conception of how the memory and rule systems interact under the blocking-and-retrieval failure hypothesis is somewhat different from fragment grammars; we defer further comments on this until our discussion of dual- versus single-mechanism models in the next section. The blocking-and-retrieval failure hypothesis differs from fragment grammar in a second way. Under the blocking-and-retrieval failure hypothesis, blocking of the regular rule by irregulars is handled by an explicit blocking principle, which states that when a stored form is available it will be used preferentially to a general rule. In fragment grammars, the preference for stored forms (when they are available) is not an explicit principle, but rather arises as an inference from the input data. We explore this in greater detail in Section 5.5.3 where we discuss the probabilistic elsewhere condition.

Yang (2002) described an approach that employs rules for all inflectional classes, as in classical linguistic theories, but enhances these rules with probabilistic application. Yang also assumed that the system posses a default regular +/d/ rule, which applies when more specific rules are either unavailable or probabilistically fail to apply. There are three important differences between the Yang (2002) model and fragment grammars. First, Yang made use of an explicit notion of defaultness but did not specify how the learner decides which rule is the default (but see Yang, 2005, 2010). Fragment grammars do not make use of an explicit notion of defaultness, but do provide a learning theory that can infer where default-like behavior should occur. In fact, the fundamental inference performed by fragment grammars is to determine which parts of input representations should be computed productively and which parts should be retrieved whole from memory. Defaultness arises as a natural consequence of the pattern of storage and reuse in systems like the English past (see discussion of the probabilistic elsewhere condition in Section 5.5.3 below).

Second, Yang (2002) did not define a specific way in which his learner estimates the probability that rules should apply, although it is clear from his discussion that these parameters should be inferred from the frequency distribution in the input data (amongst other factors). By contrast, fragment grammars provide a mathematically precise specification of how the input frequency distribution gives rise to inferences about productivity and reuse. Nevertheless, Yang's model is similar in spirit to the present work. It may be the case that the mechanism of compilation of frequently reused subtrees can be integrated with Yang's theory to provide an account of default learning (In later work, Yang, 2005, 2010; provided a theory of how to learn defaults, which is similar to fragment grammars and which we discuss in the next chapter).

A third and final difference between fragment grammars and Yang's model is that his model allows for only a single default rule. All other rules are allowed to apply only when their input conditions are matched. For irregular rules, input conditions include the set of irregular stems to which each rule is allowed to apply. While it is possible for these rules to probabilistically fail to apply, allowing for overregularization errors, it is not possible for these rules to accidentally apply when their input conditions are not met, and thus the model cannot generate erroneous irregular forms for existing stems (e.g., *bite*/*bote*). Although such errors are rare, they do occur, and, thus, providing a mechanism that allows for such overirregularization errors may be a desideratum for a model of the English past.

The model of Taatgen and Anderson (2002) is formulated in the ACT-R framework (Anderson and Lebiere, 1998), and makes use of the ACT-R distinction between stored declarative facts and procedural production rules. Similar to the fragment grammars model, Taatgen and Anderson's model also provides mechanisms to compile frequently co-occurring sequences of computations into single rules (for details, see Taatgen and Lee, 2003). Competition is resolved by choosing the rule with the highest expected outcome—a numeric value that is a composite measure of several factors including the prediction accuracy of the rule and its utility.

Taatgen and Anderson's model is similar to fragment grammars in several respects. First, the representations used for the past tense are similar to those used here: they do not take into account phonological structure, and they treat past forms as being pre-segmented into stems and inflections. Second, the model makes use of mechanisms that compile frequent sequences of rules into single reusable units. Third, and most importantly, like fragment grammars, this model explicitly treats the problem of past-tense inflection as inference from the input data by optimizing the expected outcome of production rules. As a result, like fragment grammars, it was able to learn the overall pattern of defaultness and blocking exhibited by the English past system.

Nevertheless, Taatgen and Anderson's model also differs from fragment grammars in a number of ways. First, unlike the ACT-R framework, in fragment grammars there is no a priori distinction between facts stored in memory and production rules operating on these facts. Instead, stored items are simply frequently occurring sequences of computational steps. Second, in fragment grammars competition is quantified using the probability of stored fragments (and rules in the underlying computational system). As we discussed in Section 1.2.6, this means that fragment grammars can be understood as optimizing a prediction trade-off, balancing the weight of probability assigned to reusing existing forms against the the possibility of generating novel forms in the future. By contrast, the expected outcome in the ACT-R framework is a more complex value, which takes into account not only prediction accuracy but also resource-based quantities such as the cost of production rules and the utility of goals. A third difference between Taatgen and Anderson's model and fragment grammars is that, in their model, compilation of frequently co-occurring sequences of production rules is handled by a special, distinct mechanism. By contrast, for fragment grammars, compilation is an intrinsic part of the process of generating observations.

Fourth and finally, unlike fragment grammars, the Taatgen and Anderson model was able to predict a global period of early correct learning. This was the result of their inclusion of a `do-nothing` production rule, which did not inflect verb stems. Early in learning, their model produced uninflected verb forms at high rates because the `do-nothing` rule had lower a priori cost than the other production rules in their system. Although it had low cost, the `do-nothing` rule also had poor accuracy in predicting the past-tense forms of stems. By contrast, other rules were able to predict the past tense of stems with higher accuracy, despite their higher cost. Therefore, as more data was observed, these other rules began to have higher expected outcome, and over time, became preferred. This happened first for high-frequency irregular forms, leading to a period of global early correct performance. Eventually the model discovered the regular rule and began correctly inflecting low-frequency and novel regular stems as well as overregularizing some lower-frequency irregular stems. In the simulations in this chapter, our representations did not include the possibility of such a `do-nothing` strategy. The results of Taatgen and Anderson suggest that its inclusion may provide fragment grammars with a mechanism whereby it can reliably produce a global period of early correct learning.

5.5.2 Single- versus Dual-Mechanism Models

One of the most important debates in the literature surrounding the English past is the question of how many different mechanisms or modules of grammar are required to account for the ways in which regular and irregular inflections generalize. The issue revolves around the architecture of dual-mechanism theories, such as those of Pinker and Prince (1988) and Marcus et al. (1992), which posit that productive, regular generalizations and less productive, irregular generalizations are handled by separate and distinct subsystems with fundamentally different architectures. Generalization of regulars is handled by a rule-based component of grammar, similar to traditional linguistic theories, whereas generalization of irregulars is handled by an associative memory, similar to connectionist or analogical models. By contrast, connectionist/analogical theories and many probabilistic, rule-based theories have proposed architectures where all generalization is handled by a single mechanism or component of grammar.

In this chapter and the last, we have primarily focused on the question of how various models have handled competition between regular and irregular inflections. Fragment grammars and the ACT-R model of Taatgen and Anderson (2002) resolve this competition by assigning rules (or rule combinations) a numeric value, and (implicitly) comparing all forms simultaneously based on their values. The model of Yang (2002) and the blocking-and-retrieval failure hypothesis of Marcus et al. (1992) take a different approach, based (implicitly) on the ordering of generalizations. Under the blocking-and-retrieval failure hypothesis, the blocking principle states that stored forms take precedence over productive rules. In Yang's model, all forms are instead computed by rule, but irregular rules are given a chance to apply before the default (see also Yang, 2005, 2010). Although they provide different mechanism for resolving competition, all four of these theories are similar in the way they handle productivity.

Nevertheless, these theories differ in how they model the representation and generalization of low-productivity forms (or rules), such as the irregulars. Under dual-mechanism approaches, generalization of irregulars is driven by similarity in associative memory. Under probabilistic, rule-base approaches, generalization of irregulars is handled by the same sort of phonological rules that describe the regular +/d/ generalization (see, for instance, Albright and Hayes, 2003; Yang, 2002).

Empirical arguments in favor of a dual-mechanism over single-mechanism approach have been largely based on three kinds of evidence: (i) differences in productivity between regular and irregular inflectional classes, and differences in how similarity and frequency affect the generalization of these classes, (ii) differences in the availability of regulars and irregulars as input to other morphological processes such as conversion from one category to another and compounding, (iii) differences in the vulnerability of regulars and irregulars to biological conditions such as strokes, genetic impairments, etc. (See, Pinker, 1999; Ullman, 2001).

The implications of biological evidence for the kinds of mathematical models considered in this book are poorly understood, and we leave this complex topic to future research. The problems associated with the second argument, differences between the availability of regulars and irregulars to other morphological processes, can be handled, in principle, by any theory that allows irregular forms to be stored (or otherwise marked) and treated as single units in memory—a property shared by all of the theories we have discussed here.

Arguments based on differences in productivity and sensitivity to similarity and frequency have mostly revolved around the fact, discussed in the previous chapter, that irregulars tend to form family resemblance categories (also called *variegated similarity* by Albright and Hayes, 2003). In a rule-based approach, all inputs to a rule need to be similar to one another in the same way in order for a rule to apply. In a family resemblance category, items can be similar to one another along multiple dimensions without having any single, defining characteristic. Many theorists have argued that this kind of structure is the hallmark of associative or analogical learning mechanisms (e.g., Albright and Hayes, 2003; Pinker and Prince, 1996; Prasada and Pinker, 1993; Ullman, 1999). Because of this, such family resemblance structure has been offered as evidence that irregulars are handled by associative memory and not by rules (see especially, Pinker and Prince, 1996; for detailed arguments).

However, as emphasized, for example, by Albright and Hayes (2003) and Yang (2002), probabilistic rule-based models that allow for mixtures of rules of differing degrees of abstraction and productivity can also explain family resemblance categories. It appears that difference in productivity, and sensitivity to similarity and frequency can be modeled in a variety of different computational frameworks, and these facts alone do not adjudicate between different models.

Importantly, the fragment grammars model (as well as the model of Taatgen and Anderson, 2002) is agnostic about these aspects of the single- verus dual-mechanism debate. To understand this, consider the representation for verb forms used in this chapter (as well as those of Taatgen and Anderson, 2002). Verb forms are represented as combinations of two parts: `STEM` and `INFLECTION`. The `INFLECTION` constituent is an abstraction used to represent both the operations that produces regular forms (i.e., `CONCATENATE: +/d/`) and the operations that produce irregular forms (e.g., `RETRIEVE SPECIFIC-FORM:` *went*, or `CHANGE STEM VOWEL:` /ɪ/→/æ/). However, since the phonological mechanisms that implement these operations are not defined, it would be possible for these phonological operations to be implemented using either associative/analogical mechanisms, phonological rules like those of traditional linguistic theories, or some mixture of both.

This flexibility means that fragment grammars are not committed, a priori, to any particular mechanism for phonological rule discovery and opens up the possibility of simulations that compare different approaches—all embedded in the framework-level theory of productivity and reuse advocated in this book.

5.5.3 The Probabilistic Elsewhere Condition

In the last chapter, we reviewed how most models of the past tense have made use of one variant or another of the elsewhere condition (also known Pāṇini's principle, pre-emption, the subset principle, or the blocking principle; Anderson, 1969; Andrews, 1982; Aronoff, 1976; Halle, 1997; Kiparsky, 1973; Pinker, 1984; Stump, 2001b). This principle states that more specific generalizations should be preferred to more general ones when both can apply. In fragment grammars, the elsewhere condition is not stated as an independently stipulated principle, but arises as a natural consequence of the way in which storage and computation are modeled together with the general laws of probabilistic inference. In this section, we discuss why this is the case.

In his original formulation, Kiparsky (1973) formalized the elsewhere condition in terms of a subset relation between the inputs of rules: when rule r_1 and r_2 both apply to an input, and the set of all possible inputs for r_2 is a subset of the set of all possible inputs of r_1 (i.e, $\text{Inputs}(r_2) \subset \text{Inputs}(r_1)$), then r_2 should be preferred to r_1.

For non-probabilistic models, an independently stated version of the elsewhere condition is necessary because such models do not provide any other quantification of competition between generalizations. However, in a setting where rules are formalized as probability distributions over forms, as in fragment grammars, a generalized version of the elsewhere condition arises as a consequence of principles of probabilistic inference. Consider two rules r_1 and r_2, and suppose that the set of forms to which r_1 assigns non-zero probability is a (proper) superset of the set of forms to which r_2 assignes non-zero probability. Because probability distributions must sum to one (i.e., the law of conservation of belief) r_1 will, on average, assign lower probability to forms than r_2. There can, of course, be individual forms that are more probable under r_1 than r_2, but these will be more probable at the expense of making other forms less probable.

Suppose that we observe some verb form that could have been generated by both r_1 and r_2. To determine the posterior probability of attributing the verb form to r_1 or r_2, we use probabilistic conditioning. Recall that conditioning can be understood in terms of a two-step algorithm (see Section 2.1.1). First, all rules that cannot generate the form in question are removed from consideration. Second, the probability distribution over r_1 and r_2 is renormalized to sum to one. Importantly, conditioning can never change the relative probabilities of the two rules.

Because r_1 could generate a greater number of verb forms than r_2, it must assign lower probability to each of those forms, on average. This relative preference for r_2 for most forms will be preserved by conditioning on a specific form. Thus, when trying to attribute a form to one rule or another, on average, rules generating fewer forms will be preferred to rules generating a greater number of forms. This is a probabilistic variant of Kiparsky's elsewhere condition, and does not depend on anything more than the assumption that rules define distributions over forms and that inference is based on probabilistic conditioning.

Fragment grammars implement a specific case of this general phenomenon. Consider how a fragment grammar decides to inflect a verb stem like `SING`. Since the regular rule is a default and can combine with any stem, it is available to form the past of `SING` (i.e., *singed*). However, because the regular rule also applies to many other stems, the distribution it defines cannot assign a high probability to any single inflected form. This is illustrated schematically in Figure 5.15, which shows in black outline the distribution over inflected forms associated with the regular rule.

If a fragment grammar has successfully learned the correct past form of `SING`, it will have also stored a tree like the one below.

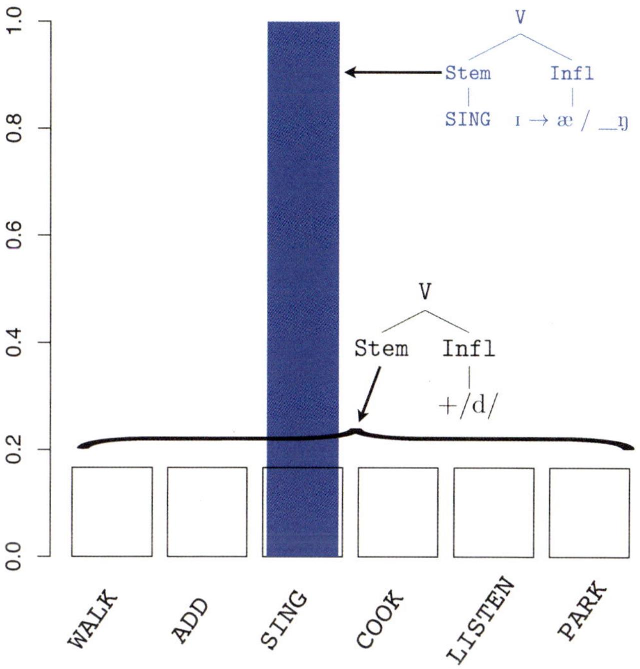

Figure 5.15

Competition Between the Regular Rule and a Stored Form: This figure shows schematic versions of the distributions associated with a productive rule (black) and a stored irregular past-tense item (blue). The productive rule must spread its probability over many forms; a stored structure concentrates its probability on a single form. Therefore, when these two possibilities compete, the stored form will usually win.

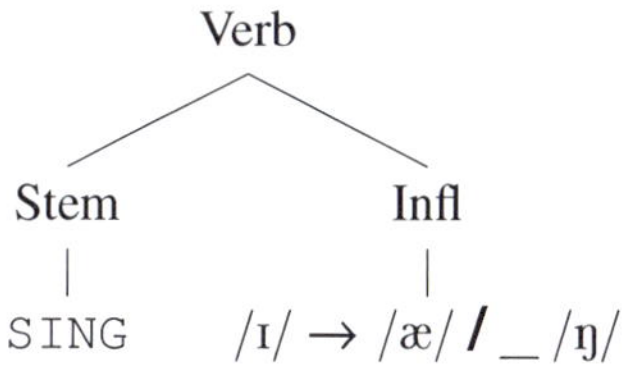

This tree can also be thought of as a rule—one whose distribution puts all of its probability on a single outcome, *sang*. This is shown in blue in Figure 5.15.[8] The bars in Figure 5.15 represent the prior probabilities of the two rules being

8. Note that the fragment grammar model will also likely predict many other possibilities that are not represented in Figure 5.15. For example, it might have a small amount of probability associated with the possibility that the past of SING is formed using the rule: /ɪ/ → /ʌ/ / (i.e., *sung*, as a past

used to inflect various different stems. The conditional probability of each rule being used to inflect a specific stem, like SANG, is computed using probabilistic conditioning. With respect to Figure 5.15, this amounts to considering only the height of the two bars over the stem SANG, and renormalizing these two probabilities so that they sum to one. Because the regular rule put probability on more stems before conditioning, and, therefore, assigned less probability to the stem SANG than the irregular rule, it will be dispreferred after conditioning. Generally speaking, fragment grammars will prefer to used stored tree fragments over abstract rules when the former are available for an input stem.

There are several advantages to this account of the elsewhere condition and blocking behavior. First, other than assuming a system based on probabilistic inference with basic mechanisms of storage and computation, fragment grammars do not need to make further stipulations to exhibit elsewhere-like behavior. Second, as reviewed in the previous chapter, blocking is an inherently probabilistic phenomenon. Empirical facts, such as the existence of doublets, overregularization/overirregularization by both children and adults, and the generalization of both regular and irregular inflections to novel stems, show that multiple generalizations compete, at different rates, to give rise to past-tense forms. Furthermore, the robustness of blocking depends on frequency. In the fragment grammars model, a more frequent stored form will tend to have higher probability than other stored forms or rules, and thus block more robustly. In morphological systems that are less dichotomous than the English past, this may prove crucial. For example, Rainer (1988) gives extensive evidence of frequency-dependent blocking behavior in both German and Italian showing that blocking is closely modulated by frequency. Likewise, Kiparsky (1982b) points out that in English derivational morphology blocking can only be considered a general tendency.

Third, this account may generalize to other cases of blocking that have been examined in the literature. In the English past, blocking occurs between a stored word form and a productive rule. However, blocking can also occur between two productive rules (Rainer, 1988; van Marle, 1986). For example, Rainer (1988) discusses cases of formation of abstract German nouns with the endings *-ität*, *-heit*, and *-ie*. He argues that although each of these affixes has a circumscribed domain of application, all three are productive and able to form novel words. The domains of application of the three affixes are nested: *-heit*

---

tense form, not a past participle form). However, if it has correctly learned the past of SING, this probability will be close to zero.

can apply to stems with final stress, *-ität* can apply to stems with final stress that are also marked as "learnèd," and *-ie* can apply to stems that are learnèd, stress final and end in *-phil*. In this system, the elsewhere condition still applies. For a given stem, the most specific rule that can be used blocks the others. This generalization follows naturally when we consider rules as distributions over forms.

There are also cases of blocking between stored word forms and syntactic structure. Di Sciullo and Williams (1987) and Poser (1992) discuss the case of the English comparative construction; in this construction forms such as *smarter* can block phrasal structures such as *more smart*. Poser (1992) discusses similar cases in Japanese and Basque. McCloskey and Hale (1983) discuss a case in Irish where the availability of a verb form that is synthetically marked for person blocks the use of a phrase combining an unmarked verb and a personal pronoun. Again, elsewhere-like competition between specific word forms and more general syntactic constructions follows naturally from the view that treats rules and stored forms as distributions, and derives blocking from principles of probabilistic inference.

Fourth and finally, this account may integrate with theories of competition between linguistic expressions based on pragmatic or communicative considerations. McCawley (1978) examines cases where a lexical item competes with a phrasal construction. He argues that Gricean principles imply that, in the default case, a speaker should choose the least complex way of expressing something. Horn (1984) extends and refines this theory, and identifies this idea with Zipf's *principle of least effort* (Zipf, 1949).[9] In both Horn and McCawley's theories, speakers prefer to use less complex forms when such forms are available; however, in neither theory is the relevant notion of complexity defined. In the current framework, blocking arises from a precise notion of complexity: the number of computational steps (random choices) required to generate a form. Using the regular rule to generate a form requires two computational steps (choose the rule; choose the stem). Generating an irregular requires a single step (choose the whole form). It is this fact that gives rise to the differences in how each kind of generalization distributes probability over forms, and, ultimately why irregulars block regulars. Thus, this account of

9. Both McCawley (1978) and Horn (1984) are concerned with explaining why the use of a more complex expression in a context where a simpler expression is available triggers certain implicatures for the hearer, rather than blocking, per se.

blocking may provide a formalization of complexity that can be used in more general, pragmatic accounts of competition between linguistic expressions.

5.6 Conclusion

In this chapter, we have presented the results of two sets of simulations on the English past tense. These results show that fragment grammars, the inference-based model, can correctly capture the relevant patterns of productivity and reuse, accounting for the sharp dichotomy between the highly generalizable regular rule and the irregular inflectional classes. Furthermore, the model can capture the correct pattern from just the distribution of morphemes in the data. In the next two chapters, we consider a very different morphological system: English derivational morphology. Unlike the past tense, this system exhibits a broad range of levels of productivity, wide variability in phonological regularity and semantic transparency, and rules with highly overlapping domains of applicability.

6 English Derivational Morphology: Productivity, Processing, and Ordering

6.1 Introduction

This chapter reviews the literature on productivity and reuse in English derivational morphology. English derivational morphology presents an interesting domain in which to examine questions of productivity and reuse for three reasons. First, in the history of generative grammar, it was here that the problem of controlling the productivity of generative processes was first highlighted as an important question, and English derivational morphemes like *-ness*, *-ity*, and *-th* are still prime motivating examples for the study of productivity (Aronoff, 1976; Bauer, 2001; Halle, 1973; Jackendoff, 1975).

Second, English derivational morphology presents a much greater diversity of phenomena than the English past tense, which we examined in the preceding two chapters. In particular, it provides examples of morphological processes occupying a wide variety of levels of productivity, from extremely productive suffixes such as *-ness*, to affixes with contextually dependent productivity such as *-ity*, to unproductive affixes, such as *-th*. This latter aspect of the system—its complex pattern of productive and unproductive word-formation processes—provides an important counterpoint to the sharp dichotomy of the English past-tense system.

Third, and finally, English derivational morphology provides a domain where it is possible to study productivity for forms with hierarchical structure and multiple interacting parts. Derivational processes produce words with multiple levels of embedded hierarchical structure (e.g., *agree*, *disagree*, *disagreeable*, *disagreeability*). They even permit recursive structure (e.g., *missile defense*, *anti-missile defense*, *anti-anti-missile defense*, etc.). While our study of the English past was focused on the productivity of single affixes, derivational morphology will allow us to study more complicated patterns of storage and composition. How should a word like *disagreeability* be analyzed—[*disagreeability*], *dis*+[*agree*]+[*able*]+[*ity*], [*disagreeable*]+[*ity*], [*disagree*]+[*ability*], etc.—and how productive are each of the parts in such an analysis? In this chapter and the next, we will examine phenomena such as affix ordering, which arise only when we consider more complex structures such as these.

This chapter is divided into three parts. First, we give an overview of derivational morphology in English, contrasting it with the inflectional system we studied in the preceding two chapters. Second, we review the empirical literature on derivational morphology. In particular, we organize this discussion around seven classes of empirical phenomena: base selectivity, phonological complexity and regularity, semantic transparency, frequency, productivity, processing, and, finally, morpheme combination and ordering. In the final part of the chapter, we review theories of three of these classes of phenomena: theories of productivity, theories of morphological processing, and theories of affix ordering. In the next chapter, we examine the ability of our models to make predictions in these three domains.

6.2 Derivational Morphology

Although there is debate about whether derivational and inflectional morphology should be handled by different formal machinery in linguistic theories (see e.g., Spencer, 1991; Stump, 2001a), the two kinds of morphology are nearly always distinguished on a descriptive level. Stump (2001a) offered the following rule-of-thumb criteria for distinguishing between the two.

1. **Change in lexical meaning and/or part of speech**: Inflectional morphology generally preserves the meaning of a stem and the part-of-speech of a word. For example, the forms produced by inflecting verbal stems are (usually) still verbs (see Chapter 4). Furthermore, these inflections all preserve the core lexical meaning and standard argument structure restrictions of the corresponding verb stem. Derivational processes, however, tend to change the category and meaning of their inputs (e.g., *work* [`verb`] → *worker* [`noun`, one who `verbs`], *bland* [`adjective`] → *blandness* [`noun`, the quality of being `adjective`]).
2. **Syntactic determination**: Inflectional structure is often syntactically determined. For example, English perfect constructions, which use forms of the verb *have* (e.g., *have gone*, *has eaten*, etc.), require that the main verb occur in the past participle form. The form of the meaning-bearing element is completely determined by the syntactic context. English derivational structure is not syntactically specified to the same degree. Instead, derivational processes can be employed when an alteration of meaning or part-of-speech is required for semantic or pragmatic reasons. As a consequence, inflectional morphology is also often obligatory whereas derivational morphology is typically optional.
3. **Productivity**: In general, inflectional structure is more productive than derivational structure. This generalization is subject to important caveats,

however. There are many highly productive derivational processes (see below), and, as seen in Chapter 4, some inflectional processes have very low productivity.

4. **Semantic regularity**: Usually, inflectional structure is highly semantically regular or transparent, while derivational structure is less so. For instance, English past-tense allomorphs have a precisely specified meaning with respect to tense and aspect, while the precise meaning of English derivational affixes, such as *-th*, can vary in ways that are not strictly compositional.
5. **Closure**: In general, the addition of inflectional morphology to a word means that further derivational structure cannot be added. While the agentive suffix *-er* can be added to a verb stem in English (e.g., *work/worker*) it cannot be added to an inflected verb (e.g., *worked/*workeder*). The converse is not true; for example, derivationally derived nouns are easily pluralized (e.g., *worker/workers*).

Another difference between inflectional and derivational morphology is that while inflectional morphology can often be organized into paradigms, as we discussed in Chapter 4, derivational morphology typically cannot. There are a few cases where English derivational morphology seems to show a ***quasi-paradigmatic*** structure. For instance, verbs ending in *-ate* often drop this morpheme when forming an adjective in *-able* (e.g., *calculate/calculable*; *graduate/graduable* rather than *caculatable/graduatable*). However, in general, paradigmatic structure is far less evident in derivational morphology.

As already mentioned, another very important way in which English derivational and inflectional morphology differ is the degree of hierarchical embedding and recursivity they permit. There is no upper bound on the size of words that can be generated by derivational processes (e.g., *agree*, *disagree*, *disagreeable*, *disagreeability*). The existence of complex embedded structures in derivational morphology makes it necessary to introduce some additional terminology. When discussing inflectional morphology in Chapters 4 and 5, we distinguished only between a stem and the inflectional process that applied to that stem. In contrast, in discussing derivational morphology, it is necessary to distinguish between a stem, a ***base***, and the derivational processes applying to bases. A stem refers to a morpheme without internal morphological structure. A base refers to a word or morpheme to which a derivational process applies—whether or not it has internal structure.[1]

Finally, another important phenomenon in derivational morphology is the existence of **bound stems**. While many English stems can stand on their own as words (e.g., *agree/agreeable*), others cannot (e.g., **vive/revive/survive*). The latter class of stems are referred to as bound stems, while the former are referred to as **free stems**.

In this section, we described some of the critical differences between derivational morphology and inflectional morphology. Next, we present a detailed review of known empirical phenomena in the domain of derivational morphology, with an emphasis on English.

6.3 Empirical Phenomena in Derivational Morphology

The following sections will present a detailed review of empirical phenomena in derivational morphology. Although we will focus on English, we will also discuss examples from other languages where needed to illustrate certain points. This review will be organized into the following eight sections.

1. **Base Selectivity:** English derivational affixes are characterized by a large number of selectional restrictions—categorial, phonological, and semantic—governing the bases with which they can combine. They vary widely in the restrictiveness of those selectional constraints.
2. **Phonological Complexity and Regularity:** English derivational processes also vary both in the degree of complexity of the phonological changes which they trigger, and the regularity (or systematicity) with which those changes apply across the lexicon. Some affixes, such as *-ness* or adverbial *-ly*, are phonologically simple, usually just concatenating to their base. Other affixes, such as *-ity* or *-ive*, trigger more complex phonological changes in their output word forms. For example, *-ity* triggers a stress shift on the base (e.g., *nórmal/normálity*). These changes vary in their regularity. For example, stress shift is triggered on every applicable base suffixed by *-ity*, while other changes are more idiosyncratic. For example, the final stem

1. Some theories use a slightly different terminology, where a *root* corresponds to our notion of stem—having no internal structure—whereas a stem corresponds to what we call a base—anything that can accept an affix.

consonant in a small, idiosyncratic set of bases (e.g., *persuade*) changes when suffixed with *-ive* (e.g., *persuasive*).

3. **Semantic Transparency and Systematicity:** English derivational affixes vary in their semantic transparency. Some affixes, such as *-ness* or adverbial *-ly* have highly predictable meanings. Other affixes, such as *-th*, often have meanings that are not entirely predictable from their combination with a base. Again, the systematicity of these different meanings varies across the lexicon.
4. **Frequency Distribution of Forms:** English derivational affixes also vary in their frequency distributions. Highly productive suffixes, such as *-ness*, tend to be characterized by frequency distributions with a large number of low-frequency forms. Less productive affixes such as *-th* tend to have frequency distributions with greater variance, often skewed towards high-frequency forms.
5. **Productivity:** Derivational affixes vary widely in their ability to form new words. Some affixes, such as *-ness* or (adverbial) *-ly*, are very productive and can be applied to form new words easily (e.g., *pine-scentedness*); other affixes, such as *-th* (e.g., *width*, *length*), rarely form novel words.
6. **Processing:** For existing words, there is variation in how different affixes are processed. Furthermore, this variation is highly correlated with their productivity. High-productivity affixes, like *-ness*, tend to be composed with bases on the fly, while low productivity affixes, such as *-th* tend to be retrieved together with their bases as whole words.[2]
7. **Combinations of Morphemes and Ordering:** English derivational morphemes also co-occur with other morphemes in highly constrained ways. Some affixes can only attach to a restricted set of stems (*sur-* in *survive*). Some suffixes are sensitive to the presence of other affixes in a base (e.g., nominalizations of verbs ending in *-ize* are formed with the suffix combination *-ation*; *nominalization*). Certain combinations of suffixes generalize productively (e.g., *-ability* in *googleability*). As a result of these co-occurrence restrictions, the number of affix sequences appearing in actual English words is far smaller than the number of logically possible sequences.

2. This property is sometimes referred to as the *parsability* or *(de)composability* of an affix (see, e.g., Hay, 2001, 2003; Hay and Baayen, 2002).

As we discussed in Section 1.2.3.1, all of the properties above are highly correlated. In general, more productive affixes tend to be composed more often during processing, show fewer selectional restrictions on their bases, exhibit more phonological regularity and semantic transparency, co-occur with more stems, and tend to appear outside of other structure in sequences of multiple affixes. Less productive affixes, by contrast, tend to be decomposed less often, are less phonologically regular and less semantically transparent, attach to fewer stems, and tend to appear inside of more productive affixes in morphologically complex forms. In the following sections, we discuss each of these phenomena in more detail.

6.3.1 Base Selectivity

The distribution of English derivational affixes is governed by a number of selectional restrictions on bases and affixes. The most obvious kind of restriction is sensitivity to the morphological (or syntactic) category of bases. Such constraints are known as ***categorial restrictions***. For example, the the suffix *-ness* typically attaches to adjectival bases, and always produces nouns. We will use the notation *-ness*:`Adj>N` to signify such categorial restrictions. Affixes vary widely in their categorial structure, and there are many affixes with the same phonological form that have different categorial structure. For instance, English has several different suffixes that are pronounced *-ly*. These include one with category `Adj>Adv` (e.g., *quick/quickly*, *stupid/stupidly*) and another with category `N>Adj` (e.g., *man/manly*).

Many affixes also have phonological restrictions on the bases to which they can be attached (see, Bauer et al., 2013). For example, the affix *-en*:`Adj>V` (e.g., *red/redden*, *short/shorten*) strongly prefers monosyllabic bases ending in an obstruent (see, e.g., Fabb, 1988; Plag, 2003; for discussion of these restrictions). The noun-forming suffix *-al*:`V>N` (e.g., *surprise/surprisal*, *refuse/refusal*, *arrive/arrival*—not to be confused with the adjective-forming suffix *-al*:`N>Adj`, e.g., *topic/topical*) prefers bases with final stress (e.g., *enter/*enteral* Plag, 2003). There may also be restrictions on word formation that are sensitive to more complex aspects of phonological structure. For example, some English affixes appear to be sensitive to whether their base is of Latinate or Germanic origin. The suffix *-al*:`N>Adj` (e.g., *development/developmental*, *distribution/distributional*) prefers to attach to bases with Latinate phonotactics. Although it is not clear whether these restrictions are driven primarily by the phonotactics of the base or by the presence of other

Latinate morphemes in the base, these preferences are psychologically real, and are present in novel-word generalization experiments (see, Anshen and Aronoff, 1981; Aronoff and Schvaneveldt, 1978; Cutler, 1980; Fabb, 1988; Gropen et al., 1989; for various evidence and arguments about such generalizations).

Word formation is also governed by numerous semantic restrictions on affixation (Bauer et al., 2013). For example, the agentive affix *-er*:`V>N` (e.g., *baker*) prefers to attach to verbs that have a subject with *agent* as its thematic role. The suffix *-ee*:`V>N` (e.g., *employ/employee*) is restricted to refer to sentient entities (Plag, 2003).

Another class of constraints make reference to selectional restrictions on the presence of particular morphemes in bases.[3] These ***morpheme-specific selectional restrictions*** represent a somewhat different phenomenon than restrictions on broader semantic or phonological categories of bases. Because this phenomenon is intimately related to the problem of affix ordering, and these together are a core explanandum of our theory of productivity and reuse, we consider them separately when we discuss morpheme combination and ordering in Section 6.3.7.

Importantly, there is wide variability in the restrictiveness of selectional constraints, and therefore wide variability in the number of bases available for affixation across different derivational suffixes and prefixes. An obvious correlate of this fact is that affixes with tighter selectional restrictions tend to be less productive than affixes with broader selectional restrictions (see, e.g., Hay and Plag, 2004).

Another important empirical observation is that most selectional restrictions appear to be violable preferences, rather than hard constraints. For example, there are exceptions to the categorial restrictions of affixes. The suffix *-ness* is highly regular and productive and typically attaches to adjectives to form nouns. However, in the case of the word *forgiveness*, it appears to take *forgive*, a verb, as a base. Likewise, the agentive suffix, *-er*, combines with both nouns (e.g., *slaver*) and verbs (e.g., *baker*). The problem of violable categorial restrictions is further complicated by the wide-spread availability of ***zero-derivation*** (also known as ***conversion***), in which the category of a lexical item can change without any overt morphological marking (e.g., *The committee will table the resolution—The resolution is tableable*). Many phonological and semantic selectional restrictions also exhibit exceptions. For example, *awaken*

3. See Section 6.4.3.2

may violate the restriction that *-en* attaches to monosyllables (unless *a-* is analyzed as a prefix). Verbs that do not have agent subjects can also be affixed with *-er*:V>N (e.g., *sinker*, as in the baseball pitch, whose subject is probably a theme).

6.3.2 Phonological Complexity and Regularity

Affixes vary considerably in their phonological complexity and regularity. Some affixes, such as *-ness*, are phonologically simple—they combine with bases via concatenation and cause no further phonological changes. By contrast, other affixes are phonologically more complex. For example, *-ity* triggers a number of phonological changes in addition to concatenating to a base: it shifts stress (e.g., *nórmal*/*normálity*); it sometimes changes the quality of base vowels (e.g., *inane*/*inanity*, /eɪ/→/æ/), and it sometimes changes the final consonant of the base (e.g., *electric*/*electricity*, /k/→/s/)).

The phonological changes triggered by affixation also vary in the systematicity with which they apply to different forms in the lexicon. Some changes, such as stress shift, are always triggered when applicable. Some changes only apply in the context of other specific affixes, but do so systematically within that restricted context. For example, all words ending in *-ic* change /k/ to /s/ when combined with *-ity* (e.g., *atomic*/*atomicity*, *electric*/*electricity*). Other changes are associated idiosyncratically with particular stems or sets of stems. For example, a set of Latinate stems ending in /d/ changes this final consonant to /s/ before the affix *-ive* (e.g., *pervade*/*pervasive*, *persuade*/*persuasive*, *invade*/*invasive*, etc.).

An important empirical generalization is that derivational morphemes sometimes share aspects of their phonological structure. While morphemes vary in their phonological complexity/regularity, this variation isn't distributed evenly throughout the derivational system. Instead, some—generally less productive—affixes appear to cluster into groups sharing properties such as the base-transformations they trigger or the set of stems with which they can be composed (Chomsky and Halle, 1968; Giegerich, 1999; Halle and Mohanan, 1985; Kiparsky, 1982b,c; Mohanan, 1986; Plag, 1996, 2002, 2003; Siegel, 1974). For example, the Latinate suffixes *-ic*, *-ity*, and *-ive* trigger similar transformations on the bases with which they combine, such as stress-shift (e.g., *átom*/*atómic*, *nórmal*/*normálity*) or consonant sibilantization (e.g., *centric*/*centricity*, *persuade*/*persuasive*). Likewise, certain suffixes attach to overlapping sets of bound stems forming quasi-paradigms (e.g., *mechanist/mechanism*, *eucharist/eucharism*).

6.3.3 Semantic Transparency and Systematicity

Affixes also vary in their semantic transparency. Words such as *curiosity*, *ability*, and *ethnicity* have two different meanings. On one hand, they can be synonymous with *curiousness*, *ableness*, and *ethnicness*—referring to the general quality of possessing the corresponding adjective. On the other hand, they can also mean a particular, specialized, countable instance of this quality—an albino rhinoceros is a *curiosity*; juggling is an *ability*; Italian is an *ethnicity*. Importantly, formations using *-ness* cannot have these more specialized meanings.

Again, cases of semantic transparency vary in their systematicity. The ability of *-ity* to refer to countable instances of the general quality seems to apply to most words using this suffix. However, complex, derived words can also often accrue completely idiosyncratic meanings. For example, *orientation* refers to the act of orienting (e.g., *We used a compass for orientation*), an introduction to some new surroundings (e.g., *During the first week on the job, he attended an orientation*), and the state of being oriented (e.g., *we lost our orientation in the woods*). Another example, *prohibition*, can be used in American English to refer to both a general act of prohibiting, and the specific historical period from 1919-1933 when various constitutional and statutory mechanisms were in place that prohibited the distribution and sale of alcohol.

6.3.4 The Frequency Distribution of Forms

As we discussed in Section 1.2.6, productive and unproductive affixes vary sharply in their frequency distribution over word forms. Because productive affixes like *-ness* often give rise to new words, their frequency distributions tend to have large numbers of low frequency forms—that is, productive affixes give rise to **large number of rare event distributions** (Baayen, 2001; Khmaladze, 1987). As a result, they also often exhibit a large number of distinct word forms (i.e., high type frequency). In contrast, less productive affixes, such as *-th*, mostly occur in existing forms, and therefore tend to appear in words across a variety of frequencies, often with a bias towards more frequent items.

An important tool for visualizing the frequency distribution of affixes is the **frequency spectrum**. Below, we reproduce the frequency spectra for *-ness*, *-ity*, and *-th* from Figure 1.2.6 in Section 1.2.6. The *x*-axis of each plot represents (log) word frequency. The height of the bars represents the densities of frequency bins with size proportional to the width of the bar, while the red lines are a smoothed kernel density estimate of each word frequency.

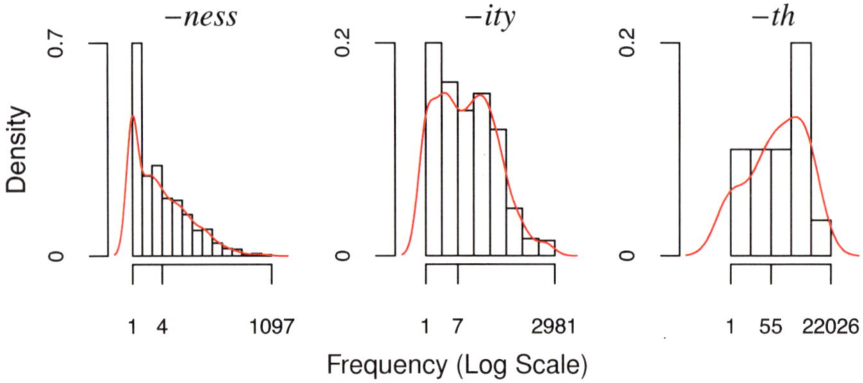

Figure 6.1
Frequency Spectra for *–ness*, *–ity*, and *–th*

The far left of Figure 6.1 shows the frequency spectrum for *-ness*. As discussed above, because *-ness* is highly productive, the spectrum is steeply peaked at low frequencies, indicating the large number of low-frequency words using this suffix. The distribution falls off quickly towards the higher frequency ranges. The spectra for *-ity* and *-th* are quite different. These suffixes have a smaller proportion of words with low frequency and a higher proportion of words with high frequency.

The differences in frequency distribution between productive and unproductive affixes can be understood in terms of a trade-off. A given word token can either be novel, or it can be a reuse of an existing word, but it cannot be both. Because productivity is exemplified by the use of an affix to derive new word forms, it is the shape of the word frequency distribution for a given affix that is most relevant in making inferences about the affix's productivity. As a result, theories of morphological productivity based only on the absolute type or token frequencies of words or affixes will fail to make the correct predictions in some cases (Baayen, 1992, 1993, 1994, 2001, 2006; Baayen and Lieber, 1991; Baayen and Renouf, 1996; Barnwell, 2010; Bauer, 1992, 2001, 2005; Du and Zhang, 2010; Fernández-Domínguez et al., 2007; Gaeta and Ricca, 2006; Hay and Baayen, 2002; Plag, 2004; van Marle, 1992; Yang, 2005).

For example, Bybee (1995), proposes that the type frequency (also known as the *morphological family size*) of an affix should predict its morphological productivity. However, because type frequency does not take into account the token frequency of individual words using an affix, it makes incorrect predictions in some cases. Baayen (2006) pointed out, for example, that the Dutch verbal prefix *ver-* has greater type frequency than the agentive suffix *-ster*, but that the former is judged less productive than the latter by speakers. He argued that this is because most types using *ver-* have a high token frequency, detracting from the morpheme's overall productivity. A similar point is made by Aronoff (1983), who suggests that high frequency forms containing an affix should actually detract from the productivity of the affix. In the next chapter, we will consider the case of the suffix *-ion* which occurs in a very large number of English words—in fact, having a type frequency greater than the productive suffix *-ness*. Nevertheless, *-ion* is not productive in isolation, and this is due to a relative lack of low frequency words which contain the suffix (see, Section 7.3.1.3). Intuitively, the importance of token frequency in these cases arises because a productive affix should frequently give rise to new forms. Therefore, a putatively productive affix that does not occur in many low (token) frequency words is highly suspicious. A related criticism of type frequency as a predictor of probability was made by Marcus et al. (1995) who argued that the most generalizable way of forming nominal plurals in German—via suffixation with *-s*—has the lowest frequency of any plural allomorphs, both in terms of types and tokens. In short, viable distributional predictors of productivity must take into account the distribution of tokens over types.

6.3.5 Productivity

Derivational morphemes vary greatly in their productivity. There is ample anecdotal evidence that English speakers are able to use some affixes (such as *-ness*, *-ly*, *-y* and *-er*) to form new words. In some cases, these intuitions are supported by experimental results. For example, in her famous "wug" study, Berko (1958) showed that agentive *-er* could be generalized to novel verbs by both adults and children. In a study of a speech corpus, MacKay (1979) found that speakers spontaneously generalize a number of derivational affixes to create novel forms. In studies of spontaneous production of affixed forms, Anshen and Aronoff (1988) and Baayen (1994) found that participants generate more novel forms for productive affixes than they do for unproductive affixes (for English and Dutch affixes, respectively; see, also, Anshen and Aronoff, 1981; Aronoff and Schvaneveldt, 1978; Cutler, 1980).[4]

The psychological reality of productivity distinctions is further reinforced by the way in which speakers sometimes fail to differentiate between potential and existing words for highly productive affixes (Aronoff, 1983). For example, Aronoff and Schvaneveldt (1978) and Anshen and Aronoff (1988) found that, for such affixes, people have difficulty distinguishing words that they are unlikely to have previously encountered from words that they are very likely to already know. Productivity also effects people's judgments in word-frequency estimation tasks. Mahowald et al. (under review) found that when asked to estimate the number of words in a hypothetical text that matched certain letter frames, such as `_ _ _ _ n e s s`, participants systematically overestimated the frequency of words consistent with frames containing productive suffixes. Furthermore, they overestimated frames with more productive suffixes at a higher rate than frames with less productive suffixes.

An important theoretical question is whether differences in productivity can be reduced to differences in the phonological, semantic, and morphosyntactic selectional restrictions of individual morphemes (Bauer, 1992). Since affixes with stricter selectional restrictions necessarily select for smaller sets of bases, all else being equal, they will be used to generate fewer novel words than affixes with broader selectional restrictions. However, native speakers exhibit gradient acceptability judgments even in cases where two words have been formed from the same base, with the same meaning, using different affixes. These cases show that speakers must represent some gradient information that influences generalization judgments beyond the selectional restrictions on affixes.

For example, Aronoff and Schvaneveldt (1978), Cutler (1980), and Anshen and Aronoff (1981) reported a series of experiments examining the formation of abstract nouns ending in *-ity*:`Adj>N` and *-ness*:`Adj>N`, from nonsense bases ending in *-ible* (e.g., *remortible*) and *-ive* (e.g., *depulsive*). These experiments demonstrated that native speakers prefer to form abstract nouns with *-ness* when the base ends with *-ive* (i.e., *depulsiveness*>*depulsivity*), but with

4. The dependent variable in production experiments such as Anshen and Aronoff (1988) and Baayen (1994) is the number of words generated for each suffix that are not present in some reference dictionary or corpus.

-ity when the base ends with *-ible* (i.e., *remortibility*>*remortibleness*). Crucially, despite the differences in acceptability ratings, both the *-ity* and *-ness* forms are judged to be acceptable English words. Since these experiments compared judgments for identical bases, and since all forms are judged acceptable, it cannot be the case that differences in the rate of generalization of *-ity* and *-ness* can be reduced purely to availability or unavailability of bases as input.[5]

Further evidence for this conclusion comes from Bauer (1992) who studied the affix combination *-lily* (i.e., adjectival *-ly*:`N>Adj` followed by adverbial *-ly*:`Adj>Adv`; e.g., *manlily*, *man +ly +ly*). Bauer showed that there is variability in readiness with which *-ly*:`Adj>Adv` attaches to different classes of words ending in adjectival *-ly*:`N>Adj`. Furthermore, he argued that this variability goes above and beyond selectional constraints on its distribution of *-ly*:`Adj>Adv`.

Cases such as these demonstrate that an account of productivity for English derivational morphology must provide quantitative mechanisms for resolving competition between multiple ways of forming new words. As was the case with the English past tense, competition in derivational morphology is often characterized by the elsewhere condition which states that more specific forms or rules are preferred to less specific forms or rules (Anderson, 1969; Andrews, 1982, 1990; Aronoff, 1976; Bresnan, 2001; Embick and Marantz, 2008; Halle and Marantz, 1993; Kiparsky, 1973, 1982b; Noyer and Harley, 1999; Plag, 2003; Rainer, 1988; Stump, 2001b; van Marle, 1986). However, as pointed out by Kiparsky (1982b), in English derivational morphology the elsewhere condition must be viewed only as a tendency due to the large number of doublets, words like *toleration* and *tolerance*, which have the same base and meaning, but are derived from different morphemes.

6.3.6 Processing

A fundamental question in psycholinguistics is whether morphologically complex words are retrieved whole from memory or composed on the fly. The

5. This isn't to say that differential base availability cannot play a role in productivity judgments. It is just to say that productivity cannot be reduced purely to differential rates of word formation predicted by differing selectional restrictions. Of course, if speakers represent meta-information about a word-formation process, such as the number of bases available to the process, this might influence how easily they use it to form new words. However, this idea goes beyond the simple hypothesis that differential rates of generation of novel forms is driven solely by the number of bases available.

results of over fifty years of intense research on this question have definitively shown that both composition and retrieval processes are employed. For example, one case that has been studied extensively is English prefixation. Using a lexical decision task, Taft and Forster (1975) showed that bound stems from truly prefixed words (e.g., *juvenate* from *rejuvenate*) take longer for subjects to reject than pseudo-stems (e.g., *pertoire* from *repertoire*). They concluded that prefixed words are represented in terms of their parts. Further evidence for this conclusion comes from experiments using different manipulations (e.g., prefixed non-words versus unprefixed words) and other experimental paradigms (e.g., primed lexical decision; see, e.g., Forster and Azuma, 2000; Taft et al., 1986). However, there is also evidence for the lexical storage of prefixed forms. For example, in an identity priming task Stanners et al. (1979b) showed that prefixed words with bound stems elicit weaker priming of their stems than prefixed words with free stems, suggesting that words with bound stems are represented as wholes.

These results show that there is pervasive variability in word composition and retrieval. This variability extends down even to the level of individual words Hay (2001).[6] Although there is a broad consensus in the literature that lexical processing mixes composition and retrieval, the factors that predict each mode of computation are numerous and complex, and there are systematic relationships with frequency, semantic transparency, and phonological regularity.

It has long been known that both base/stem frequencies and whole-form frequencies play a role in morphological processing. High-frequency words are more likely to be retrieved, while low-frequency words are more likely to be composed. This fact has been demonstrated with a wide variety of experimental methods, including lexical decision, identity priming, speeded naming, and free reading. (e.g., Alegre and Gordon, 1999; Baayen et al., 1997b, 2003, 2007; Beck, 1997; Bertram et al., 2000b; Bradley, 1980; Burani and Caramazza, 1987; Burani et al., 1984; Clahsen et al., 2004; Colé et al., 1989, 1997; Gordon and Alegre, 1999; Manelis and Tharp, 1977; Meunier and Segui, 1999; Moscoso del Prado Martín et al., 2005; Prasada et al., 1990; Rubenstein et al., 1970; Sereno and Jongman, 1997; Taft, 1979; Vannest and Boland,

6. The model presented in this book allows variability to extend even down to the level of individual uses of individual words. In fact, the fragment grammars model makes crucial use of this fact. It is the average rates of parsing that give rise to differences in productivity and processing in the model. Allowing this kind of variation may be important for modeling differences between experimental tasks. For example, Andrews (1986) shows that decomposition of morphologically complex words is modulated by the inclusion of compound words in tests.

1999). While most studies have focused on the processing of simple stem-affix combinations, more recently it has been shown that the frequencies of even deeply-embedded word constituents can have effects on processing latencies (Kuperman, 2008).

As was the case for the English past tense, processing latencies for high-frequency forms are often better predicted by their whole-form frequencies, indicating that high-frequency forms are stored as wholes in memory, while latencies for low-frequency forms are often better predicted by the frequencies of their parts, indicating that these forms are composed during processing. Using a lexical decision task, Colé et al. (1997) showed that the processing of morphologically complex words in French exhibits an interaction between base and whole-form frequencies. When the frequency of the base is less than the frequency of the whole form, latencies are better predicted by the base frequency. However, when whole-form frequency is greater than the frequency of the base, latencies are better predicted by the whole-form frequency. Using similar French materials, Meunier and Segui (1999) showed that only low-frequency suffixed words prime their stems in an identity priming task.

In general, whether word parts are decomposed or a word is retrieved whole from memory appears to depend on the ratio between part- and whole-form frequencies. This generalization has been termed the *relative frequency hypothesis* by Hay and colleagues (see, Section 6.4.2.2). Hay (2001) reported the results of an experimental study supporting the relative frequency hypothesis. Participants were asked to rate the complexity of various English derived word forms. The study matched words using identical affixes differing in whether their whole-form frequency was greater than or less than their base frequency. Hay found that words for which the base frequency was greater than the whole-form frequency were rated as more complex than the corresponding matched words for which the opposite frequency relation held.

While it appears that the ratio of part- and whole-word frequencies is a major determinant of decomposition/retrieval, there has been some debate in the literature about which kinds of part-word frequencies are most appropriate in predicting processing latencies. Historically, most studies have made use of the token frequency (i.e., total number of occurrences) of bases or stems to predict the processing of low-frequency forms. However, a number of more recent studies have emphasized the use of type frequency, sometimes called the morphological family size of a base or affix—the number of distinct words that the base or affix appears in. For example, Bertram et al. (2000a) found that morphological family size of bases predicts the processing time of complex

Dutch words and that this effect is driven by semantically transparent family members, suggesting that morphological family size is specifically linked to regular and productive composition. Similarly, Moscoso del Prado Martín et al. (2005) demonstrated an effect of morphological family size for both Dutch and Hebrew, and further showed that this effect is based on shared morphological structure, rather than just shared semantics. In fact, the morphological family size of a base even predicts processing time when the base is presented in isolation, suggesting that the size of a word's morphological family is important even when other affixes are not present (Baayen et al., 2006; Schreuder and Baayen, 1997).

Baayen et al. (2007) argued that only morphological family size—not base token frequency—is relevant for predicting the processing times of low-frequency words. In a study of low-frequency English words, they demonstrated that, while whole-form token frequency and morphological family size both have a facilitatory effect on lexical decision latencies, base token frequency has no effect when these (and other) factors are controlled by inclusion in a linear model. This result should be interpreted with caution. A linear model assumes that the relationship between variables, such as the (log) morphological family size and (log) base token frequency, can be captured as a linear combination of their values. However, in the present case, this assumption is clearly wrong. The type frequency of a base is the number of distinct words in which a base was used, and, therefore, provides a lower bound on the token frequency of the base. This relationship cannot be expressed as a linear combination. For low-frequency forms, where both family size and base token frequency are small, like those studied by Baayen et al. (2007), base type and token frequencies are especially confounded.

Moving on from frequency-based predictors of processing, numerous experimental studies have shown that decomposition and retrieval are also correlated with phonological complexity, phonological regularity, and semantic transparency . Generally, affixes triggering more complex phonological changes tend to be decomposed less often than affixes triggering less complex changes (MacKay, 1978). Words that are more phonologically irregular tend to decomposed less often than more regular words (Derwing and Baker, 1979; Marslen-Wilson et al., 1994; Vannest and Boland, 1999; Wheeler and Schumsky, 1980), and words that are less semantically transparent tend to be decomposed less often than more transparent words (Bradley, 1980; Derwing and Baker, 1979; Feldman and Soltano, 1999; Marslen-Wilson et al., 1994).

Finally, there is a close relationship between productivity and processing. Affixes which tend to give rise a greater number of novel words also tend to be decomposed more often during the processing of existing words (see, e.g., Frauenfelder and Schreuder, 1991). Most models of morphology unify productivity and processing by assuming that both the productive generation of new words and the decomposition of existing words are the result of a single underlying mechanism—such as rules with variables. While this assumption is near universal in the literature and is favored by considerations of parsimony, it is important to emphasize that it is a hypothesis. It is possible that generalization of novel forms and composition of existing forms are handled by different underlying mechanisms or representations.

Unfortunately, it is very difficult to directly test the hypothesis that composition of existing and novel forms share the same mechanism. Such a test would require comparing empirical data on generalization rates for a number of affixes (e.g., wug tests or naturalness ratings) with some measure of composition of those affixes in existing words. At present, this is challenging for two reasons. First, wide coverage databases of affix generalization data do not yet exist. Second, because composition and retrieval are not directly observable properties of words or affixes—in experimental work, they are inferred from patterns of processing latencies—a measure of composition will necessarily be a derived, model-based quantity, and there is, as yet, no general consensus on the correct approach.

Nevertheless, there have been a small number of studies that have tried to establish a relationship between composition and productivity. For example, Baayen (1993) and Hay and Baayen (2002) compared the predictions of several models of morphological processing with a number of corpus-based estimators of productivity (e.g., Baayen's $\mathcal{P}^*$, see Section 6.4.1.2). They showed that affixes with higher productivity scores were more likely to be composed by the processing model. Because these studies compared derived, model-based measures of composition and productivity, they cannot be taken as providing direct evidence that the same representations underlie the two phenomena. Nevertheless, the models of processing and the estimators of productivity used in these studies were defined independently and based on very different assumptions and mathematical tools. Therefore, it is encouraging that these distinct approaches converge in the way that would be expected if the same mechanisms underlie composition during processing and productivity.

6.3.7 Morpheme Combination and Ordering

English derivational morphology also exhibits strong preferences on the co-occurrence and order of morphemes. It is well established that certain suffixes can only attach to certain bases. For example, *-th* only appears on a restricted set of stems (e.g., *warm*, *wide*, etc.). Similarly, bound stems only occur with a restricted set of affixes, but never on their own (e.g., *-vive* in *survive*). Such preferences extend to sequences of more than two morphemes. Nominalizations of verbs ending in *-ize* are predominantly formed with the ending *-ation* (e.g., *nominalize/nominalization*), despite the availability of many other suffixes that can be used to form abstract nouns from verbs (e.g.,*-ment*, *-age*, *-al*), some of which are preferred to *-ation* in other contexts. Furthermore, to form an adjective from such an abstract noun ending in *-ization*, the suffix *-al* (*nominalizational*) is used in preference to, for example, *-ary* (*inflationary*). Aronoff and Schvaneveldt (1978) gave several other examples, noting that *-ousness* is preferred to *-osity* (e.g., *humorousness* v. *humorosity*) and *-iveness* is preferred to *-ivity* (e.g., *restrictiveness* v. *restrictivity*), while *-ility* is preferred to *-ileness* (e.g., *servility* v. *servileness*), and *-icity* is preferred to *-icness* (e.g., *concentricity* v. *concentricness*).

Such preferences for combinations of morphemes are pervasive in English derivational morphology and lead to a long-standing empirical puzzle known as the **affix-ordering problem**. The affix ordering problem arises from the observation that the number of affix sequences that actually appear in English words is much smaller than the number of logically possible sequences. A number of empirical studies have quantified this phenomenon. For example, in a study of 43 English suffixes, Fabb (1988) showed that there are 663 pairs of suffixes that are possible when categorial restrictions are taken into account. However, only 50 of these combinations are actually attested in his sample.

The pattern of possible and impossible affix sequences has been the subject of much theoretical work in the literature (see, also, Section 6.4.3). Hay and colleagues have observed that an important empirical generalization holds in this domain: more productive affixes tend to occur outside of less productive affixes (Hay, 2002, 2003; Hay and Plag, 2004; Manova, 2010; Plag and Baayen, 2009; Zirkel, 2010). We will call this fact the **productivity-and-ordering generalization**. The productivity-and-ordering generalization follows as a consequence of the mental representations underlying morphological systems that allow both the lexical storage of (morphologically complex) words and the productive composition of (sequences of) affixes with bases. As we have emphasized throughout the book, particular combinations of morphemes are sometimes stored together in memory (e.g., *warmth*) and sometimes productively computed (e.g., *pine-scentedness*). Therefore, the distribution of possible sequences of affixes can at least be partially explained by an appeal to three factors: (i) the reuse of stored, morphologically complex words, (ii) the productive use of individual affixes and combinations of affixes and, (iii) the assumption that the set of stored, morphologically complex words is the result of historical patterns of storage and productive composition.

It is possible for a word token to be generated by productively attaching a morpheme outside of a stored morpheme sequence, but the opposite is logically impossible (for concatenative affixation). Therefore, because unproductive morphemes mostly occur in existing words (by definition) and productive morphemes can attach to such existing words, morphemes that tend to give rise to a greater number of new words also tend to appear more often outside of morphemes that give rise to fewer new words.

Furthermore, if we make the standard theoretical assumption that the same mechanisms underlie the productive generalization of new words and the composition of existing words during processing, then the productivity-and-ordering generalization also predicts that there should be a close relationship between affix ordering and decomposition/retrieval during processing. Affixes that tend to be decomposed during processing tend to occur outside of affixes that tend to be retrieved as parts of stored words. In support of such a theoretical position, Hay and Plag (2004) and Plag and Baayen (2009) showed that ordering relationships between affixes correlate with a number of theoretical measures of productivity and composability (see Section 6.4.3.3).

Given the other correlations with productivity that we have discussed in this chapter, the productivity-and-ordering generalization also predicts that affixes that tend to appear closer to word stems should also tend to be more phonologically complex, less phonologically regular, and less semantically transparent. These predictions are supported by the empirical literature (Aronoff, 1976; Aronoff and Sridhar, 1983; Chomsky and Halle, 1968; Fabb, 1988; Giegerich, 1999; Halle and Mohanan, 1985; Hay, 2002; Hay and Plag, 2004; Kiparsky, 1982b,c; Manova and Aronoff, 2010; Mohanan, 1986; Plag, 1996, 2002, 2003; Plag and Baayen, 2009; Siegel, 1974).

While the productivity-and-ordering generalization holds generally across English words, there are numerous exceptions. Some of these can be explained by an appeal to word-by-word variation in decomposition and retrieval—in particular, by the composition of a (generally) less productive affix with a stored base containing a (generally) more productive affix. For example, Plag (2002) discussed the pair of suffixes *-ion* and *-al*. The suffix *-al* more freely combines with bases than the suffix *-ion*, which tends to occur mostly in stored forms (e.g., *confession*). As a result, in most English words containing both suffixes, *-ion* appears before *-al* (e.g., *confess**ional***). However, this ordering is reversed in words such as *coloni**al**iz**ation*** (Plag, 2002). Examining a set of such exceptional words, Hay and Plag (2004) found that the base (e.g., *colonial*), was more frequent than its own base (i.e., *colony*), which predicts that *colonial* should therefore be stored as a whole in memory (see Section 6.3.6). Furthermore, unlike most instances of *-ion*, the suffix combination *-ate⁀ion* (i.e., *-ation*) is semi-productive in English. Thus, words like *colonialization* are explained by appealing to word-by-word patterns of storage and computation. In this case, the semi-productive suffix combination *-ation* combines with the stored base, *colonial*, leading to the unexpected ordering between *-al* and *-ion*. In another example, Hay (2002) found that *-ment* forms that are likely to be stored as wholes (e.g., *government*) are rated by English speakers as more acceptable when suffixed with *-al* (e.g., *governmental*) than *-ment* forms that are likely to be composed (e.g., *wonderment*/*wondermental*).

A more difficult set of exceptions to the productivity-and-ordering generalization are the case of **paradoxical suffix combinations**. Paradoxical suffix combinations arise when a generally unproductive and/or phonologically complex suffix, such as *-ity*, generalizes productively in the context of another morpheme. A prototypical example of this phenomenon is the suffix combination *-ability*. For the most part, the suffix *-ness* is used to productively derive new abstract nouns from adjectives in English (*pine-scentedness*), rather than the suffix *-ity* (**pine-scentedity*). However, Anshen and Aronoff (1981) showed that for bases ending in *-able*[7] (e.g., *remortible*), speakers prefer the abstract noun ending in *-ity* (i.e., *remortibility*) to the abstract noun ending in *-ness* (i.e., *remortibleness*), although both forms are acceptable (see, also, Aronoff and Schvaneveldt, 1978; Cutler, 1980; Embick and Marantz, 2008). Crucially, since the bases used in this experiment were novel, this preference cannot be reduced to word-by-word variation in storage and retrieval, unlike the *-al*/*-ion* examples discussed above. Such cases appear to require a mechanism whereby *-ity* can be sensitive to the presence of *-able* in a base, even if the base itself is composed (i.e., a morpheme-specific selectional restriction). Equivalently, these cases can be handled by a mechanism whereby *-ability* can generalize

7. Anshen and Aronoff (1981) actually studied the variant of *-able*: *-ible*.

as a single unit. Other examples of such paradoxical suffix combinations that have been mentioned in the literature include *-ation*, *-ility*, *-icity*, and *-mental* (Anshen and Aronoff, 1981; Aronoff and Schvaneveldt, 1978; Cutler, 1980; Embick and Marantz, 2008; Hay, 2003; Plag, 2002).

6.3.8 Summary

In this section, we have reviewed the empirical literature on English derivational morphology, focusing on seven classes of phenomena. Derivational affixes vary in the kinds of selectional restrictions they exhibit, and the restrictiveness of those restrictions. They vary in their phonological complexity and regularity, and in the semantic transparency of words in which they appear. Furthermore, they vary in the frequency distributions of forms in which they appear and in their productivity. One important generalization that arose in this context is that more productive affixes are characterized by frequency distributions with a greater number of low frequency forms.

Affixes also vary in whether they are composed or retrieved as parts of words during processing. This variation in decomposition/retrieval is also related to frequencies: words that have a higher part- to whole-form frequency ratio are more likely to be decomposed during processing. This fact is codified in Hay's relative frequency hypothesis. There is also a tight relationship between decomposition and productivity; more productive affixes tend to be composed more often during processing. This fact can be explained by models that assume that a single mechanism underlies both the productive generation of new words and the decomposition of existing words.

Finally, affixes exhibit strong constraints on which other morphemes they can co-occur with, and this leads to restrictions on the set of possible affix sequences. An important generalization in this domain is the productivity-and-ordering generalization which observes that more productive affixes tend to occur outside of less productive affixes. However, while the productivity-and-ordering generalization holds in general for English derived words, there are numerous exceptions. The most difficult class of such exceptions is the class of paradoxical suffix combinations—cases where a generally unproductive suffix, such as *-ity*, is able to generalize productively in the context of another suffix, such as *-able*, even for novel forms.

In addition to the relationships between the phenomena just discussed, we have also emphasized that more productive affixes tend to be less phonologically complex, more phonologically regular, and more semantically transparent, while less productive affixes tend to be more complex, less regular, and less transparent.

6.4 Theoretical Approaches

In this section, we review the theoretical literature on English derivational morphology. There have been a large number of theories proposed to account for the structure of English derived words, focusing on different subsets of the phenomena we reviewed in the previous section. In the next chapter, we will show how the model proposed in this book—fragment grammars—can explain and unify several of these empirical domains, including differences in productivity between individual affixes, differences in the frequency distribution of words associated with different affixes, differences in composition and retrieval during the processing of existing words, and constraints on suffix combination and ordering. With this in mind, our review below is divided into three parts. First, we consider models of productivity of individual affixes. Second, we discuss models of composition and retrieval during processing. Third, and finally, we discuss theories of affix combination and ordering.

6.4.1 Models of Productivity

As we discussed in Section 1.2, the problem of accounting for the differential productivity of grammatical rules has been at the heart of linguistic theory since the early days of generative grammar. The fundamental tension arises from the need to simultaneously account for productive linguistic processes, such as *-ness* affixation, context-dependent productive processes such as *-ity* affixation, and unproductive processes such as *-th* affixation. In that section, we reviewed how, in response to this problem, the field has gradually adopted a heterogenous account of linguistic structure allowing the storage and computation of a variety of structures with varying degrees of abstraction.

Throughout this book, we have also argued that any account of productivity must provide a way of resolving competition when several rules can apply simultaneously (see Sections 4.4.2 and 6.3.5). Furthermore, because there are cases where several rules can apply, but at different rates, an account of competition resolution should explain such quantitative variation. Finally, we have reviewed a large amount of evidence that productivity is intimately related to the frequency distribution of whole- and part-word forms. In this section, we review quantitative theories of productivity that base their predictions (at least partially) on these distributional aspects of the input data.

As we reviewed in Sections 6.3.4 and 6.3.5, since more productive processes give rise to greater numbers of novel forms, their frequency distributions are characterized by a large number of low-frequency words (Baayen, 1992, 1993, 1994, 2001, 2006; Baayen and Lieber, 1991; Baayen and Renouf, 1996; Barnwell, 2010; Bauer, 1992, 2001, 2005; Du and Zhang, 2010; Fernández-Domínguez et al., 2007; Gaeta and Ricca, 2006; Hay and Baayen, 2002; Plag, 2004; van Marle, 1992; Yang, 2005). The frequency distributions of less productive processes exhibit greater variance in word frequencies, often with a bias towards higher-frequency forms. Since a word can be either novel or an instance of the reuse of an existing form, but not both, the relationship between frequency and productivity represents a trade-off. As we argued above, this means that any viable distributional theory of productivity must take into account the distribution of word tokens over types, for both productively derived and stored forms, rather than the absolute type or token frequency of an affix.

Below, we review theories that attempt to capture such a trade-off between productivity and reuse, integrating both token and type information in their predictions. For the most part, these theories focus on predicting the productivity of individual affixes—we consider theories of morpheme combination and ordering in Section 6.4.3. The theories of individual affix productivity we discuss below fall into two classes. The first are statistical estimators intended to predict productivity from the frequency distribution of words and part-words in a corpus sample. We will focus in particular on Baayen's statistical estimators, $\mathcal{P}^{**}$, $\mathcal{P}^{*}$, and $\mathcal{P}$, discussing the theoretical underpinnings of these measures and their relationship to more explicitly grammatical or cognitive models.

The second set of models we examine are those theories deriving productivity predictions from cognitively-driven assumptions about the structure of words. We will focus in particular on the optimization-based theory of Yang (2005, 2010), arguing that, while this theory provides an interesting optimization-based perspective on productivity and has reasonable empirical coverage, it fails to capture some important phenomena in the domain of productivity and reuse. Before reviewing these two classes of theories, we first give a careful formal definition of productivity in terms of several abstract probabilistic quantities.

6.4.1.1 Defining Productivity

Although there are many different notions of morphological productivity (see Bauer, 2001; for detailed discussion), in this book, we define productivity as the psychological phenomenon whereby morphological processes may give rise to new word forms. In order to clarify how this phenomenon can be quantified, we begin by defining three abstract quantities that measure the rate at which different morphological processes can give rise to new words: $\Pi_{S,r}$ (***absolute productivity***), $\Pi^{*}_{S,r}$ (***between-rule productivity***), and $\Pi^{\circ}_{S,r}$ (***within-rule productivity***).

These quantities are inspired by Baayen's three hapax-based productivity statistics introduced in the following section (Baayen, 1992, 2001, 2003). Baayen's statistics are meant to estimate three probabilities, but are defined in terms of a specific statistical estimation technique (Good-Turing estimation) for these quantities. Here we generalize Baayen's definitions to apply to any arbitrary model of word structure which provides some distribution $P(w)$ over the set of possible words in a language.

Consider the set of all the possible words $\mathcal{W}$ in some language. By this, we mean the infinite set consisting of all simple stems together with all morphologically complex words that can be built by applying each morphological rule $r \in \mathcal{R}$ in the language to all stems where the rule is applicable, and then again, recursively, to all resulting bases, and so on. In other words, $\mathcal{W}$ is the closure of the set of stems under $\mathcal{R}$. For example, if the set of stems was {*clue*}, and the set of rules consisted of affixation of the morphemes {-*less*:`N>Adj`, -*ness*:`Adj>N`}, then $\mathcal{W}$ would be {*clue*, *clueless*, *cluelessness*,*cluelessnessless*, *cluelessnesslessness*,...}.

Suppose that we have access to a model of word structure which allows us to estimate the total (i.e., marginal) probability $P(w)$ of any word $w \in \mathcal{W}$. By this, we mean the probability that the model assigns to word w averaging over other, extraneous factors.[8] Consider a particular rule $r \in \mathcal{R}$ and let $\delta_r[w]$ be the function that returns one if the word w was the output of rule r, for example, δ_{ness} `Adj` > `N`$[clue] = 0$ and δ_{ness} `Adj` > `N`$[cluelessness] = 1$. Note that $\delta_r[w] = 1$ only when w is the direct output of rule r, that is, when r

8. A model, and thus $P(w)$, might (or might not) reflect any factors that influence the causal processes underlying the production and comprehension of words: phonotactic well-formedness, phonological structure, morphological structure, frequency, syntactic distribution, meaning, stylistic and sociolinguistic properties, practical utility across situations, communicative efficiency, variation amongst individuals, etc.

was the most recent rule applied to the base. Amongst other things, this means that if r added a suffix, it will be the last suffix on the word. The total probability of all words in a language that are the output of rule r, Π_r, is given by Equation 6.1. Note that $\delta_r[w]$ has the effect of multiplying words not output from r by 0 in the sum.

$$\Pi_r = \sum_{w \in \mathcal{W}} P(w) \cdot \delta_r[w] \tag{6.1}$$

Now consider some finite sample of words from $S \subset \mathcal{W}$ and let $\delta_{\setminus S}[w]$ be the function that returns 1 if $w \notin S$ and 0 otherwise. The set S can refer to any finite set of distinct words from $\mathcal{W}$. It might represent the set of word types in some text corpus, or the set of words in a dictionary, or all of the word types a speaker has either spoken or comprehended at some point in their lives. The function $\delta_{\setminus S}[w]$ indicates whether a word is novel with respect to this set. The quantity Π_S defined in Equation 6.2 gives the total probability associated with words that are new with respect to the finite set S, that is, the total probability of all words not in that set. Note that $\delta_{\setminus S}[w]$ has the effect of multiplying the probability of words in S by 0 in the sum.

$$\Pi_S = \sum_{w \in \mathcal{W}} P(w) \cdot \delta_{\setminus S}[w] \tag{6.2}$$

It is important to stress that, because it involves (by definition) unobserved events, the quantity Π_S will always be defined by some model projected from a set of observed words. In particular, a model that provides an approximation to the distribution over words, $P(w)$, will typically include a component that handles the generation of new words from parts (e.g., *pine-scented*+*-ness*:`Adj>N`). Various aspects of this model of word generation will have to be learned or estimated from a sample of word forms. In our discussion below, we will always assume that the sample of word forms used to estimate a model that approximates $P(w)$ is also the set S against which novelty is defined. In particular, our discussions below will focus on how models approximating $P(w)$ project from a corpus of observed words to a distribution over new word: $S \mapsto \Pi_S$.

Our first abstract measure of the productivity, $\Pi_{S,r}$, which we call the *absolute productivity*, is defined in Equation 6.3 below.

$$\Pi_{S,r} = \sum_{w \in \mathcal{W}} P(w) \cdot \delta_r[w] \cdot \delta_{\setminus S}[w] \tag{6.3}$$

$\Pi_{S,r}$ is just the total probability that a word is new with respect to the input sample S and that it was produced as the output of rule r. In a sense, $\Pi_{S,r}$ is the simplest possible measure of the productivity of a rule r, since it is just the conditional probability that a word w is new and produced by r: $P(w|w \notin S \wedge \exists x \ r(x) = w)$. However, this quantity is sometimes difficult to interpret because it measures productivity with respect to all possible words. In other words, it asks: if we chose a word at random from $P(w)$, how likely is it to be a productive use of r?

It is often more intuitive to measure productivity relative to more restricted subsets of possible words. For example, we might wish to measure the productivity of a rule relative to all the words generated using that rule: how likely is it that r gives rise to a new word? Alternatively, we may wish to measure the productivity of a rule relative to other new words: if we are given a new word, how likely is it to be built using r? There are two conditional probabilities which can be defined in terms of $\Pi_{S,r}$ (absolute productivity) and which address these questions (Baayen, 1992, 2001, 2003).

The first, $\Pi^*_{S,r}$, is called the ***between-rule productivity*** and focuses on the probability that a particular rule r will be used given that a novel word is the result.

$$\Pi^*_{S,r} = \frac{\sum_{w \in \mathcal{W}} P(w) \cdot \delta_r[w] \cdot \delta_{\setminus S}[w]}{\sum_{w \in \mathcal{W}} P(w)\delta_{\setminus S}[w]} = \frac{\Pi_{S,r}}{\Pi_S} = \frac{\Pi_{S,r}}{\sum_{r' \in \mathcal{R}} \Pi_{S,r'}} \tag{6.4}$$

$\Pi^*_{S,r}$ is the total probability of all novel words produced by r, relative to all possible novel words. In other words, it is the conditional probability that a rule was used to generate a word, given that the word is novel. This quantity can be interpreted as the proportion of overall vocabulary growth at some moment in time that is attributable to a rule r (see, Baayen, 2001). $\Pi^*_{S,r}$ is appropriate for comparing productivity between different word-formation processes, such as word formation with *-ity*:`Adj>N` versus word-formation with *-ness*:`Adj>N`.

Note that $\Pi^*_{S,r}$ captures less information than $\Pi_{S,r}$. In particular, $\Pi^*_{S,r}$ does not take into account the ***base rate*** of novelty—the overall rate with which new words are being added to the language. Also note that Π_S, the denominator in our definition of $\Pi^*_{S,r}$, is a constant with respect to different rules

r. Thus, if we are only concerned with the relative values of $\Pi^{*}_{S,r}$, for example if we are performing a correlation between this quantity and experimental data, then we are free to use the joint probability $\Pi_{S,r}$, instead. Finally, the last term in the equation above, which replaces the denominator Π_S with the sum $\sum_{r' \in \mathcal{R}} \Pi_{S,r'}$ is included to emphasize that many models of productivity only examine the relative productivity of morphological rules (e.g., *-ity*:`Adj>N` versus *-ness*:`Adj>N`), and do not consider other ways of forming new words that would otherwise be included in the distribution over all words $P(w)$ (e.g., neologisms, compounding, acronymization, etc.).

A third quantity that can be derived from $\Pi_{S,r}$ measures productivity with respect to a particular rule, and is called $\Pi^{\circ}_{S,r}$ or the ***within-rule productivity***. It is defined in Equation 6.5.

$$\Pi^{\circ}_{S,r} = \frac{\sum_{w \in \mathcal{W}} P(w) \cdot \delta_r[w] \cdot \delta_{\setminus S}[w]}{\sum_{w \in \mathcal{W}} P(w)\delta_r[w]} = \frac{\Pi_{S,r}}{\Pi_r} \tag{6.5}$$

Π°_{r} is the total probability of all novel words produced by rule r relative to all words produced by r. In other words, it is the conditional probability that a word is novel, given that it was generated by rule r. Intuitively, this quantity measures the amount of probability associated with r that is reserved to generate novel forms, rather than assigned to existing forms. It can be interpreted as the rate of growth in the vocabulary of words using rule r. Unlike $\Pi^{*}_{S,r}$ the denominator is not constant with respect to rules, and we must compute an approximation to it if we wish to compare across different rules.

All of the theories that we examine in the following sections define a distribution over words $P(w)$ based ultimately on just distributional information. Therefore, these models only take into account aspects of productivity reflected in the frequency distribution of words and part-words in a sample, and necessarily leave out such important factors as the meaning and pragmatic utility of different affixes. We now turn to these models.

6.4.1.2 Statistical Estimators of Productivity

In this section, we review statistical estimators designed to predict productivity directly from the frequency distribution of forms in corpus samples. The simplest kind of statistical estimators of productivity are those based on the absolute token or type frequencies of words and part words in a corpus, and there have been various proposals in the literature based on these quantities (e.g., Bybee, 1995). However, as we discussed above, any predictor that takes into account only type or token frequency will produce incorrect predictions in some cases (see Section 6.3.4). A viable predictor of morphological productivity must take into account the distribution of tokens over types. There have been several proposals in the literature that have integrated type- and token-frequency information to produce more accurate productivity estimates.

Barnwell (2010) defined an estimator of productivity that is the product of two quantities, the log frequency of whole word forms appearing with a particular affix and the entropy of the set of forms appearing with that affix. The productivity of a rule, like affixation with *-ness*:`Adj>N`, is computed by taking the product of the logarithm of the sum frequency of all words ending in *-ness*, and the entropy of the set of bases that appear before *-ness* in a corpus. Since entropy is maximized when a distribution is uniform, and furthermore is greater for distributions with greater numbers of outcomes, this measure assigns the highest productivity to rules used across many, frequent words. Although this measure is likely to be better than raw type frequency, it still suffers from being unable to take into account the fact that high-frequency forms can detract from the productivity of a rule (Aronoff, 1983; Baayen, 2006; Marcus et al., 1995; see, also, Section 6.3.4). Furthermore, this measure does not have a clear interpretation in terms of the probabilistic measures of productivity that we defined in the previous section.[9]

By far the most thoroughly studied and widely used measures of productivity are Baayen's $\mathcal{P}^{**}$, $\mathcal{P}^{*}$ and $\mathcal{P}$. These three statistics can be understood as approximating the abstract measures of productivity we defined in Section 6.4.1.1. In particular, $\mathcal{P}^{**}$ estimates the quantity $\Pi_{S,r}$ (absolute productivity), $\mathcal{P}^{*}$ estimates $\Pi^{*}_{S,r}$ (between-rule productivity), and $\mathcal{P}$ estimates $\Pi^{\circ}_{S,r}$ (within-rule productivity; Baayen, 1992, 1993, 1994, 2001, 2003, 2006; Baayen and Lieber, 1991; Baayen and Renouf, 1996; Du and Zhang, 2010; Hay and Baayen, 2002; van Marle, 1992).

Because productive affixes frequently give rise to new words, their word frequency distributions typically contain a large number of low frequency forms. This observation suggests that a reasonable estimate of the productivity of an

9. Although Barnwell's measure uses information-theoretic quantities, it does not have a clear interpretation in terms of the standard operations of probability theory: marginalization and conditioning. As a result, the measure does not seem to correspond to an estimate of the entropy, relative entropy, mutual information, etc. of any probability distribution(s) associated with a corpus.

affix might be based on the number or proportion of low-frequency words associated with the affix. Baayen's statistics are based on this idea. In particular, all three quantities estimate the probability of unobserved (i.e., novel) words using the proportion of word types which appear only once in a corpus. In the literature, words that occur only once are known as ***hapax legomena***, or ***hapaxes*** for short.

Imagine that we have some corpus S of N words together with their frequencies. Let $V(1,N)$ be the number of words appearing in corpus S with frequency 1—the hapax legomena. Let $\mathbb{E}[V(1,N)]$ be the expectation of $V(1,N)$. Baayen proposes that an estimate of the probability of observing a novel word type with respect to sample S (i.e., the quantity Π_S in Section 6.4.1.1), can be given by the following expression:

$$\Pi_S \approx \frac{\mathbb{E}[V(1,N)]}{N} \tag{6.6}$$

In other words, the probability of novel words is estimated using the proportion of observed words which occur only once in a sample. As we discussed in Section 6.4.1.1, the simplest abstract measure of productivity is $\Pi_{S,r}$ (absolute productivity), the total probability of all words that use rule r and are novel with respect to S. Using the logic just described to estimate Π_S, Baayen defines an estimator for $\Pi_{S,r}$ (absolute productivity) which he calls $\mathcal{P}^{**}$ or the ***unconditioned productivity***.

Let $\mathbb{E}[V_r(1,N)]$ be the expected number of hapaxes that are the output of a morphological rule r in corpus S of size N. $\mathcal{P}^{**}$ is defined in Equation 6.7.

$$\Pi_{S,r} \approx \mathcal{P}^{**} = \frac{\mathbb{E}[V_r(1,N)]}{N} \tag{6.7}$$

$\mathcal{P}^{**}$ is just the proportion of words using rule r and occurring only once in a sample. In the last section, we discussed the fact that it is often more useful to measure productivity relative to certain subsets of words, rather than all possible words. In particular, we introduced the conditional probabilities $\Pi^{\circ}_{S,r}$ (within-rule productivity), which measures the probability of novel words using r relative to all novel words, and $\Pi^{*}_{S,r}$ (within-rule productivity), which measures the probability of novel words using r relative to all words using r. Baayen defines two quantities that estimate these conditional probabilities. The first, which he calls $\mathcal{P}^{*}$, the ***expanding productivity***, or the ***hapax-conditioned productivity*** is an estimator for $\Pi^{*}_{S,r}$ (within-rule productivity), and is defined in Equation 6.8.***expanding productivity***

$$\Pi^*_{S,r} = \frac{\Pi_{S,r}}{\Pi_S} \approx \mathcal{P}^* = \frac{\mathbb{E}[V_r(1,N)]}{N} \bigg/ \frac{\mathbb{E}[V(1,N)]}{N} = \frac{\mathbb{E}[V_r(1,N)]}{\mathbb{E}[V(1,N)]} \tag{6.8}$$

In other words, $\mathcal{P}^*$ is just the proportion of all hapaxes in the corpus that are attributed to rule r, out of all hapaxes in the corpus. Like the quantity it estimates, $\Pi^*_{S,r}$ (between-rule productivity), $\mathcal{P}^*$ is useful for comparing the differential productivity of various affixes.

The second quantity, known as $\mathcal{P}$, the ***category-conditioned degree of productivity***, or ***potential productivity*** is an estimator for $\Pi^\circ_{S,r}$ (within-rule productivity). Let $\mathbb{E}[V_r(N)]$ be the expected frequency of all words in the corpus that make use of rule r. Then, the category-conditioned degree of productivity is given by Equation 6.9.

$$\Pi^\circ_{S,r} = \frac{\Pi_{S,r}}{\Pi_r} \approx \mathcal{P} = \frac{\mathbb{E}[V_r(1,N)]}{N} \bigg/ \frac{\mathbb{E}[V_r(N)]}{N} = \frac{\mathbb{E}[V_r(1,N)]}{\mathbb{E}[V_r(N)]} \tag{6.9}$$

In other words, $\mathcal{P}$ is the proportion of all words in the corpus that are attributable to r and appear only once, out of all words attributable to r. Like the quantity which it estimates, $\Pi^\circ_{S,r}$, $\mathcal{P}$ measures the growth rate of the set of words which are derived using r.

In the next chapter, we will compare values of $\mathcal{P}^*$ and $\mathcal{P}$ computed from corpora with estimates of $\Pi^*_{S,r}$ (within-rule productivity) and $\Pi^\circ_{S,r}$ (within-rule productivity) computed against the models studied in the book. It is therefore appropriate to say a few words about the theoretical interpretation and empirical validity of Baayen's statistics.

The intuition underlying Baayen's productivity measures is that, because productive processes give rise to a large number of low-frequency forms, a reasonable estimate of productivity can be based on the proportion of low-frequency forms in some sample. This intuition can be rigorously justified under several different sets of mathematical assumptions. Baayen (2001) describes two such formal derivations of the category-conditioned degree of productivity, $\mathcal{P}$.

The first derivation is based on a simplified model of word sampling. Under this model, it is assumed that words are sampled according to a categorical distribution from finite lexicon (i.e., the distribution over words is specified by a probability vector of length K, where K is the number of words in the lexicon). Given this assumption, it is possible to derive an expression for the expected number of different word types, $\mathbb{E}[V(N)] \leq K$, which will be observed in a sample of size N. Productivity can be defined as the rate of change in the expected number of observed words as the sample size, N, increases. Using a Poisson approximation to the expression for $\mathbb{E}[V(N)]$, Baayen shows that this rate of change is (approximately) equal to $\mathcal{P}$.

It is useful to consider the ways in which this derivation clarifies the simplifications inherent in the $\mathcal{P}$ statistic. $\mathcal{P}$ is an estimator for $\Pi^{\circ}_{S,r}$ (within-rule productivity). Recall that $\Pi^{\circ}_{S,r}$ (within-rule productivity) was defined in terms of the abstract probability distribution over words, $P(w)$, which captures all possible factors that influence the causal processes underlying word production and comprehension—phonotactic well-formedness, morphological structure, likelihood across syntactic contexts, meaning, pragmatic utility, etc. Obviously, any practical model that approximates $P(w)$ must simplify away some of these factors. Baayen's derivation, for instance, assumes that the causal processes underlying word use can be summarized using a fixed, finite-length vector of probabilities over words. It is useful to contrast this assumption with the assumptions of theories that explicitly model the grammatical and cognitive processes underlying word formation, such as the fragment grammars model.

Unlike Baayen's approach, fragment grammars provide an explicit model of the hierarchical, recursive processes underlying productivity and reuse in morphology. To the degree that this system is a more realistic model of morphology, we should expect it to provide a more veridical approximation to $P(w)$ than Baayen's approach. Thus, we should also expect approximations to $\Pi^{\circ}_{S,r}$ (within-rule productivity) based on fragment grammars to be more accurate than those based on $\mathcal{P}$.

What if $\mathcal{P}$ and fragment grammars make convergent predictions? Since the assumptions underlying fragment grammars are more realistic than those underlying the derivation of $\mathcal{P}$, when both theories make similar predictions for some case, we can conclude that $\mathcal{P}$ provides an approximate, high-level description of the behavior of the fragment grammars model, for that case. Put another way, fragment grammars' predictions about productivity are based on a complex, global optimization balancing a trade-off between productivity and reuse and potentially considering exponentially many possible stored and/or computed word fragments. If, in certain cases, the predictions of this complex optimization scheme coincide with $\mathcal{P}$, then we can conclude that inference in fragment grammars is approximated in those cases by considering the proportion of hapaxes associated with an affix. In this sense, a cognitive and linguistic model like fragment grammars may provide a justification for Baayen's statistics.

Baayen (2001) also describes a second mathematical derivation for $\mathcal{P}$, based on the theory of ***Good-Turing estimation***. Good-Turing estimation is a well-known statistical technique for estimating the true frequency of an event in some underlying population based on its frequency in a sample from that population. Suppose that we take a sample of size N from an underlying population with K possible event types. If we estimate the probability of each event type based on its sample frequency, m—that is, if we use m/N as an estimate of the probability of the event—we will tend to systematically underestimate the probabilities of low probability event types. In particular, we will assign probability zero to any event type that failed to make it into the sample. Roughly, Good-Turing is based on the idea that frequency estimates should be corrected by a factor that takes into account the number of different event types that occur at frequency $m+1$.[10] In particular, Good-Turing states that the corrected estimate of frequency zero events should be $\mathbb{E}[V(1,N)]$, the number of events that occurred exactly once. Turning this frequency into a probability by normalizing by the sample size N, gives $\mathbb{E}[V(1,N)]/N$, which is the definition of $\mathcal{P}$.

Good-Turing estimation is widely used and has proven especially useful for smoothing natural language frequency distributions (i.e., accounting for probability mass belonging to words unobserved in a training sample; see, e.g., Chen and Goodman, 1998; Goldwater et al., 2006; Teh, 2006). Although there are many mathematical derivations of Good-Turing estimation (see e.g., Gale and Sampson, 1995; Orlitsky et al., 2003), it is beyond the scope of the current discussion to examine these in detail. However, we emphasize that the various derivations all make simplifying assumptions about the nature of frequency distributions. In particular, they do not explicitly capture the generative processes underlying word formation that are inherent in models like fragment grammars. Therefore, to the degree that fragment grammars and $\mathcal{P}$ make similar predictions with regard to productivity, we can again interpret Good-Turing estimation as an approximate, high-level description of the behavior of the model.

Turning to the empirical status of Baayen's statistics, there are a number of reasons to believe that they correctly capture at least some aspects of morphological productivity. First, Baayen's measures have been the subject of a

10. Precisely, Good-Turing states that the corrected estimate of the frequency for an event that occurred m times in the sample is given by $(m+1)\frac{\mathbb{E}[V(m+1,N)]}{\mathbb{E}[V(m,N)]}$.

long discussion in the literature, and, in many cases, seem to fit with the intuition of linguists (Baayen, 1993, 2006; Baayen and Lieber, 1991; van Marle, 1992). Second, Baayen's estimators also replicate across corpora. For example, Baayen and Renouf (1996) showed that estimates of $\mathcal{P}$ and $\mathcal{P}^*$ from two different corpora achieve similar (although not identical) results. Third, there is some empirical evidence that these measures correlate with estimates of productivity from experimental data. Baayen (1994) presented a production experiment in which speakers were asked to generate as many forms as possible using particular Dutch affixes. Because participants were provided with a specific affix and then asked to generate forms, the number of words produced for each affix can be seen as an estimator of $\Pi^{\circ}_{S,r}$ (within-rule productivity). Baayen found that the category-conditioned degree of productivity, $\mathcal{P}$—which also estimates $\Pi^{\circ}_{S,r}$ (within-rule productivity)—correlated well with experimental results. Finally, as we will discuss in greater detail below, Baayen's quantities correlate with a number of other factors known to be related to productivity, such as measures of composition/retrieval during processing, and ordering.

6.4.1.3 Cognitive Models of Productivity

In the preceding section, we discussed corpus-based estimators of productivity. In this section we turn to models of productivity based on explicitly cognitive theories of word formation.

One early proposal was given by Aronoff (1976). The bulk of Aronoff (1976) is devoted to outlining a rule-based theory of word formation. However, Aronoff also argued that this theory of word formation could be used to provide an estimate of the productivity of individual linguistic rules. In particular, he proposed that the productivity of a rule should be given by the ratio of existing word types using a rule to the number of possible word types using that rule, where the set of possible words is defined by his theory of word formation. Under the assumption that all potential and existing words are equiprobable (i.e., that $P(w) \propto 1$ for all w), Aronoff's proposal is equal to one minus the within-rule productivity, that is $1 - \Pi^{\circ}_{S,r}$. Although Aronoff motivated this measure in terms of his grammatical theory of possible words, it is not clear how it could be approximated in practice. Furthermore, the assumption that all words are equally probable does not allow this measure to take into account the various kinds of frequency effects we discussed earlier in the chapter.

A more recent cognitive theory of productivity was provided by Yang (2005, 2010). Yang's proposal was based on a serial search model of morphological processing. Consider a speaker who wishes to generate a morphologically complex word from a given base. In Yang's model, this computation is implemented by a linguistic rule consisting of three components. The first component is the rule's *structural description*, which describes the set of selectional restrictions on bases to which the rule can apply. Second, a rule has an optional *default structural change*, which specifies a default transformation to apply to a base to produce an output word form. The default structural change provides a mechanism whereby rules can productively apply to new bases. However, the default structural change only applies to bases that are not listed in the third component of a rule: the (possibly empty) *exception list*. Bases listed in the exception list do not undergo the default structural change, but instead have a different output specification. These bases may be associated with specific stored word forms, or may undergo transformations that are distinct from the default structural change.

To generate an output form given a base, a rule first checks if the base satisfies its structural description. If it does, the rule then checks the base against each element of the exception list, which is ordered in terms of the decreasing token frequency with which that exception has applied in the past (i.e., the frequency of the output form associated with that exception). As a result, more frequent exceptions will be matched before less frequent exceptions. If the base matches an element of the exception list, the corresponding output form is computed. Only when no exceptions match, does the rule apply its default structural change. Under Yang's model, a rule is productive whenever it provides a default structural change. Since the default structural change is optional, a rule can consist merely of a list of memorized exceptions, in which case the rule is unproductive.

Yang proposed an optimization criterion for determining when a rule will have a default structural change called the *tolerance principle*. The tolerance principle states that a default will be favored when the expected time required to apply the rule with the default case is less than the expected time without the default. Yang assumed that checking each exception takes one time step. Therefore, when the default applies to a base (i.e., when the base is not listed in the exception list), the time taken will be the total number of elements in the exception list. However, when a base matches an exception, the time taken will be the number of bases that came before it in the exception list. Because exceptions are ordered in terms of decreasing token frequency, more frequent exceptions will be processed faster than less frequent exceptions. The tolerance principle states that a rule will have a default structural change when providing such a default leads to faster expected processing time, averaging over all inputs. Thus, whether a rule is productive will depend on the frequency distribution of words in the input data.

Yang derived a mathematical estimate of the threshold at which defaults would be favored. Based on the assumption that word frequency distributions are Zipfian, Yang showed that when the number of exceptions M is less than the total number of words N divided by the logarithm of N, a productive default case will be favored; that is, a productive default will be favored when $M < \frac{N}{ln(N)}$.

Like the models considered in the last section, the tolerance principle implicitly integrates both type and token frequency in determining whether a rule should be productive, and is thus able to capture certain aspects of the tradeoff between productivity and reuse. To see this, consider how the principle balances the token frequency of individual exceptions against the type frequency of words generated by the rule. When a rule generates a large number of forms with common morphological structure (e.g., a large number of words ending in *-ness*), a default structural change will be favored. Without the default, all words would have to be listed explicitly, and therefore the average time spent searching the exception list would be very high. By contrast, when a rule is only used to produce a small number of high-frequency words, there will be no advantage to including a default structural change since the exception list will be short and, on average, a base will be matched quickly.

Yang's tolerance principle provides an interesting, optimization-based perspective on productivity. However, there are a number of empirical phenomena related to productivity and reuse that the model appears to have difficulty with. First, as we discussed above, there is empirical evidence for the storage of at least some high-frequency words containing suffixes that are regular and productive in other contexts (e.g., Baayen et al., 1997b; Bertram et al., 1999; Sereno and Jongman, 1997). However, Yang's model does not have a mechanism that can predict the storage of forms that can otherwise be handled by a default structural change. Storing any form in the exception list can only increase the average processing time for a rule. Therefore, if a form can be handled by the default, it always should. It is possible to imagine an extension of Yang's model that allowed the storage of high-frequency regulars by associating a cost with the default structural change. If this cost were high enough, the average processing time of a rule might be minimized by storing very high-frequency regulars in the exception list, where they could be processed quickly. Although Yang did argue that rule frequency should play a role in processing time, the details of how rule frequency should relate to the processing cost of the default were not spelled out for his proposal.

A second problem with Yang's model is that it doesn't provide a mechanism allowing for quantitative variability in morphological marking. For example, as we discussed in Section 6.3.7, English speakers prefer to nominalize adjectival bases ending in *-able* with *-ity* (e.g., *agreeability*). However, the more productive nominal suffix *-ness* (e.g., *agreeableness*) is also acceptable, although less preferred in this context. Yang (2010) suggested that linguistic rules subserving the same function (such as *-ity* and *-ness* suffixation) should be strictly ordered in terms of the elsewhere condition—more specific rules are applied before general rules. This account does not allow both *agreeability* and *agreeableness* to be acceptable forms, nor does it provide predictions of the differential acceptability of the two forms. Again, a small extension of Yang's model could address this issue. In earlier work, Yang (2002) suggested that the input conditions to rules might be matched probabilistically so that forms could occasionally fail to be handled by a rule, even when they matched its input conditions. Such a probabilistic matching mechanism would allow the system to generate both *agreeability* and *agreeableness* at different rates. Details of this proposal remain to be worked out.

A third and final problem with Yang's model is that it predicts that the productivity of a rule will depend on the interaction of the token frequencies of exceptional forms, and the type frequency of non-exceptional forms, but will not depend on the token frequency of forms handled by the default. However, as we discussed in Section 6.3.4, even highly type-frequent affixes can be unproductive if the token frequency of words containing them is sufficiently high (Baayen, 2006). To illustrate, consider a case where a single suffix occurs in 1000 words with no exceptions. The tolerance principle predicts that all 1000 forms should be handled by the default structural change of a single rule with no listed exceptions. This remains true regardless of the token frequency of each of the 1000 forms. However, empirical evidence suggests that increases in the token frequency of words should detract from the productivity of their parts, and, if the all 1000 words become sufficiently frequent, the affix should cease to be productive (Aronoff, 1983; Baayen, 2006; Marcus et al., 1995). It not clear how to capture such effects in Yang's model as stands.

6.4.1.4 Summary

In this section, we have reviewed models of the morphological productivity of individual affixes. In order to clarify the problem, we first defined several abstract probabilistic measures of productivity. The fundamental idea behind these measures is that productivity can be defined with respect to an abstract distribution over words, $P(w)$. Since this distribution can be approximated by a variety of empirical techniques and formal models, our definitions of productivity are valid under many different theoretical approaches.

We then discussed two classes of models of productivity: corpus-based estimators, and cognitively-driven models of morphological knowledge. Consistent with our remarks earlier in the chapter, all of the models considered here make use of both the token frequency of individual words and the type frequencies of word parts.

Our discussion of corpus-based estimators of productivity focused on Baayen's $\mathcal{P}^{**}$, $\mathcal{P}^{*}$, and $\mathcal{P}$. These statistics are based on the idea that the probability of novel words (which do not appear in a sample, by definition) can be estimated by the proportion of low-frequency words in the sample. We examined the mathematical underpinnings of these statistics and discussed the extremely simplified models of $P(w)$ that they are based on. We therefore surmised that models that make more realistic assumptions about word generation should provide better estimates of productivity. Furthermore, to the degree that a more realistic model makes similar predictions to Baayen's measure, we can take Baayen's statistics as a high-level, approximate description of the behavior of the model in the cases where it applies.

Finally, we discussed several models of productivity which are based on cognitive theories of morphology. We focused especially on Yang's tolerance principle which provides an interesting optimization-based perspective on productivity. However, we also identified several places where Yang's model cannot handle known empirical phenomena without further extensions.

6.4.2 Models of Morphological Processing

As we discussed in Section 6.3.6, a central question in psycholinguistics is whether morphologically complex words are decomposed during processing or retrieved whole from memory. Five decades of intense research into the question has definitively shown that storage and computation are properties that vary from word to word and rule to rule, with complex mixtures of both in many cases. In this section, we review processing models that allow for such mixtures of computation and storage. As we also discussed above, a standard assumption in models of lexical processing is that the decomposition of existing forms and the productive generation of new forms share common psychological mechanisms (see, Frauenfelder and Schreuder, 1991). With this in mind, our review will focus on how processing models have been used to make predictions about productivity.

Our review is divided into two parts. In the next section, we discuss dual-route models of processing latency from psycholinguistics. There are two points that will emerge from this discussion. First, some models of lexical processing provide (indirect) support for the hypothesis that the decomposition of existing words and the productive generation of new words share the same psychological mechanisms. Second, implicit in some dual-route models is the assumption that composition/retrieval do not depend directly on the surface frequency of forms, but, rather, on the frequency of the mental computations employed by speakers when using language.

In the second part of our review, we discuss Hay's relative frequency hypothesis, which states that whether a word is composed or retrieved depends on the ratio of the frequencies of the word's parts to the frequency of the word as a whole. Under the hypothesis that decomposition of existing forms and the productive generation of new forms share psychological mechanisms, the relative frequency hypothesis predicts that the productivity of individual affixes should also be related to the part-word/whole-word frequency ratio.

6.4.2.1 Dual-Route Models of Lexical Processing

In psycholinguists, a number of computational models have been proposed that are designed to explain the pattern of processing times which arise in lexical decision, priming, and other experimental paradigms. An important class of such models are known as ***activation-based, dual-route*** models (e.g., Baayen and Schreuder, 1999, 2000; Bybee, 1988, 1995; Chialant and Caramazza, 1995; Hay, 2003; Laudana and Burani, 1985; Marslen-Wilson et al., 1997; Morton, 1969, 1970; Schreuder and Baayen, 1995; Taft, 1994). In these models, it is assumed that there are two routes, composition and whole-form retrieval, which compete to analyze (or generate) morphologically complex word forms. Such dual-route models have internal units representing both whole words (e.g., *happiness*), and individual stems (e.g., *happy*) and affixes (e.g., *-ness*). Each unit has an associated real-valued activation, which determines the speed or likelihood that the unit will be accessed during the processing of a word. The dynamics of activation of these units during processing determines the outcome of the competition between whole word and compositional processing routes. Changes in activation levels are used to predict processing times and, therefore, such models are often described using the temporal metaphor of a race between competing derivations of a word. After processing a word, the resting activation of the (set of) node(s) that "win" the race is increased, allowing these models to learn patterns of composition and retrieval from training data.[11]

Although there are a large number of dual-route models of lexical processing in the literature (see, Hay, 2003; for a review of many), only a small number of studies have explicitly addressed their relationship to productivity. A pair of such studies, Baayen (1993) and Hay and Baayen (2002), related the predictions of two specific processing models to Baayen's statistical estimators of productivity ($\mathcal{P}$ [potential productivity] and $\mathcal{P}^*$ [expanding productivity], see, Section 6.4.1.2). Baayen (1993) defined a statistic, called $\mathcal{A}$, which measures the composability of a morpheme (e.g., *-ness*:`Adj>N`) with reference to a given model of lexical processing $\mathcal{M}$.[12] The idea behind $\mathcal{A}$ is that a measure of composition for a morpheme can be derived from the total number of times that model $\mathcal{M}$ composes a word using the morpheme—in other words, the total frequency of all words built using that morpheme via composition. Mathematically, the composition score $\mathcal{A}$ for a rule (or morpheme) r can be defined as follows. Let $\#(w)$ be the frequency of word w from the set of all words considered by the model $\mathcal{W}$. Let $\delta_{+_{\mathcal{M},r}}[w]$ be the function that returns one if and only if the model $\mathcal{M}$ derives w by the active application of r to some base. The notation $+_{\mathcal{M},r}$ is meant to signify productive composition, $+$, of a word using rule r under model $\mathcal{M}$. $\delta_{+_{\mathcal{M},r}}[w]$ will be zero either when w is not

11. Activation-based, dual-route models typically conceive of the competition between different processing routes as occurring in parallel. There has been a second class of models in the psycholinguistics literature that treats lexical search, composition, and retrieval as serial processes with search ordered by the frequency of words and part words (e.g., Forster, 1976, 1979, 1981; Forster and Bednall, 1976; Murray and Forster, 2004; Rubenstein et al., 1970). However, Townsend (1974) argues that parallel and serial models can simulate one another, and therefore the two classes of models are equivalent (see, also, Anderson, 1990).

12. Baayen (1993) introduced $\mathcal{A}$ as a measure of productivity. However, since this statistic actually measures the rate of composition of existing words under a model of lexical processing, it is better understood as a measure of composability. Baayen's identification of $\mathcal{A}$ with productivity is an example of the standard assumption that composition of existing forms and productive creation of new forms use the same underlying mechanism.

the output of r or when w can be derived using r, but instead was retrieved as a whole from memory. $\mathcal{A}$ is defined in Equation 6.10.

$$\mathcal{A} = \sum_{w \in \mathcal{W}} \#(w) \cdot \delta_{+\mathcal{M},r}[w] \tag{6.10}$$

In other words, $\mathcal{A}$ is just the sum frequency of all words which use r compositionally.[13] Note an important feature of this quantity: it does not directly use the corpus frequencies of words containing rule r; instead, it uses the frequency with which r has been used to generate words compositionally in the past. Composition and retrieval are not directly observable properties of word forms; rather, they are properties of the mental computations that language users perform to produce and comprehend words. Thus, Baayen's $\mathcal{A}$ reflects an implicit hypothesis: composition and retrieval do not depend on the surface frequencies of words and part-words, but, rather, on the frequencies of the mental computations underlying language use. As we shall discuss below, the frequencies of words and word parts have an influence on whether a form is retrieved or composed. However, Baayen's hypothesis suggests that the connection between these surface frequencies and lexical processing is indirect and mediated by internal mental representations.

Baayen (1993) computed values of $\mathcal{A}$ using a dual-route model in which competition between whole-form retrieval and composition is resolved with reference to a global frequency threshold θ. When a word has a frequency greater than θ, the model retrieves it as a whole from memory. When the word's frequency is less than θ, the model composes it from its constituent morphemes (i.e., $\delta_{+\mathcal{M},r}[w] = 1$ if and only if $\#(w) < \theta$).

Importantly, Baayen (1993) examined the relationship between $\mathcal{A}$ and various statistical estimators of productivity, such as his $\mathcal{P}^*$, concluding that $\mathcal{A}$ provides more accurate predictions than the corpus-based estimators. Baayen speculated that this is likely because $\mathcal{A}$ is based on a cognitive model of lexical

13. Our definition of $\mathcal{A}$ generalizes the definition given in Baayen (1993) to apply to an arbitrary model of lexical processing, $\mathcal{M}$. Baayen (1993), instead, defined $\mathcal{A}$ with respect to the specific model he introduced in that paper. However, as discussed in the main text, Hay and Baayen (2002) calculated $\mathcal{A}$ against a different model. Therefore, we introduce this generalized definition, which can capture both uses of the statistic.

processing, in contrast with quantities such as $\mathcal{P}^*$, which are simply corpus-based statistics. Because $\mathcal{A}$ is based on a more realistic model of word processing, it can factor in properties, such as the semantic transparency of words, that are inaccessible to corpus-based estimators.

Hay and Baayen (2002), however, called into question the empirical adequacy of the assumption that the frequency threshold of composition, θ, should be fixed globally for all words. Adopting Hay's relative frequency hypothesis (discussed below), they argued that θ should vary from word to word and be related to the ratio of the base frequency of a word to its whole-form frequency. When this ratio is high, the word should be more likely to be composed; when it is low, it should be more likely to be retrieved as a whole from memory. Furthermore, they showed that a more recent dual-route model of lexical processing, MATCHECK (Baayen and Schreuder, 2000; Baayen et al., 2000), implements this principle, producing composition/retrieval decisions for words related to their base/whole-form frequency ratios. Hay and Baayen (2002) computed values of $\mathcal{A}$ for a set of English affixes using the composition/retrieval decisions made by the MATCHECK model to derive word-specific values for $\delta_{+\mathcal{M},r}[w]$. As with Baayen's earlier study, the $\mathcal{A}$ values computed by Hay and Baayen (2002) correlated with a number of corpus-based productivity estimators.[14] Crucially, they argued that their variant of $\mathcal{A}$, which relied on base/whole-form frequency ratios, produced better productivity predictions than the fixed-frequency-threshold variant of $\mathcal{A}$ studied in (Baayen, 1993).

There are three points to take from this discussion. First, the studies reviewed above provide (indirect) evidence that the composition of existing forms and the productive generation of new forms share common psychological mechanisms. The quantity $\mathcal{A}$ measures the degree to which an affix is expected to be decomposed in existing words (under a particular model). By contrast, corpus-based estimators of productivity, such as $\mathcal{P}^*$, estimate the probability that a new word will be generated using an affix. If these two quantities are related, it must be because they share some underlying mechanism. Of course, the strength of this conclusion depends on the correctness of the model used to compute $\mathcal{A}$ and the appropriateness of the sampling assumptions used to derive $\mathcal{P}^*$. Nevertheless, the results are encouraging.

14. They also showed that corpus-based productivity estimators correlated with a variant of $\mathcal{A}$, $\mathcal{A}'$, which was computed using the number of composed word types, rather than the total token frequency of composed words (i.e., $\mathcal{A}' = \sum_{w \in \mathcal{W}} 1 \cdot \delta_{+M,r}[w]$). Note that the notation $\mathcal{A}'$ is ours.

The second important point in this discussion is that Baayen's definition of $\mathcal{A}$ involves the implicit hypothesis that it is not the surface frequency of forms that matters in lexical processing, but, rather, the frequency of mind-internal computations. This perspective is shared by all of the models studied in this book, and we will address it in more detail in the next chapter. Third and finally, Hay and Baayen (2002) argued that models of lexical processing should base composition/retrieval decisions for complex word forms on the relative frequencies of words and their parts, rather than on absolute frequencies. We discuss this idea, Hay's relative frequency hypothesis, in the next section.

6.4.2.2 The Relative Frequency Hypothesis

In the preceding section, we discussed the study of Hay and Baayen (2002), which demonstrated a correlation between several statistical estimators of productivity and the composability scores of affixes derived from a model of lexical processing. Crucially, the model of lexical processing considered in that study made composition/retrieval decisions based on the ratio of base to whole-form frequencies of words, rather than any absolute frequencies of words or word-parts. As we mentioned, this is an instance of Hay's relative frequency hypothesis (Hay, 2001).

The relative frequency hypothesis proposes that whether a word is composed or retrieved depends on the ratio between the frequency of its base and the frequency of the word itself. Since the base frequency of a word *w* is the sum of the whole-form frequencies of all words that include that base (including the base without further morphological structure), when the base/whole-form frequency ratio is low, then the base tends to occur most often as part of the word *w*, rather than in other words. Under these conditions, the relative frequency hypothesis predicts that the word is likely to be stored and retrieved in its entirety. When the base/whole-form frequency of a word is high, the base must occur frequently in other words (or on its own), and the relative frequency hypothesis predicts that composition of that base with affixes is more likely.

Critically, Hay argued that although many researchers have stressed the importance of absolute frequencies, when examined in detail, their models predict that composition/retrieval decisions actually depend on the ratio between part-form and whole-form frequencies for individual words.

> While the proponents of many of these models emphasize the role of surface frequency, examination of the models themselves reveals that they predict an interaction between the surface frequency of the complex form, and the frequency of the parts. Maximally

> decomposable forms should be those which are much less frequent than the parts they contain. Non-decomposable forms should be those which are more frequent than the parts they contain (Hay, 2003; p. 60).

Hay highlighted two important corollary predictions of the relative frequency hypothesis. First, whether a word is parsed or retrieved as a whole is a property that should vary from word to word depending on the precise distributions of the word and its parts (Di Sciullo and Williams, 1987; make a similar point). Second, assuming the decomposition of existing words and the generation of new words share a common mechanism, the productivity of a morpheme should depend on the frequency with which it is composed across existing words.[15]

As we reviewed in Section 6.3.6, there is a wealth of experimental evidence showing an interaction between whole and part-word frequencies during processing. This evidence is consistent with the relative frequency hypothesis. Highly frequent words with less frequent parts tend to exhibit processing latencies best predicted by their whole-form frequencies, while less frequent words with more frequent parts tend to exhibit processing latencies better predicted by the frequency or morphological family size of their bases and affixes. Furthermore, both corollary predictions of the relative frequency hypothesis are supported in the literature. Composition and retrieval vary on a word-by-word basis, and the overall composability of a morpheme is closely related to its productivity.

There are two additional points to make before moving on. First, while the relative frequency hypothesis provides a general guideline for predicting composition and retrieval during lexical processing, it does not completely specify how the frequencies of all the components of morphologically complex words should be factored into a model. For example, in complex words involving more than a single stem and affix, like *agreeability*, the relative frequency hypothesis does not provide a specific recipe for determining how the frequencies of *agree*, *-able*, *-ity*, *agreeable*, and *-ability* should interact. A more general model of composition and retrieval would ideally explain how all such quantities affect processing. Second, the relative frequency hypothesis emphasizes that decisions about composition and retrieval should depend on a

15. More accurately, Hay proposed that the baseline rate of composition for an affix should be related to the frequency with which it is composed across existing words. However, if baseline rate of composition is interpreted as average composability, then this strikes us as truism. We believe that our formulation is the intended prediction.

comparison of different frequencies (i.e., relative frequencies), rather than on absolute frequency values. As we will argue later, the dependence of composition/retrieval decisions on such relative values is a necessary property of any fully probabilistic model that allows the storage of a heterogeneous collection of structures and estimates the probabilities of such structures proportional to their frequencies of use. We elaborate on both these points in Section 7.3.2.2.1.

6.4.2.3 Summary

In this section, we have reviewed models of lexical processing, focusing on those models that allow a mixture of computation and storage, and highlighting the relationship between the decomposition of existing words and the productive generation of novel words. We showed that some dual-route models of lexical processing provide support for the hypothesis that decomposition and productivity are subserved by the same psychological mechanisms. We also discussed the hypothesis, implicit in some dual-route models and in the models considered in the next chapter, that composition and retrieval do not depend on the surface frequency of words and word parts, but, rather, on the frequency of mental computations underlying surface forms.

We then discussed Hay's relative frequency hypothesis, which states that whether a word is composed or retrieved depends on the ratio of part-word to whole-word frequencies. Under the assumption that productivity and decomposition share common mechanisms, this also implies that productivity will depend on these ratios. In particular, it will be related to the number of words in which an affix is composed. Finally, we observed that a version of the relative frequency hypothesis, in fact, follows as a necessary consequence of the assumptions made in certain classes of probabilistic models of word structure. We revisit these observations in Section 7.3.2.2.1.

6.4.3 Models of Morpheme Combination and Ordering

As we reviewed in Section 6.3.7, English derivational morphology exhibits strong restrictions on the co-occurrence and order of morphemes that can appear together in a word. Because derivational morphology can give rise to multiply-embedded, hierarchical structures, producing words with sequences of multiple affixes (e.g., *disagreeability*), a long-standing theoretical problem in the field has been providing an account that explains which morphemes can appear together in such complex words, and in what order.

The set of attested affix sequences is at least partially explained by the combination of three factors: (i) the storage of morphologically complex words, (ii) the productive combination of (sequences of) prefixes and suffixes with stored bases and (iii) the assumption that the set of stored, morphologically complex words is the result of historical patterns of productivity and reuse. Since all else being equal a productive affix can attach to a stored, morphologically complex base containing an unproductive affix, these factors give rise to the important empirical generalization that more productive affixes tend to occur outside of less productive affixes (the productivity-and-ordering generalization).

While all theories of ordering have appealed to the three factors above to some degree, three additional empirical considerations have also influenced the development of the field. First, word-formation processes vary in their phonological, semantic, and morphosyntactic selectional restrictions, and such selectional restrictions influence the distributions of attested morpheme sequences. Second, while English derivational morphemes vary in their phonological complexity/regularity, semantic transparency, and productivity, this variation isn't distributed evenly throughout the system. Instead, some less productive affixes cluster into groups that share phonological structure. Third, since more productive affixes also tend to be decomposed more often during processing, more decomposable affixes tend to appear outside of less decomposable affixes.

We discuss three (classes of) theories of affix ordering: *stratal* theories, *selectional-restriction based* theories, and Hay's *complexity-based ordering*. Stratal theories of ordering are built around the existence of phonological clusters of affixes. They propose that the grammar of word-formation groups affixes into a number of ordered levels or *strata* sharing phonological structure. By contrast, selectional-restriction based theories attempt to reduce all ordering generalizations to affix-specific selectional constraints, while complexity-based ordering grounds ordering generalizations in a theory of lexical processing. We describe these theories and their ability to capture the relevant empirical phenomena below.

6.4.3.1 Stratal Theories of Affix Ordering

One very influential theory of affix co-occurrence and ordering is known as the *level-ordering generalization*. The level-ordering generalization has its origins in Chomsky and Halle's detailed treatment of English phonology in *The*

Sound Pattern of English (Chomsky and Halle, 1968). As we noted in Sections 6.3.1 and 6.3.2, there are substantial differences between English affixes in their phonological selectional restrictions and the complexity/regularity of the phonological changes they trigger. For example, while some affixes simply concatenate to bases (e.g., *-ness*: *happy/happiness*), others trigger stress shift (e.g., *-ic*: *átom/atómic*) or consonant sibilantization (e.g., *-ity*: /k/ → /s/ in *atomic/atomicity*) on their bases. Furthermore, certain affixes cluster together triggering similar phonological changes. For example, a number of Latinate suffixes such as *-ic* and *-ity* trigger similar stress-shifts on bases to which they attach. To account for facts like these, Chomsky and Halle (1968) proposed that different English affixes create different kinds of boundaries within a word. Productive and phonologically simple/regular affixes, such as *-ness*, create **word boundaries**. Unproductive and phonologically complex/irregular affixes, such as *-ity*, create **formative boundaries**.

Under the theory of Chomsky and Halle (1968), the type of boundary that an affix creates determines the kinds of phonological changes that can be triggered by an affix. Certain changes, such as stress-shift, can be triggered by affixes that create formative boundaries, but not by affixes that create word boundaries. The level-ordering generalization results from the observation, first made by Siegel (1974), that the distinction between word- and formative-boundary affixation also partially explains the ordering of affixes. In particular, when both formative-boundary affixes and word-boundary affixes occur in a single word, the former generally appear outside of the latter. Siegel (1974) also noted the correlation between the two kinds of affixes and productivity: word-boundary affixes tend to be more productive than formative-boundary affixes.

To explain these facts, Siegel (1974) proposed a theory of morphology that divided the process of word generation into a number of ordered levels or strata.[16] Formative-boundary affixes are joined with bases in the first level of the system; only after derivation in this stratum is complete, are word-boundary affixes concatenated onto the resulting forms in the second level. Other properties, such as semantic transparency and the availability of different affixes for productive generalization, can also be associated with different strata. Stratal theories of the lexicon reached their most complete state of development in **lexical phonology**, in which a variety of models, with varying numbers of

16. Note that stratal theories, including Siegel's, are lexicalist. They use the term lexicon to refer to a generative component of the language system independent of other rule systems such as the system governing phrase structure.

strata, were proposed (Halle and Mohanan, 1985; Kiparsky, 1982a,b,c, 1985; Mohanan, 1986).[17]

Stratal theories provide a (partial) explanation for why morphological rules sometimes share phonological structure, such as triggering stress-shift (although they still need to specify which morphemes trigger which transformations), but, like other theories of morphology, they also allow individual morphemes to be marked with additional selectional restrictions—including restrictions on the set of bases with which a morpheme can combine. This mechanism can be used to specify, for instance, that an unproductive morpheme like *-th*, can only attach to a small set of stems (*warm*, *wide*, etc.). Such restrictions allow stratal theories to model the lexical storage of morphologically complex words.

However, despite these advantages, stratal theories suffer from several empirical shortcomings. First, under these theories, it appears that some affixes must be assigned to multiple levels. For example, consider the combination of the affix *-able* and the verb *compare*. There are two acceptable pronunciations of the word, one of which includes a stress shift and vowel reduction (i.e., *cómparable*) and one of which does not (i.e., *compárable*; Plag, 1996). The former pronunciation indicates that the affix must reside in an earlier level of the lexicon, while the latter indicates that it should reside in a later level. Giegerich (1999) provides extensive examples of such multiple level assignment problems and argues that they represent a pervasive and decisive challenge to (affix-centered) stratal theories.

Another problem with level-ordering is that it gives rise to **bracketing paradoxes** (Aronoff and Sridhar, 1983; Strauss, 1982; Williams, 1981). For example, the suffix *-ity*, which typically attaches to adjectives, is usually assigned to the first lexical level, while the prefix *un-*, which also selects for adjectives, is usually assigned to a higher level. Consider the word *ungrammaticality*. Since *-ity* triggers a stress shift on the base *grammátical*, *grammaticálity* must be output by a level before *un-* is prefixed, which gives rise to the bracketing: [*un-* [[*grammatical*] *-ity*]]. However, this violates the selectional restrictions for *un-*, since *grammaticality* is a noun, and, by hypothesis, *un-* only

17. For example, Kiparsky (1982a,b,c, 1985), articulated a theory with three strata of (morpho)phonological rules. The first level handled formative-boundary derivational affixes (e.g., *-ity*) and irregular inflections such as the irregular past tense forms of English verbs (e.g., *sing/sang*). The second level handled word-boundary derivational affixes (e.g., *-ness*) and compounding. The third level handled regular inflection such as the English past tense +/d/ rule. Halle and Mohanan (1985) and Mohanan (1986) proposed a theory with four levels, which moved compounding to level three and regular inflection into level four.

attaches to adjectives or verbs. Furthermore, this bracketing implies the incorrect meaning, "not the state of being grammatical," as opposed to the correct meaning, "the state of being ungrammatical." However, the other possibility, [[*un-* [*grammatical*]] *-ity*], violates the assumption that *un-* is assigned to a higher stratum than *-ity*. Thus, either bracketing violates a prediction of the level-ordering generalization.[18]

A more serious problem with the level-ordering generalization is that it predicts the existence of sequences of affixes that are not attested (i.e., it over-generates), while simultaneously failing to predict the existence of sequences of affixes that are attested (i.e., it under-generates). Undergeneration occurs in cases where an affix that should be assigned to a lower level based on phonological considerations appears after an affix that should be assigned to a higher level. For example, many English words involve the paradoxical suffix combinations: *-istic*, *-ableity*, and *-mental*. In all three cases, the outermost affix triggers stress shift, and therefore is typically assigned to a lower stratum than the innermost affix. However, this predicts that these affix combinations should be impossible (Aronoff and Sridhar, 1983; Fabb, 1988; Giegerich, 1999; Hay, 2002). Even more problematic is the degree to which stratal theories over-generate, predicting affix sequences that are unattested. In an influential study of forty-three English affixes, Fabb (1988) showed that there are 663 pairs that are possible when categorial restrictions are taken into account. Taking into account several additional phonological and semantic selectional restrictions reduces this number to 614, and the addition of level-ordering restrictions further reduces this number to 459. However, Fabb (1988) showed that only fifty of these 459 predicted combinations actually occur in his sample. The level-ordering generalization does not rule out a sufficient number of affix combinations.

These problems with the level-ordering generalization have led researchers to gradually retreat from the theory. The following sections describe two alternative approaches that have been proposed in the literature: selectional-restriction-based theories and complexity-based ordering.[19]

18. Most examples of bracketing paradoxes in the literature concern cases, like this, of interaction between prefixes and suffixes. One possible solution is to assume that level ordering applies differently to the two kinds of affixes (Strauss, 1982). This solution, however, still leaves open the question of how the correct meaning of such forms is computed.

19. A third approach was provided by Giegerich (1999), who argued that lexical strata should be organized around bases rather then affixes, as in the traditional approach. The intuition behind this approach is that certain affixes (e.g., *-ity*) seem to be selected by bases ending in certain other suffixes (e.g., *-able*). Organizing lexical levels around bases accounts for such facts.

6.4.3.2 Selectional-Restriction-Based Theories

An alternate approach to stratal theories is to reject the idea of lexical levels altogether and, instead, propose that ordering effects are an epiphenomenon of the large number of affix-specific selectional restrictions characterizing derivational morphology (see Section 6.3.1). In his paper criticizing the overgeneration of level-ordering theories, Fabb (1988) proposed an alternative theory of ordering based on selectional restrictions. He argued that affixes can be categorized into four classes, each of which has different affix-specific selectional constraints. Class I consists of those suffixes that select for morphologically simple stems, never attaching to an affixed word (e.g., Fabb claims that *-ion* is in this class). Class II consists of suffixes which have selectional restrictions that specify other affixes that are allowed to be present in the bases to which they can attach (e.g., Fabb claims that *-ary*:`N>Adj` only attaches to *-ion*:`V>N`). Class III are those fully productive affixes (e.g., *-ness*) which can attach to any base. Class IV contains a residue of six remaining problematic cases that do not seem to fit into any of the earlier classes.[20]

Fabb's theory dispenses with the notion of lexical strata, explaining the distribution of affix sequences primarily by appealing to two kinds of selectional restrictions: selectivity for morphologically simple bases and selectivity for the presence of specific morphemes in bases (i.e., morpheme-specific selectional restrictions). However, as pointed out by Plag (1996), Fabb's classification scheme also has significant problems of both over- and undergeneration. To address these issues, Plag (2002, 2003) developed a selectional-restriction-based theory of ordering which includes a much larger number of (potentially overlapping) selectional restrictions with morphosyntactic, phonological, and semantic content.

If individual morphemes can be marked with morpheme-specific selectional restrictions, then selectional-restriction-based theories can model the lexical storage of morphologically complex words, like *warmth*. Importantly, this mechanism allows these theories to account for the generalizability of paradoxical suffix combinations, like *-ability*, which were problematic for stratal theories. Furthermore, if the components of morphological rules—such as the phonological changes they trigger or the sets of bases to which they can

20. Fabb (1988) also discusses the verbal affix *-en* separately, pointing out the fact, mentioned earlier, that it has strong selectional restrictions, preferring bases that are monosyllabic and end in an obstruent.

attach—can be shared between rules, then selectional-restriction-based theories can also provide mechanisms that explain why certain affixes form phonological clusters (e.g., *-ic* and *-ity* both triggering stress shift).

Hay and Plag (2004) examined the ability of Plag's selectional-restriction based theory to explain the attested orderings of a set of fifteen English suffixes. They found that while selectional restrictions were able to rule out most unattested sequences of affixes, there were a handful of combinations that were predicted to be possible, but did not appear in their corpus sample (e.g., *-nessless*). Hay and Plag (2004) argue that these cases can be ruled out by an appeal to Hay's theory of complexity-based ordering. We turn to this theory in the next section.

6.4.3.3 Complexity-Based Ordering

As we discussed Section 6.3.7, Hay and colleagues have drawn attention to the important empirical generalization that more productive affixes tend to appear outside of less productive affixes (the productivity-and-ordering generalization Hay, 2002, 2003; Hay and Plag, 2004). In that section, we discussed how the productivity-and-ordering generalization follows as a consequence of the fact that words can be generated both by the composition of productive affixes with stored bases and the lexical storage of morphologically complex words. Hay's theory of complexity-based ordering provides a somewhat different account of this generalization, deriving it from a model of lexical processing.

Complexity-based ordering is formulated in terms of a dual-route model of word processing. To process a word form, a language user must correctly identify its subparts and perform any composition operations that are necessary to derive the whole form. Complexity-based ordering is built on the assumption that the processor is biased to preferentially identify the base of a word before identifying and composing productive affixes (see, also, Cutler et al., 1985).[21] Under a dual-route model, a morphologically complex word may be stored as a whole or be built using some combination of retrieval and composition, and, therefore, often has several ambiguous derivations. Since the processor prefers

21. The origins of this bias are not entirely clear. One possibility is that this bias arises from the needs of semantic composition. Morphologically complex stored bases can accumulate non-transparent meanings. Since identification of semantically idiosyncratic units logically precedes regular semantic composition of those units, the processor may be biased towards eagerly searching for the maximal possible stored units before applying regular semantic rules.

to identify bases before composing affixes, and since processing proceeds left-to-right, words that provide earlier cues to base-segmentation should be easier to process.

To understand how segmentation cues relate to ordering predictions, consider a word of the form `stem -suff1 -suff2`, and, furthermore, assume that `stem -suff1` is a stored base. One cue to segmentation is the probability of phoneme transitions between morphemes. If the transition between the final phoneme of the `stem` and the first phoneme of `-suff1` is less probable than the transition between `-suff1` and `-suff2`, this may cause the processor to delay recognizing that `stem -suff1` is a stored unit. On the other hand, if the second transition is more probable than the first, recognizing that `stem -suff1` is a single unit will be easier. Based on such considerations, complexity-based ordering predicts that suffixes will tend to be ordered in terms of decreasing juncture probability. Furthermore, since productive suffixes, in general, occur with a larger variety of bases than stored suffixes, they tend to produce lower probability phoneme transitions and therefore tend to occur later in words.

In addition to the probability of phoneme transitions, Hay also explores another cue to segmentation: the frequency of words and word-parts. Complexity-based ordering proposes that the word processor takes into account the frequency of the candidate morphemes that it discovers over the course of left-to-right processing. Derivations containing more frequent units are considered to be more likely than derivations containing less frequent units. Building on the relative frequency hypothesis (see Section 6.4.2.2), complexity-based ordering proposes that it is not the absolute frequency of a morpheme that matters, but rather the frequency of the morpheme relative to morphemes in competing derivations, for example, the relative frequency of bases and whole-forms.

Hay and colleagues have given several kinds of evidence in support of complexity-based ordering. Some of this evidence is consistent only with Hay's processing-based account (Hay, 2002, 2003; Hay and Plag, 2004; Manova, 2010; Manova and Aronoff, 2010; Plag and Baayen, 2009; Zirkel, 2010). For example, Hay (2003) argued that differences in segmentability and transparency between prefixes and suffixes can be explained by an appeal to processing biases. Prefixes tend to be less segmentable and to exhibit more idiosyncratic and less transparent semantics than suffixes. Hay argued that this follows from the bias to segment bases before composed affixes and the left-to-right nature of processing. Since prefixes occur before bases, words that store prefixes together with their bases will be more processable than words that compose prefixes.

However, most of the evidence cited in support of complexity-based ordering is consistent with both Hay's processing-based account and with a more general account based on the pattern of productivity and reuse in derivational morphology. For example, complexity-based ordering is able to recover differences between suffixes assigned to level 1 and level 2 under stratal theories of ordering. The relative frequency hypothesis predicts that when the base to whole-form frequency ratio of a word is low, then the word should tend to be retrieved whole from memory. The overall rate of decomposition for a suffix can be computed using the frequency-weighted average of its base to whole-form ratio across words in which it appears. Hay (2002) showed that the average ratio for level 1 suffixes is lower than the average ratio for level 2 suffixes—predicting that level 1 suffixes are decomposed less often than level 2 suffixes. A similar analysis showed that complexity-based ordering also recovers differences between the four classes of suffixes proposed by Fabb (1988). Hay and Plag (2004) and Plag and Baayen (2009) offered an even finer-grained classification of suffixes, showing that English derivational suffixes can be arranged into a hierarchy where suffixes that appear higher in the hierarchy rarely occur in words before suffixes that are ranked lower in the hierarchy, and furthermore, that a suffix's position in the hierarchy correlates with theoretical measures of productivity and composability.

These studies showed that suffixes can be classified in terms of order and that these classifications correlate with the productivity of suffixes, and therefore are consistent with the productivity-and-ordering generalization. However, this evidence is ambiguous with regard to the origins of these phenomena. They may be the result of biases in favor of ease of processing; alternatively, it may result from the pattern of storage and computation in English derivational morphology (or both).

As we discussed in Section 6.3.7, an appeal to word-by-word variability in decomposition/retrieval can also explain cases of ordering that are anomalous with regard to the productivity-and-ordering generalization. For example, the productivity-and-ordering generalization predicts that the suffix *-al* should tend to appear after the suffix *-ion* because the former is generally more productive than the latter. While this generalization holds in most cases, there are exceptions such as *coloni***al***izat***ion**. This set of exceptions, however, can be explained if the base *colonial* is stored in memory and affixed with the semi-productive combination *-ation*.

As pointed out by Hay (2003), complexity-based ordering has difficulties accounting for the *paradoxical suffix combinations* we discussed in Section 6.3.7. These are cases where an unproductive suffix, such as *-ity* can be used productively when it occurs after a more productive suffix like *-able*. Hay (2003; p. 158) discussed *-ability* and several other cases (*-istic*, *-mental*, and *-ation*), pointing out that the frequency distribution of the suffixes involved in these combinations suggests that the final suffix should be less composable than the initial suffix, and therefore that the combination should be disfavored under processing by complexity-based ordering. Moreover, since combinations such as *-ability* can be used on completely novel bases, these examples cannot be explained by appealing to the storage of bases that include the inner suffix. In the next chapter, we show how the approach to productivity and reuse proposed in this book can explain these cases.

Although complexity-based ordering grounds affix ordering generalizations in lexical processing, it does not preclude the possibility of affix-specific selectional restrictions. As we discussed in the last section, Hay and Plag (2004) explored the relationship between the predictions of complexity-based ordering and the predictions of Plag's selectional-restriction based theory. They found that the two theories were often complementary. Complexity-based ordering ruled out a number of affix combinations that were predicted to be possible by the selectional-restriction-based theory, but did not appear in their corpus sample. Likewise, selectional restrictions ruled out a number of unattested suffix combinations that complexity-based ordering predicted should be allowed. More surprisingly, however, they also found that for many combinations of suffixes both theories made the same predictions. This finding implies that less productive/decomposable suffixes also tend to have tighter selectional restrictions—a candidate synergy of learning (see Section 1.2.3.1).

6.4.3.4 Summary

In this section, we reviewed theories of morpheme combination and ordering. We first observed that the set of attested affix sequences in English words can be explained by an appeal to three factors: (i) the storage of morphologically complex words, (ii) the productive combination of (sequences of) prefixes and suffixes with stored bases, and (iii) the assumption that the set of stored, morphologically complex words is the result of historical patterns of productivity and reuse. This combination of factors also gives rise to the productivity-and-ordering generalization, which states that more productive affixes tend to occur outside of less productive affixes in morphologically complex words (Hay, 2002, 2003; Hay and Plag, 2004).

We then examined three classes of theories of affix ordering: stratal theories, selectional-restriction based theories, and Hay's complexity-based ordering. Stratal theories attempt to explain ordering generalizations by dividing affixes into a number of ordered levels or strata, however, they face decisive empirical challenges that have led researchers to alternative approaches. Selectional-restriction-based theories of ordering attempt to reduce all ordering generalizations to a large number of affix-specific selectional contraints. One especially important class of selectional restrictions are those that make reference to the presence of specific morphemes in bases. Such morpheme-specific selectional restrictions allow theories to model the lexical storage of certain stems and affix combinations (e.g., *warmth*) as well as generalizable, paradoxical suffix combinations, such *-ability*. In the next chapter, we will show how the framework presented in this book can be understood as a model of the learning of such morpheme-specific selectional restrictions.

Finally, we considered Hay's theory of complexity-based ordering, which attempts to provide a somewhat different explanation of the productivity-and-ordering generalization than the one given above. Instead of explaining the tendency for productive affixes to appear outside of less productive affixes in terms of computation and storage, complexity-based ordering argues that the tendency arises from the needs of efficient processing. However, as we reviewed above, the majority of the evidence which Hay and colleagues adduce is consistent with both accounts. We also noted that, like stratal theories, complexity-based ordering has difficulties accounting for one particular class of exceptions to the productivity-and-ordering generalization: paradoxical suffix combinations. In the next chapter, we will show how the fragment grammars model explains such cases.

6.5 Conclusion

In this chapter, we have discussed the empirical and theoretical literatures on English derivational morphology. Although we examined many empirical findings and theoretical models, the majority of our discussion has focused on three phenomena and their theoretical explanations.

First, we discussed differences in productivity between individual affixes, and the relationship between productivity and the frequency distribution of words using those affixes. English affixes vary in their productivity, with more productive affixes giving rise to larger numbers of low-frequency words. This fact motivates Baayen's corpus-based estimators of productivity $\mathcal{P}^*$ and $\mathcal{P}$. In the next chapter, we will show that the fragment grammars model in some sense justifies the intuition behind these statistics.

Second, we discussed differences in decomposition/retrieval between words and part-words. Whether a word is decomposed or retrieved whole appears to be a function of the ratio of its part-word to whole-word frequency, a generalization Hay has termed the relative frequency hypothesis. However, as we discussed above, under many models of lexical processing the relevant frequencies are not the raw, surface frequencies of words and word parts, but rather the frequencies of the mental computations in which morphemes have been composed or retrieved as part of a larger stored unit. Furthermore, under the hypothesis that the decomposition of existing words is related to the productive generation of new words, the relative frequency hypothesis also predicts that the productivity of affixes should be related to the number of words in which they are decomposed. In the next chapter, we will show why probabilistic models like fragment grammars necessarily predict that the relative frequency hypothesis should hold, but make additional predictions beyond those of the relative frequency hypothesis.

Finally, we considered the empirical phenomena of morpheme combination and ordering. Out of all the logically possible sequences of English morphemes, only a small proportion are attested. Part of the structure in the set of attested affix sequences can be explained by the productivity-and-ordering generalization, which states that more productive affixes tend to appear outside of less productive affixes. As we have argued, the productivity-and-ordering generalization arises as a natural consequence in systems that mix productive affixation and the lexical storage of morphologically complex words. However, we also considered a number of classes of exceptions to the productivity-and-ordering generalization, the most difficult of which is the case of paradoxical suffix combinations, like *-ability*. Although this phenomenon presents difficulties for several of the theories of affix ordering we discussed above, it can be handled by theories that allow morpheme-specific selectional restrictions. Since morpheme-specific restrictions can also model the lexical storage of words including unproductive affixes (e.g., *warmth*), this suggests that the two cases might be unified by a theory of productivity and reuse. In the next chapter, we will show how fragment grammars provide this unification, predicting the productivity-and-ordering generalization, explaining exceptional cases, and providing a learning theory for morpheme-specific selectional restrictions.

7 English Derivational Morphology: Simulations

7.1 Introduction

This chapter reports simulation results examining three classes of empirical phenomena from English derivational morphology—productivity, processing, and morpheme combination and ordering—for a set of 338 English suffixes. Although there is a substantial body of work focussing on these phenomena (see Chapter 6), quantitative modeling in this domain remains difficult for several reasons. First, as we will describe in the next section, construction of an appropriate training corpus presents a number of challenges. Second, and more critically, there are no broad-coverage datasets available to directly test model predictions. Because, in English, there is a clear notion of the correct past-tense form for existing verb stems, the performance of the five models studied in the book could be evaluated directly in the domain of the past tense (see Chapter 5). English derivational morphology, by contrast, offers few such clear-cut cases. Instead, it presents wide variability in the acceptability of different affixed forms, including sharp preferences (e.g., *furiousness* v. **furiosity*), weaker preferences (e.g., *expectation*/?*expectedness*), and many doublets (e.g., *tolerance* v. *toleration*; Kiparsky, 1982b). The appropriate data to evaluate such a system—for example, databases of naturalness judgments or wug tests—do not yet exist.

Throughout this chapter we again compare the performance of the five models introduced in Chapters 1 and 2: (i) our inference-based model, Fragment Grammars (FG), (ii) the full-parsing model, implemented with Dirichlet-Multinomial PCFGs (DMPCFG), (iii) the full-listing model, implemented with Maximum a posteriori Adaptor Grammars (MAG), and two versions of the exemplar-based approach, (iv) Data-Oriented Parsing 1 (DOP1), and (v) Data-Oriented Parsing: Equal-Node estimator (ENDOP). In the first section of this chapter, we explain the construction of the training corpus used for these simulations, highlighting the difficulties involved in providing morphological analyses for large numbers of English words, and describe the simulations that were run in order to perform the analyses below.

In the second part of the chapter, we turn to our simulation results and analyses. We first examine the ability of the models to predict the productivity of individual English suffixes. Because there are no broad-coverage databases available for evaluating the productivity of individual English suffixes, it is not possible to perform a large-scale quantitative evaluation of our models in this domain. Instead, we will examine a number of specific case studies, and demonstrate that all of the models besides the FG (inference-based) model make a number of highly implausible predictions. We also perform an analysis demonstrating that, for the case of individual suffixes, the predictions of the fragment grammars model are highly convergent with predictions based on Baayen's corpus-based estimators of productivity $\mathcal{P}$ and $\mathcal{P}^*$. This result both sheds light on the nature of the inference performed by the FG (inference-based) model and provides further theoretical justification for Baayen's statistics.

The second set of simulation results concern the decomposition and retrieval of existing forms. We start from Hay's relative frequency hypothesis, which states that whether a word is composed or retrieved during processing should depend on the ratio of part-word to whole-word frequencies. Three of the models, FG (inference-based), DOP1 (exemplar-based), and ENDOP (exemplar-based) produce predictions about composition and retrieval that are consistent with a version of the relative frequency hypothesis. In fact, we argue that this must be the case for any probabilistic model of word structure that allows the storage of arbitrary tree fragments, and that estimates the probability of those fragments using their frequencies. Although such probabilistic models are consistent with the relative frequency hypothesis, they go beyond it in terms of predictive power. Furthermore, although both exemplar-based models (DOP1 and ENDOP) and the FG (inference-based) model are consistent with the relative frequency hypothesis, there is an important difference between the former and the latter. While the two exemplar-based models (DOP1 and ENDOP) predict that many words and affixes are ambiguous with respect to composition and retrieval, the FG (inference-based) model predicts a sharp dichotomy between the two derivational routes, preferring to treat words and affixes as nearly always composed or nearly always stored as part of more complex structures. We discuss the implication of this phenomenon for theories of productivity and reuse.

In the final set of empirical analyses, we turn to the phenomenon of morpheme combination and ordering. We first examine the ability of the five models to capture the frequency distribution of suffix pairs in a corpus compiled by Plag and Baayen (2009). In this evaluation, the MAG (full-listing) model best predicts the data, with the FG (inference-based) model coming a close second. However, in a second analysis, we show that the performance of the MAG (full-listing) model results from its memorization of all input word forms in the training set. When we consider the ability of each of the models to predict the generalizable structure in the set of suffix pairs, the FG (inference-based) model outperforms the others.

Our next set of analyses examine the generalizable structure in the set of attested suffix sequences in more detail. An important empirical observation in the domain of affix ordering is the productivity-and-ordering generalization, which states that more productive affixes tend to occur outside of less productive affixes in morphologically complex words (Hay, 2002, 2003; Hay and Plag, 2004). As we argued in the last chapter, the productivity-and-ordering generalization is a necessary consequence of models of morphology that allow productive affixes to attach to stored, but morphologically complex, bases. We show that, of the five models, the `FG` (inference-based) model produces predictions that are most consistent with the productivity-and-ordering generalization. Although the productivity-and-ordering generalization describes much of the structure in attested affix sequences, there are a number of exceptions to the generalization, the most difficult of which is the case of paradoxical suffix combinations. These are cases where a generally less productive suffix, such as *-ity* generalizes productively in the context of another (often more productive) suffix, like *-able*. We demonstrate that only the `FG` (inference-based) model is able to capture the generalizability of suffix combinations like *-ability*, and, in particular, is able to correctly predict the pattern of results for one pair of cases that has been studied experimentally (*-ivity*/*-iveness* versus *-ability*/*-ableness*). We also present a number of other examples of suffix combinations that are predicted to generalize as units by the `FG` (inference-based) model. These examples include many of the cases that have been discussed in the literature. Such generalizable suffix combinations can be modeled in theories that allow morpheme-specific selectional restrictions. We argue that the `FG` (inference-based) model can be seen as a learning theory for this kind of restriction.

7.2 The Training Corpus and Simulations

This section presents the necessary details to understand the simulations and analyses presented below. The first part of this section describes and justifies the underlying CFG and input representations that were used characterize the set of possible structures in English derivational morphology, and served as input to all of the models examined in this chapter. In the second part of this section, we describe the construction of our training corpus. As we mentioned in the introduction, constructing a training corpus that exposes the necessary structure for our examination of productivity and reuse presents a considerable number of challenges. We explain the process by which the corpus was constructed, and motivate the decisions made during that process. In the last part of the section, we briefly describe the simulation runs which resulted in the grammars used for the analyses later in the chapter.

7.2.1 The Starting CFG and Input Representations

W	⟶	N	
W	⟶	V	
W	⟶	Adj	
W	⟶	Adv	
N	⟶	Adj	*-ness*
N	⟶	Adj	*-ity*
N	⟶	*dog*	
	…		
V	⟶	N	*-ify*
V	⟶	Adj	*-ize*
V	⟶	*re-*	V
V	⟶	*agree*	
	…		
Adj	⟶	*dis-*	Adj
Adj	⟶	V	*-able*
Adj	⟶	N	*-ic*
Adj	⟶	*tall*	
	…		
Adv	⟶	Adj	*-ly*
Adv	⟶	*today*	
	…		

Figure 7.1
A Portion of the CFG Used in Derivational Morphology Simulations

A fragment of the underlying CFG used to define the set of possible trees in English derivational morphology is shown in Figure 7.1. The rules in this CFG were read off of the set of depth-one trees in the training corpus we describe in the next section. There are several properties to note about the model of English derivational morphology implicit in this grammar. First, affixes are treated as *heads*, which take their bases as *dependents*. This is consistent with a number of morphological theories that treat word structure as hierarchically structured and headed (e.g., Halle and Marantz, 1993; Lieber, 1980; Selkirk, 1982; Williams, 1981). Representing head-dependent structure in this way allows the grammar to capture categorial structure of affixes. For example, the morpheme *-ness*:`Adj>N` can be represented by the rule `N` ⟶ `Adj` *-ness*. Second, no attempt is made to represent any of the phonological or semantic structure associated with individual morphemes. In particular, as with the case for the past tense simulations reported in Chapter 5, none of the phonological or semantic selectional restrictions associated with particular affixes are modeled.[1] This system gives rise to derivation trees like the one shown in Figure 7.2.

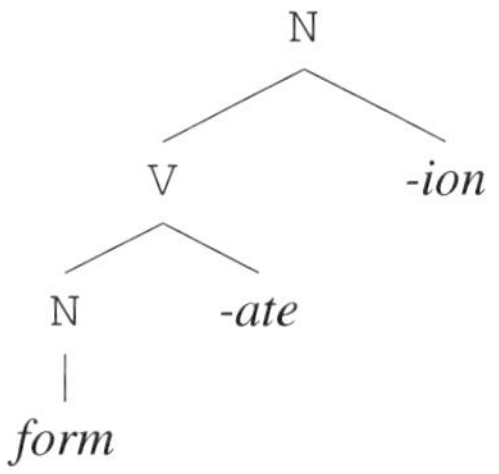

Figure 7.2

Derivation Tree: A derivation tree showing the structure of the word *formation*. Note that we have suppressed the top `W` node.

From a psychological perspective, similar remarks apply to this grammar as applied to the CFG used in the past-tense simulations reported in Chapter 5 (see, Section 5.2.1). This CFG provides the models with both more and less information than is available to the child learner. On one hand, unlike real language input, all forms are segmented into component morphemes, which are labeled with their category (or input-output categories in the case of affixes). Furthermore, the grammar exposes the head-dependent structure of affixes and bases. On the other hand, this representation fails to provide the many kinds of phonological and semantic information that are available to the child learner.

1. The grammar also lacks any representation of morpheme-specific selectional restrictions. However, such restrictions can be equivalently represented by the lexical storage of morphologically complex forms in models like fragment grammars. In fact, we argue below that fragment grammars can be seen as a model of the learning of such restrictions

Importantly, these assumptions lead to an underlying CFG that is massively over-productive. In particular, this grammar implicitly defines many more logically possible structures than are actually possible in English derivational morphology. Since the representation includes no semantics or phonology, the only sources of information available to the models to constrain the productivity of individual morphemes are the categories and pattern of co-occurrence of morphemes in the training set.

7.2.2 The Training Corpus

The goal of the analyses presented later in the chapter is to examine the extent to which the distribution of words (and word parts) in a sample of English derived forms can provide evidence of the pattern of computation and storage in the computational system which gave rise to that sample. Crucially, we wish to isolate purely distributional factors from other sources of information such as phonological regularity or semantic transparency. We also wish to focus tightly on the question of productivity and reuse, setting aside the many other difficult induction problems that the learner must solve to acquire a morphological system—for example, morpheme segmentation, allomorph identification, and the inversion of (semi-)regular phonological rules.

These goals lead to a number of desiderata for our training corpus. First, the corpus should provide a representative sample of the frequencies of co-occurrences of English derived words and morphemes, including low-frequency, productively derived forms. Second, because unsegmented input forms are stored by definition, and thus presume a solution to the problem we wish to study, words should be maximally segmented into morphemes, including those derived using unproductive morphemes such as *-th*. Third, since the occurrence of a morpheme across multiple contexts provides evidence of its productivity, phonologically different allomorphs of the same underlying morpheme should be unified. For example, in pairs such as *revolve/revolution*, the base *revolu* and the verb *revolve* should be represented as identical in the input to provide maximal evidence of the productivity of *-(t)ion*.

Note that all of these desiderata maximize the combinatorial possibilities inherent in the starting system. If a model is able to learn to constrain this combinatorial potential, this will indicate that the distribution of morphemes in the input provides strong evidence to the learner about productivity and reuse independent of other considerations (such as phonological regularity or semantic transparency).

The existing morphological dataset that most closely meets these desiderata is the CELEX database (Baayen et al., 1993), which provides a large number of segmented, morphologically analyzed English words. However, the CELEX sample still suffers from a number of serious problems for the present study. Some examples are given below.

- CELEX only provides morphological analyses for words which appeared at a frequency threshold of at least fifteen occurrences in eighteen million. Consequently, the analyzed portion of the CELEX database under-samples low-frequency, productively derived words, thus eliminating a crucial information source for the present models (see, also, Evert and Lüdeling, 2001; Hay and Baayen, 2002).
- Only bases appearing as independent words in CELEX are segmented from their affixes. In particular, bound stems are never segmented (e.g., *possible* is not segmented).
- Certain suffix combinations are represented as single units. For example, a common suffix combination in English is *-ate -ion* (i.e., *-ation*). In many cases, there is no corresponding word ending in *-ate* for an existing word ending in *-ation* (e.g., *form*/**formate*/*formation*). In such cases, CELEX typically represents the suffix combination *-ate -ion* as a single suffix *-ation*.
- Certain derivational suffixes exhibit quasi-paradigmatic structure, with systematic correspondences between a set of suffixes across morphological categories or meanings (e.g., *-ism*/*-ist*, *-ent*/*-ence*). In such cases, CELEX often analyzes a form with one ending as the base of forms that have the other endings. For example, nouns ending in *-ence* (e.g., *competence*) are analyzed with the corresponding *-ent* form (i.e., *competent*) as their base, rather than deriving both *-ent* and *-ence* forms from the bound stem: *compet*.
- Many semantically non-transparent deverbal forms are coded with verb particle constructions as their bases. For example, one analysis of the word *reliable* in CELEX gives the verb-particle construction *rely on* as the base, presumably reflecting the annotator's intuitions that *reliable* is semantically most related to this lexical entry than to the simple verb form *rely*.

To address these (and other) problems, we extended the set of morphologically analyzed forms in the CELEX database with a large number of low-frequency words, and improved and corrected the result via an iterative process that made use of a combination of heuristic parsing and hand annotation.

In addition to the set of morphologically analyzed forms, CELEX provides a separate list of all word strings that appeared in the original samples used to construct the database. This list provides a more representative sample of low-frequency, productively-derived forms. Because the majority of English derivational affixes are suffixes, and most empirical work in linguistics and psycholinguistics has focused on the subdomain of suffixation, we focused on detecting and segmenting low-frequency, suffixed forms when expanding our training set.[2] To assign analyses to these forms, we developed a heuristic segmentation algorithm, which we describe below.

This algorithm started at the end of each word and attempted to match each of the suffixes found in the original set of CELEX analyses, one at a time, in order of increasing frequency. If the algorithm matched a suffix, it stripped this suffix and recursively applied the same process to the resulting base, continuing until no further suffixes could be matched. If the final stem that resulted when no more suffixes could be matched was contained in the set of original morphological analyses, then the word was considered correctly segmented, and a tree representation of the parse was returned. Thus, the parser was conservative in the sense that it permitted only categorially well-formed morpheme combinations with known stems as their innermost component. [3]

Many of the words in CELEX are annotated with orthographic transformations that apply to a base before attaching an affix. For example, CELEX specifies that the word *curate* loses its final *e* before appending *-or* to derive the word *curator*. In the CELEX analysis for the word *curate*, the suffix *-or* is annotated with information that specifies the loss of the final *e* on the base. We made use of this information in our segmentation algorithm. When constructing the set of possible suffixes used by the algorithm, we noted the set of base transformations that appear with that suffix in analyses. During the suffix segmentation process, these base transformations were inverted, and the resulting base (e.g., with restored final *e*) was passed recursively back to the algorithm as input to the next stage of segmentation. Thus, the algorithm was able to exploit systematic spelling changes (e.g., loss of final *e*, or the change of final *y* to *i* before *-ness*).

2. Note, however, that any segmented prefixes that appeared in words in the original CELEX analyses were included in the training set.

3. This parser was implemented with the Ruby programming language (Flanagan and Matsumoto, 2008) and can be made available upon request.

When possible, the algorithm was also extended to automatically correct problems with the original set of CELEX analyses. For example, the parser segmented bound stems (e.g., *possible*⟶*poss -ible*) and suffix combinations (e.g., *-ation*⟶*-ate -ion*), collapsed across related bases (e.g., *rely* and *rely on*) and allomorphs of suffixes (e.g., *-ible*/*-uble*⟶*-able*), and represented families of related words sharing the same base (e.g., *competence*/*competent* share *compet*).

Although the segmentation algorithm resulted in a substantial extension and improvement of the CELEX sample, it was not always possible to make such corrections automatically. Moreover, the unanalyzed portion of CELEX is very noisy, containing many misspellings and non-words, which led to many incorrect analyses. To further mitigate these problems, approximately 10,000 forms were corrected by hand during iterative rounds of heuristic parsing and examination of the results. Particular attention was paid to candidate forms containing suffixes discussed in the linguistic and psycholinguistic literature (e.g., *-ness* and *-ity*).

The resulting set of morphologically analyzed words contained 338 suffixes (which we cross referenced with a list compiled from the literature), over $25,000$ word types, and over 7.2 million word tokens. The training set used for all simulations below was the subset of this corpus containing all words either (i) include one of the 338 suffixes, or (ii) include a stem appearing with one of these suffixes. A list of suffixes that appeared in at least 10 word types in the training corpus is provided in Appendix B.

7.2.3 Simulations

Trees representing the morphological analyses of word forms, along with their token frequencies, were input to all of the models. Eighty simulations using the `FG` (inference-based) model were run on the training set, each with $105,000$; $95,000$; $85,000$; $75,000$; $65,000$; $55,000$; $45,000$; $35,000$; $25,000$; and $15,000$ inputs, as described in Sections 3.3–3.5. These ran for between one and 115 sweeps through the training corpus, depending on time available on the MIT CSAIL cluster. We approximated the maximum a posteriori fragment grammar by using the highest probability approximating PCFG discovered by the search procedure described in Section 3.2. The approximating PCFGs for the other models were computed directly from the input corpus as described in Section 3.3.

7.3 Simulation Results and Analyses

Below, we present analyses of simulations performed on the training corpus described in the previous section. This section is divided into three parts. In the first, we describe analyses examining the productivity of individual English suffixes. In the second, we turn to an examination of model predictions about the composition and retrieval of existing words. In the final section, we study the ability of the models to account for the pattern of morpheme combination and ordering in English derivational morphology.

7.3.1 The Productivity of Individual English Suffixes

In this section, we investigate the productivity predictions of our five models for individual English suffixes. Productivity is the readiness with which a morphological rule, such as *-ness* affixation, can be used to generate new word forms. An ideal comparison of our five models would, therefore, test their ability to correctly predict experimental data about the generalizability of a large number of English suffixes. Unfortunately, however, broad-coverage databases of such speaker judgments are currently unavailable. In lieu of a wide-coverage, quantitative evaluation, we will present two analyses.

First, we will examine the most productive suffixes predicted by each model. Our examination of these suffixes will reveal that all of the models besides the `FG` (inference-based) model make a number of predictions that are both intuitively implausible and at odds with discussions in the literature. Second, we will compare the predictions made by our models to Baayen's hapax-based estimators of productivity $\mathcal{P}$ and $\mathcal{P}^*$ (see, Section 6.4.1.2). This analysis will show that the `FG` (inference-based) model produces predictions that converge strongly with the predictions of Baayen's statistics. This result both clarifies the nature of the inference performed by the `FG` (inference-based) model, and provides a justification for the assumptions underlying Baayen's measures.

7.3.1.1 The Five Most Productive Suffixes

Table 7.3 shows the five most productive suffixes predicted by each model, measured using approximations of $\Pi_{S,r}$ (absolute productivity)—the probability that each suffix will give rise to novel forms, relative to all possible words under each model. As we discussed in Section 6.4.1.1, $\Pi_{S,r}$ (absolute productivity) is the simplest and most basic measure of productivity, and, unlike the other quantitates defined there—$\Pi^*_{S,r}$ (between-rule productivity) and $\Pi^\circ_{S,r}$ (within-rule productivity)—it takes into account all distributional factors affecting the likelihood of new words, including the base rate of novelty in a grammar and variability in productivity across suffixes.

For simplicity, $\Pi_{S,r}$ (absolute productivity) was approximated by computing the probability that the models assigned to the combination of each suffix with a novel wug stem, which was not present in the training set.[4] As a consequence, this approximation does not take into account variability in $\Pi_{S,r}$ (absolute productivity) that arises due to differences in the availability of bases. However, the effect of base availability on productivity is significant mainly when an affix attaches only to a very small number of bases.[5] Since most of the affixes in Table 7.3 attach to major categories—like `V` (verb) and `N` (noun), which contain thousands of words, this simplification is unlikely to have a major effect.

Turning to Table 7.3, the first thing to note is that two of the most frequently-discussed productive suffixes (see, e.g., Di Sciullo and Williams, 1987; Fabb, 1988)—agentive *-er*:`V>N` and adverbial *-ly*:`Adj>Adv`—are predicted to be highly productive by all of the models. For these suffixes, the distribution of forms in the training set provides robust evidence of productivity under a variety of different theoretical assumptions about what factors should indicate generalizability. More importantly, there are several suffixes for which the models make divergent predictions, revealing important differences in the kind of evidence used by each model. We focus our discussion on three in particular: the highly productive suffix *-ness*:`Adj>N`, the suffix *-ate*:`BND>V`, which attaches only to bound stems, and the unproductive suffix *-ion*:`V>N`.

The suffix *-ness*:`Adj>N` is perhaps the most commonly-discussed productive suffix in English; however, only the `FG` (inference-based) model places it in the top five. By contrast, all four competing models predict that the unproductive suffix *-ion*:`V>N` is either the most, or second most, productive suffix in English. Although *-ion*:`V>N` does occur in the (semi-)productive suffix combination *-ation*,[6] these four models make the obviously absurd prediction that *-ion*:`V>N` should readily attach to arbitrary verbs, such as *meet*, on its own, producing forms such as **meetion* (a MEETING event).

4. The probability of the novel wug stem was assumed to be a small fixed constant.

5. For example, it has an effect on the absolute productivity of the ordinal number marker, *-th*:`Q>Q` (e.g., *fourteenth*), which is restricted to attaching to number words (category `Q`)—a small set of bases.

6. We consider this suffix combination when we examine generalizable combinations of suffixes in Section 7.3.3.4.

FG (inference-based)

Suffix	Example
-ly:Adj>Adv	*quickly*
-er:V>N	*talker*
-ness:Adj>N	*tallness*
-y:N>Adj	*mousey*
-er:N>N	*prisoner*

DMPCFG (full-parsing)

Suffix	Example
-ion:V>N	*regression*
-ly:Adj>Adv	*quickly*
-ate:BND>V	*segregate*
-ment:V>N	*development*
-er:V>N	*talker*

MAG (full-listing)

Suffix	Example
-ly:Adj>Adv	*quickly*
-ion:V>N	*regression*
-er:V>N	*talker*
-ly:V>Adv	*bitingly*
-y:N>Adj	*mousey*

DOP1 (exemplar-based)

Suffix	Example
-ion:V>N	*regression*
-er:V>N	*talker*
-ment:V>N	*development*
-ate:BND>V	*segregate*
-ly:Adj>Adv	*quickly*

ENDOP (exemplar-based)

Suffix	Example
-ion:V>N	*regression*
-ly:Adj>Adv	*quickly*
-ment:V>N	*development*
-er:V>N	*talker*
-ate:BND>V	*segregate*

Figure 7.3
The 5 Most Productive Suffixes Predicted by Each Model

This counter-intuitive prediction is also at variance with a number of facts about *-ion*:V>N, which have been discussed in the literature. First, this suffix is a classical example of a formative-boundary affix, which triggers base-specific sound changes (e.g., *transmit/transmission*; Chomsky and Halle, 1968). Since base-specific phonological structure is typically associated with less productive affixes, this provides additional evidence that the predictions of the four competing models are incorrect. Second, in a lexical decision study, Bradley (1978) found that, unlike a number of other suffixes she examined (e.g., *-ness*:Adj>N), base-frequency manipulations did not change latencies for *-ion*:V>N when whole-form frequencies were controlled, suggesting that *-ion*:V>N is not decomposed during lexical processing. Since decomposition is assumed to be correlated with productivity, this is further evidence against the prediction that *-ion*:V>N should attach to arbitrary verbs. Finally, Wheeler and Schumsky (1980) found that forms using this suffix were rated as "uncertain suffix" words in a metalinguistic classification task. Thus, its appearance alongside such clearly productive suffixes as *-er*:V>N and *-ly*:Adj>Adv in the four competing models is puzzling.

Three of the five models—DMPCFG (full-listing), DOP1 (exemplar-based), and ENDOP (exemplar-based)—also place *-ate*:BND>V amongst the top five most productive suffixes. This prediction is also highly implausible since this suffix is restricted, by its categorial definition, from attaching to anything besides bound stems.

In summary, all of the systems besides the FG (inference-based) model produce highly implausible and non-intuitive predictions concerning the productivity of certain suffixes. The competing models all fail to predict the high degree of productivity of *-ness*:Adj>N, and they all incorrectly predict that *-ion*:V>N should readily attach to arbitrary verbs. Furthermore, several of the models make the implausible prediction that the suffix *-ate*:BND>V, which attaches only to bound stems, is highly productive.

7.3.1.2 Comparison with Baayen's Corpus-Based Productivity Estimators

In this section, we compare the predictions of our five models to the predictions of Baayen's hapax-based productivity estimators $\mathcal{P}^*$ and $\mathcal{P}$ (Baayen, 1992, 1993, 1994, 2001, 2003, 2006; Baayen and Lieber, 1991; Baayen and Renouf, 1996; Du and Zhang, 2010; Hay and Baayen, 2002; van Marle, 1992). As we discussed in Section 6.4.1.2, $\mathcal{P}^*$ and $\mathcal{P}$ are estimators of the abstract conditional probabilities $\Pi^*_{S,r}$ (between-rule productivity) and $\Pi^\circ_{S,r}$ (within-rule productivity), respectively.

The conditional probabilities, $\Pi^*_{S,r}$ (between-rule productivity) and $\Pi^\circ_{S,r}$ (within-rule productivity), both measure the total probability of all words produced by affixation of a morpheme *r*, which are novel with respect to some sample of words *S*. They differ in the set of words that is taken as a reference for comparison. The first conditional probability, $\Pi^*_{S,r}$ (between-rule productivity), measures the likelihood of novel words using *r* relative to other novel words (produced by any morphological process). The second conditional probability, $\Pi^\circ_{S,r}$ (within-rule productivity), measures the likelihood of novel words using *r* relative to other words using *r* (i.e., attested or novel).

In Section 6.4.1.2, we explained that Baayen's statistics are based on the idea that, because productive affixes frequently give rise to new words, and therefore their word frequency distributions typically contain a large number of low-frequency forms, a reasonable estimate of the productivity of an affix can be based on the proportion of low-frequency words associated with the affix. In particular, Baayen's statistics estimate the productivity of an affix *r* using the proportion of word types that include *r* and appear only once in a corpus (i.e., the hapax legomena).

As we discussed in Section 6.4.1.2, there is empirical evidence that Baayen's statistics quantify at least some aspects of the intuitive notion of productivity, and, therefore, there is reason to believe that they should provide reasonably accurate predictions for the productivity of individual English suffixes. Nevertheless, because there is no "gold standard" measure of productivity nor, currently, any broad-coverage databases of native speaker generalization data, there is no way to precisely quantify the degree to which Baayen's approach accords with psychological reality, and the predictions of any of the models studied here may or may not be more accurate. In particular, as we argued in Section 6.4.1.2, the mathematical justifications for Baayen's statistics are based on highly simplified assumptions about word distributions, which do not take into account morphological structure. By contrast, all five of the models studied here characterize the distribution over words by explicitly defining possible morphological computations. To the degree that each model's account of morphological computation is more in accord with psychological reality than Baayen's theory, it may provide more accurate predictions about the productivity of individual suffixes.

Hay and Baayen (2002) reported the values of $\mathcal{P}$ and $\mathcal{P}^*$ for a large number of suffixes computed from a corpus constructed from a combination of morphological analyses from the CELEX database and additional analyses derived via automatic segmentation of word strings.[7] Table 7.1 shows the Pearson correlations between these values and estimates of $\Pi^{\circ}_{S,r}$ (within-rule productivity) and $\Pi^{*}_{S,r}$(between-rule productivity) derived from our models. Values for $\Pi^{\circ}_{S,r}$ (within-rule productivity) were computed by dividing the values of $\Pi_{S,r}$ (absolute productivity), from the last section, by an approximation to Π_r (the total probability of all words output by suffix r; see, Section 6.4.1.1). This quantity was computed by summing the probability of all possible words consisting of a single stem and suffix r. Because, $\Pi^{*}_{S,r}$ (between-rule productivity) is proportional to $\Pi_{S,r}$ (absolute productivity) when comparing across suffixes, and, because our analysis involves correlations (which are not sensitive to constant multiplicative factors), we use our estimates of $\Pi_{S,r}$ (absolute productivity) in place of $\Pi^{*}_{S,r}$ (between-rule productivity).[8]

Measure	FG	DMPCFG	MAG	DOP1	ENDOP
$\Pi^{\circ}_{S,r}$ v. $\mathcal{P}$	**0.907**	-0.0003	0.692	0.346	0.143
$\Pi^{*}_{S,r}$ v. $\mathcal{P}^*$	**0.662**	0.480	0.568	0.402	0.500

Table 7.1

Correlation with Baayen's Corpus-Based Productivity Statistics: The correlation between estimates of $\Pi^{\circ}_{S,r}$ (within-rule productivity) and $\Pi^{*}_{S,r}$ (between-rule productivity), computed against the models, and the values of Baayen's $\mathcal{P}$ and $\mathcal{P}^*$ provided in Hay and Baayen (2002).

As can be seen in Table 7.1, the FG (inference-based) model provides the best correlation between Baayen's hapax-based and our model-based estimators of $\Pi^{\circ}_{S,r}$ (within-rule productivity) and $\Pi^{*}_{S,r}$(between-rule productivity), with the MAG (full-listing) model also producing reasonable correlations. There are two points to take from these results.

7. Although both the present simulations and the values reported Hay and Baayen (2002) are based on corpora ultimately derived from CELEX, there are large differences in the samples, especially for highly-productive suffixes, because both samples were substantially extended using distinct heuristic segmentation algorithms and the training corpus used the present study also included extensive correction by hand.

8. Hay and Baayen (2002) provided only the string versions of suffixes, without input and output categories. To compare with their values, we therefore actually summed across all suffixes with the same string form.

First, although there is currently no way to quantitatively measure the psychological accuracy of the predictions of any of the models, Baayen's measures are known to accord reasonably well with intuition. Thus, the failure of the predictions of the `DMPCFG` (full-parsing), `DOP1` (exemplar-based), and `ENDOP` (exemplar-based) models to correlate highly with $\mathcal{P}$ suggests that these models are not accurately capturing the productivity of individual English suffixes. Taken together with the implausible predictions about specific suffixes made by these models in the previous section, we can conclude that they do not represent a viable theory of productivity. Second, the strongly convergent predictions made by Baayen's measures and the `FG` (inference-based) model indicate that these two theories are likely to be making use of similar information in the distribution of English word forms. We discuss the implications of this finding below.

Before discussing these results in more detail, we note one additional aspect of the data in Table 7.1. For the `FG` (inference-based) and `MAG` (full-listing) models, the correlation between $\mathcal{P}$ and $\Pi^{\circ}_{S,r}$ (within-rule productivity) is somewhat higher than the correlation between $\mathcal{P}^{*}$ and $\Pi^{*}_{S,r}$ (between-rule productivity). This is likely due to greater sampling error in estimates of $\Pi^{*}_{S,r}$ (between-rule productivity). Because $\Pi^{\circ}_{S,r}$ (within-rule productivity) is computed relative to all forms using the same affix *r*, a large part of any estimate for this value is determined by actual, attested forms, some of which are quite frequent. In any finite sample, the distribution of frequent word forms is much more stable than the distribution of low-frequency, productively-derived forms (Baayen, 2001). By contrast, $\Pi^{*}_{S,r}$ (between-rule productivity) is computed relative to all novel words, which, by definition, cannot appear in the sample, and whose probabilities therefore must be estimated from existing words. As a result, estimates of both $\Pi^{*}_{S,r}$ (between-rule productivity) and $\mathcal{P}^{*}$ will tend to be noisier and more sample-dependent.

7.3.1.3 Discussion

In the preceding sections, we compared and contrasted the predictions made by each of our five models on the productivity of individual English suffixes. First, we showed that the five most productive suffixes predicted by the `FG` (inference-based) model were more intuitively plausible than those predicted by the other four models. Only the `FG` (inference-based) model placed the highly productive suffix *-ness*:`Adj>N` in the top five, and, unlike the other four models, it did not infer that *-ion*:`V>N` was amongst the most productive suffixes of English. In our second analysis, we showed that the `FG` (inference-based) model exhibited predictions about the productivity of individual suffixes that were strongly convergent with Baayen's corpus-based estimators. Because, both the `FG` (inference-based) model and Baayen's estimators project productivity from the frequency distribution of forms in the input data, this result indicates that they are likely to be making use of similar kinds of information.

Throughout the book, we have emphasized that, since productive affixes frequently give rise to new words, their frequency distributions are characterized by a large number of low frequency forms. Baayen's hapax-based predictors are based on this fact, and we have suggested that this property is also crucial to the inferences drawn by the `FG` (inference-based) model. The convergent predictions of these two theories confirms this suggestion. To further illustrate this point, and contrast the behavior of Baayen's measures and the `FG` (inference-based) model with the other four models, it is useful to examine the frequency distribution of *-ness*:`Adj>N` and *-ion*:`V>N` in more detail.

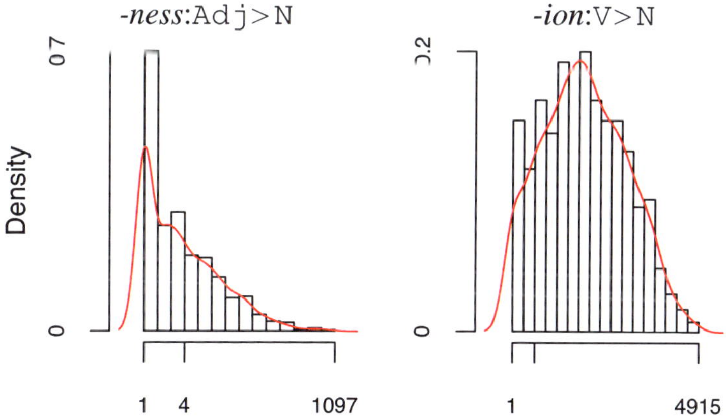

Figure 7.4
Frequency Spectra for –*ness*:`Adj>N` and –*ion*:`V>N`

Both *-ness*:`Adj>N` and *-ion*:`V>N` occur in a large number of word types in the input corpus (1024 and 1117, respectively). The suffix *-ion*:`V>N` is much more frequent, however, occurring in 162,573 word tokens, while *-ness*:`Adj>N` occurs in only 15,568. The distribution of word tokens over word types is also very different. This can be seen in Figure 7.4, which plots the frequency spectra for these two affixes (see Sections 1.2.6 and 6.3.4).

The frequency spectrum for *-ion*:V>N is sharply peaked at low-frequency forms (350 hapaxes) and declines steadily as frequency increases. By contrast, *-ion*:V>N has fewer low frequency forms (83 hapaxes). The distribution of *-ness*:Adj>N—but not *-ion*:V>N—is characteristic of a productive affix.

Baayen's statistics are defined directly in terms of the frequency of hapax forms, and we have argued that low-frequency forms are also crucial to the inferences made by the FG (inference-based) model. However, why do the other four models incorrectly infer that *-ion*:V>N is productive? The DMPCFG (full-parsing) model bases productivity inferences purely on the token frequency of suffixes, while the MAG (full-listing) model bases productivity inferences purely on the type frequency of suffixes. Because *-ion*:V>N is highly frequent in both types and tokens, these models both infer, incorrectly, that it is highly productive. The two exemplar-based models (DOP1 and ENDOP) can be understood as interpolating between type and token frequencies in different ways. Roughly speaking ENDOP (exemplar-based) tends to behave more like the DMPCFG (full-parsing) model, while DOP1 (exemplar-based) tends to behave more like the MAG (full-listing) model. Therefore, these two models also incorrectly infer that *-ion*:V>N is a productive suffix.

As we have shown, the FG (inference-based) model makes crucial use of the number of low frequency forms to infer the productivity of affixes; Baayen's statistics simply define productivity in these terms. What is the theoretical relationship between the two approaches? As we have emphasized, although Baayen's measures can be mathematically justified in several different ways, these justifications all make unrealistic simplifying assumptions about the distribution of possible words. The FG (inference-based) model, by contrast, defines the distribution over words using an explicit model of word generation that incorporates psychologically motivated mechanisms of computation and storage. It infers the pattern of productivity and reuse for a given language via a complex inference that evaluates the explanatory adequacy of an enormous number of grammars each of which consists of thousands of stored fragments of differing levels of abstraction. In particular, the inferences drawn by the model cannot be expressed by any simple mathematical expression involving type or token frequencies. These results show that, at least for the case of individual suffixes, Baayen's hapax-based approach provides easy-to-calculate predictions that approximate the complex inference performed by the FG (inference-based) model.

In Section 6.4.2.1, we discussed Baayen's $\mathcal{A}$ statistic, which measures the composability of affixes with respect to a specific, psychologically motivated model of lexical processing $\mathcal{M}$ (in the literature, several different models have been used). Baayen (1993) argued that $\mathcal{A}$ provides a better predictor of productivity than his hapax-based measures, and furthermore speculated that this is because it is based on an explicit, psychologically motivated model, whereas $\mathcal{P}$ and $\mathcal{P}^*$ are motivated by the mathematical properties of word frequency distributions. We believe that similar remarks apply to the relationship between the FG (inference-based) model and his measures. Since the FG (inference-based) model is based on a psychologically motivated theory of morphological computation, we hypothesize that in cases where its predictions differ from hapax-based predictions, the FG (inference-based) model is more likely to be correct. Future work will reveal whether this hypothesis can be supported by empirical evidence.

There is an additional theoretical consequence of the relationship between Baayen's statistics and the FG (inference-based) model. Since the hapax-based measures provide an approximate description of fragment grammar inferences, but the FG (inference-based) model is grounded in more realistic assumptions, these results can be interpreted as providing an additional theoretical justification for the use of hapaxes in estimating productivity.

7.3.2 Relative Frequency and Decomposition

In Section 6.4.2.2, we discussed Hay's relative frequency hypothesis, which states that, during processing, whether a word is decomposed or retrieved as a whole depends on the ratio of the frequencies of a word's parts to its whole-form frequency. In most published work, Hay has focused on the specific relationship between the base and whole-form frequencies of morphologically complex words. When the frequency of a base is greater than the frequency of a derived word, affixes will tend to be decomposed from the base during processing. When the frequency of the derived word is greater than the frequency of the base, the word will tend to be stored and processed as a whole.

In this section, we examine the behavior of our five models from the perspective of the relative frequency hypothesis. In the next two sections, we show that the FG (inference-based) model and the two exemplar-based models (DOP1 and ENDOP) are consistent with the relative frequency hypothesis, decomposing existing word forms at rates that depend on their base to whole-form frequency ratios. We argue, however, that this result is not surprising: any probabilistic model (including the FG [inference-based], DOP1 [exemplar-based], and ENDOP [exemplar-based] models) that allows heterogenous lexical storage and relates the probability of stored fragments to their frequency must make composition/retrieval decisions related to part-word to whole-word ratios.

Although these three models are consistent with the relative frequency hypothesis, there is an important difference between the FG (inference-based) model, on one hand, and the two exemplar-based models (DOP1 and ENDOP) on the other. While the exemplar-based models predicted that many words are ambiguous with regard to decomposition and retrieval, the FG (inference-based) model predicts a sharp dichotomy between a set of words that are nearly always decomposed and a distinct set of words that are nearly always retrieved as wholes from memory. We discuss the reasons for this difference and its implications.

7.3.2.1 Decomposition of Existing Words

In order to examine the decomposition of existing words under the five models, we performed an analysis of the set of suffixed words in our training corpus. For each suffixed form, we computed the total (marginal) probability that each model assigned to derivations of the word where the final suffix was decomposed from the base (i.e., where the word was derived using a fragment consisting of the suffix together with a variable in the base position). Because words can have multiple derivations of differing probability, this value can fall anywhere between zero and one. Figure 7.5 plots the probability that the final suffix for each word in the training corpus was decomposed, as a function of the training word's whole-form (x-axis) and base (y-axis) token frequencies. Words that tended to be retrieved as wholes (i.e., whose probability of decomposition was less than or equal to 0.5) are plotted in shades of blue, with darker shades of blue indicating higher rates of retrieval. Words that tended to be decomposed (i.e., whose probability of decomposition was greater than 0.5) are plotted in red, with deeper shades of red indicating higher rates of decomposition. Lighter blue and lighter red points are closer to the 0.5 boundary value and are, therefore, more ambiguous with regard to decomposition versus retrieval.

The plots in Figure 7.5 reveal several patterns. First, because it necessarily and fully decomposes all words, all points in the plot for the DMPCFG (full-parsing) model are strongly red. By contrast, all points in the plot for the MAG (full-listing) model are strongly blue, showing that this model retrieves existing words as wholes with probability close to 1. In the latter case, this is not a logical necessity. Although, the MAG (full-listing) model stores all attested forms,

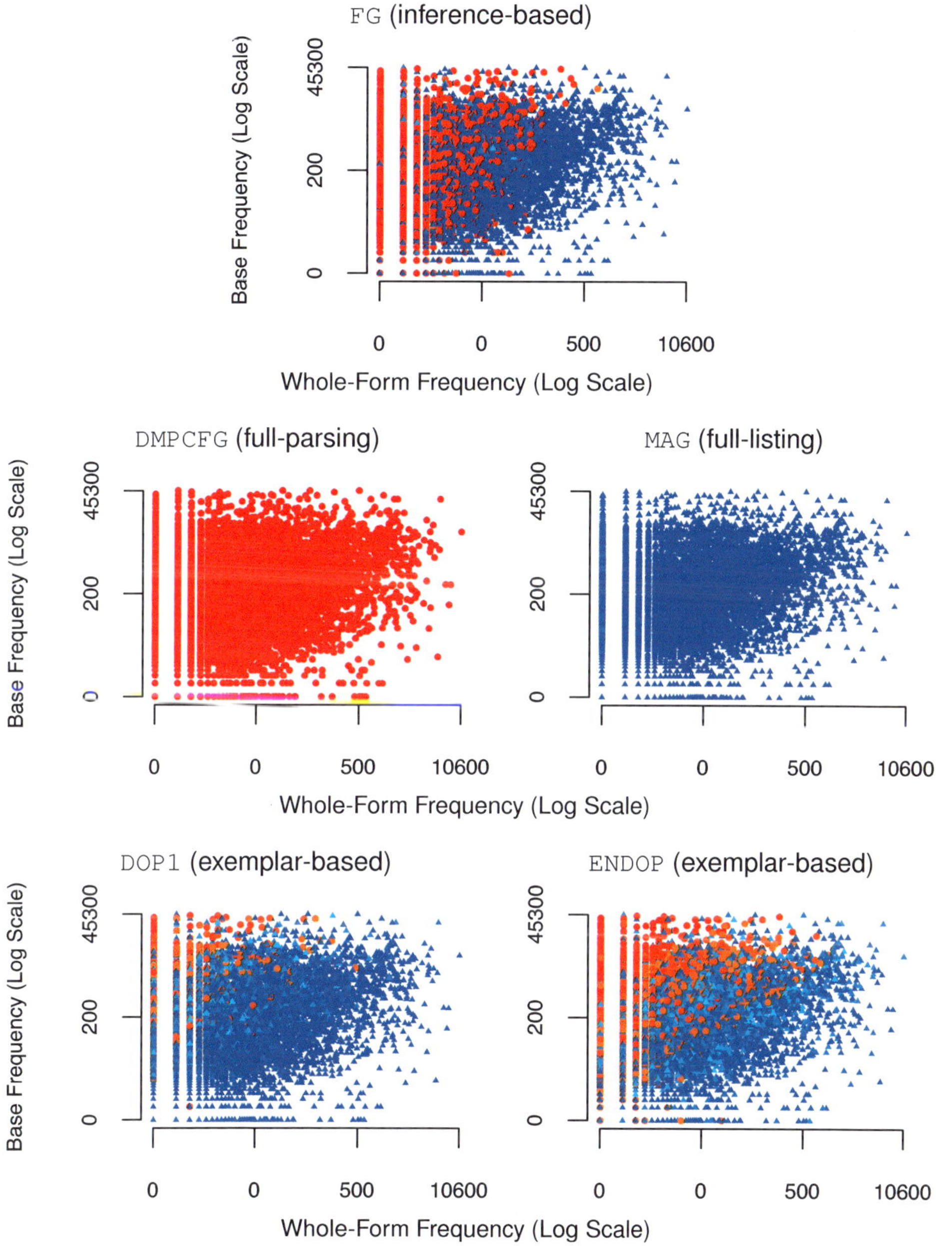

Figure 7.5

Decomposition and Retrieval of Suffixed Words: The probability of decomposition for the final suffix of words in the training set as a function of their whole-form (x-axis) and base (y-axis) log-frequencies. Deeper blue points represent words which were more likely to be retrieved as wholes. Deeper red points represent words which were more likely to be composed.

it can also compose forms from bases using its underlying PCFG. However, Figure 7.5 shows that the decomposition of existing forms is very improbable under the model.

The other three models, FG (inference-based), DOP1 (exemplar-based), and ENDOP (exemplar-based), predict a greater degree of between-word variability in decomposition. Note that all three models tend to predict more retrieval than decomposition, across all words from the training set. This observation is confirmed in Table 7.2, which shows the weighted and unweighted mean probability that existing forms are decomposed under the five models. Because the weighted mean takes into account the probability of each word, and because more probable (i.e., frequent) words are more likely to be stored as wholes, the weighted mean decomposition rate tends to be significantly lower than the unweighted mean rates. Nevertheless, even for the unweighted mean, there is a clear global preference for retrieval over decomposition in all three models. This result accords with the basic intuition that most words are memorized and that word-formation processes less often give rise to new forms than, for instance, syntactic processes.

Model	FG	DMPCFG	MAG	DOP1	ENDOP
Decomp. Prob. (Weighted Mean)	0.00165	1.0	0.000172	0.0172	0.212
Decomp. Prob. (Unweighted Mean)	0.248	1.0	0.00165	0.0813	0.243

Table 7.2

Overall Probability of Decomposition: This table shows the total probability that the final suffixes of words from the training set were decomposed by the models. The first row shows the expected probability of decomposition, weighted by the total probability of each word. The second row shows the unweighted mean value.

Crucially, the three models that allow for heterogenous lexical storage—FG (inference-based), DOP1 (exemplar-based), and ENDOP (exemplar-based)—all predict rates of decomposition consistent with the relative frequency hypothesis. To see this, consider the relationship between base and whole-form frequencies apparent in these plots. The x-axis in each plot represents the whole-form frequency of each suffixed word, while the y-axis represents the base frequency of each suffixed word. Therefore, the base frequency of individual words is greater than their whole-form frequency for points that fall above a line where $x = y$ (i.e., where whole-form frequency equals base frequency). If decomposition depended only on whole-form frequency, then red and blue points would tend to be separated by a vertical line. If decomposition depended only on base frequency, then red and blue points would tend to be separated by a horizontal line. However, for all three models, red and blue points tend to be separated by a line with positive slope, showing that decomposition is related to the ratio of base to whole-form frequency, as predicted by the relative frequency hypothesis.

Although both the FG (inference-based) model and the two exemplar-based models (DOP1 and ENDOP) are consistent with Hay's Relative Frequency Hypothesis, there is an important difference between them. While the FG (inference-based) model tends to predict that words are either decomposed with high probability (strongly red points) or retrieved with high probability (strongly blue points), the exemplar-based models predict a greater number of words that are ambiguous with regard to decomposition and retrieval (lighter blue and lighter red points).

Model	FG	DOP1	ENDOP
Decomp. Prob. *decomposed-subset* (Unweighted Mean)	0.990	0.680	0.720
Decomp. Prob. *decomposed-subset* (Weighted Mean)	0.972	0.719	0.770
Decomp. Prob. *retrieved-subset* (Unweighted Mean)	0.0010	0.0588	0.243
Decomp. Prob. *retrieved-subset* (Weighted Mean)	0.0007	0.0140	0.212

Table 7.3

Probability of Decomposition for Decomposed/Retrieved Subsets: This table shows the decomposition probability for training set words broken down into subsets according to whether they tend to be decomposed (*decomposed-subset*) or retrieved (*retrieved-subset*). The FG (inference-based) model produces a sharper dichotomy between these two subsets than the two exemplar-based models.

The data in Table 7.3 illustrate this point. This table shows the weighted and unweighted mean decomposition rates for words broken down into two subsets. The first subset, labeled *decomposed-subset*, are those words whose probability of decomposition is greater than 0.5. The second subset, labeled *retrieved-subset*, are those words whose probability of decomposition is less than 0.5. The FG (inference-based) model predicts a sharp dichotomy between these two subsets: words that tend to be decomposed, tend to be decomposed with probability close to one; words that tend to be retrieved as wholes, tend to be decomposed with probability close to zero. For the two exemplar-based models, by contrast, the dichotomy is much less pronounced. For the ENDOP (exemplar-based) model in particular the rates of decomposition and retrieval for the two subsets are near the logical mean points of 0.25 for decomposed forms and 0.75 for retrieved forms.

7.3.2.2 Summary and Discussion

In this section, we discuss the origins and consequences of two phenomena. First, we examine the relationship between the predictions of our five models and Hay's relative frequency hypothesis. Second, we examine the sharp dichotomy between decomposition and retrieval predicted by the FG (inference-based) model, compared to the greater degree of ambiguity in composition predicted by the two exemplar-based models (DOP1 and ENDOP).

7.3.2.2.1 Relationship to the Relative Frequency Hypothesis

Our analyses above demonstrated that three of the models (FG [inference-based], DOP1 [exemplar-based], and ENDOP [exemplar-based]) predicted rates of decomposition for individual words related to each word's base to whole-form frequency ratio, consistent with Hay's relative frequency hypothesis. Why do these models exhibit this behavior?

These models share three assumptions. First, they allow arbitrary fragments of structure to be stored in memory—including large, specific fragments, and small, abstract fragments. Second, they define the probability of a derivation as the product of the probabilities of the fragments used in the derivation. Third, they estimate the probability of each stored fragment using its frequency across derivations. We claim that any model that makes these three assumptions, and uses probability theory to infer whether a word should be decomposed or retrieved, must exhibit behavior consistent with the relative frequency hypothesis.

To see this, consider a word composed of two parts: a stem and a suffix. Assume that a model that shares the three assumptions above has stored the following three fragments in memory: a fragment b representing the stem (on its own), a fragment s representing the suffix (on its own), and a fragment w representing the full word form (stem and suffix). The probability of any stored fragment, x (of category A) is given by $P(x) = \frac{\#(x)}{C_A}$, where $\#(x)$ is the frequency with which fragment x is used across word derivations and C_A is a normalizing constant, which is equal to the the sum of the frequencies of all fragments with the same category as x (i.e., all fragments of category A).

In this setting, the decision to compose the word or retrieve it whole from memory is equivalent to the decision to derive the word by combining b and s, or, alternatively, to retrieve the whole-word fragment w directly from memory. If the model uses probabilistic conditioning to make this decision, then the probability that the model will compose the word using fragments b and s is given by the conditional probability $\frac{P(b)P(s)}{P(w)+P(b)P(s)}$. We can consider this conditional probability a measure of the (de)composability of the word. First, note that the composability of a word is defined as a ratio between the probabilities of fragments representing parts of the word, and a sum including the probability of the whole-word fragment. Second, recall that these probabilities are proportional to the frequencies of the fragments.

If we increase the frequency of the base fragment b while holding the frequency of other fragments constant, then the composability of the word will increase. Likewise, if we increase the frequency of the whole-form fragment w while holding the other frequencies constant, then the composability will decrease. However, if we increase or decrease the base and whole-form fragment frequencies by exactly the same amount, the composability will not change. Thus, in probabilistic models making the three assumptions discussed above, composability is related monotonically to the base-to-whole-form frequency ratio, as predicted by the relative frequency hypothesis.

However, there are several important ways in which the measure of composability defined above differs from that suggested by the relative frequency hypothesis. First, the probabilities above are defined in terms of the frequency of stored fragments, not the frequency of the surface morphemes. While fragment frequencies will, in general, be related to surface frequencies, they are not themselves directly observable in the linguistic input. This difference between the relative frequency hypothesis and the theory advocated in this book follows directly from a consideration of the problem of language acquisition. The goal of the language learner is not to characterize the statistics of corpus data, but rather to infer the properties of the mind-internal computational system that generated the data.

Similar points have been made by other authors. For example, in Section 6.4.2.1, we discussed Baayen's measure of morpheme composability, $\mathcal{A}$. This statistic is defined with respect to a psychologically motivated model of decomposition and retrieval, $\mathcal{M}$ (that is, it is parameterized by $\mathcal{M}$). Like the measure of composability discussed above, $\mathcal{A}$ is not defined in terms of the surface frequency of morphemes, but rather the frequency with which morphemes are composed by the model $\mathcal{M}$. Baayen (1993) argued that, because $\mathcal{A}$ quantifies the behavior of a psychologically motivated model of composition and retrieval, it is a superior predictor of composability to measures based purely on corpus statistics.

Another important difference between probabilistic models making these assumptions and the relative frequency hypothesis is that the former make mathematically precise predictions in many cases where the latter does not. For example, the definition of composability above predicts that an increase in the frequency of a base fragment will lead to larger changes in word composability when the suffix is highly frequent than when the suffix is less frequent. The relative frequency hypothesis doesn't provide a precise, quantitative account of this interaction. Probabilistic models can also provide mathematically precise predictions about the interactions of any number of morphemes in morphologically complex words of any size. Moreover, because each fragment probability $P(x)$ includes a normalizing constant C_A, which depends on the frequency of other forms with the same category as x, these probabilistic models also provide predictions about how morphemes that are not even present in a particular form can affect the composability of that form. For example, these models predict that the effect of a change in base frequency will be greatest when there are a small number of other bases also available to combine with a given suffix. These and many other predictions remain to be tested empirically. Nevertheless, probabilistic models like fragment grammars can be understood as providing a justification for the relative frequency hypothesis, while going significantly beyond it in predictive power.

7.3.2.2.2 The Dichotomy between Storage and Composition

Our analysis above revealed a way in which the fragment grammars model differs from the two exemplar-based models (`DOP1` and `ENDOP`). While the exemplar-based models predicted that many words should be ambiguous with regard to decomposition and retrieval, fragments grammars predict a sharp dichotomy between these possibilities—individual words are nearly always decomposed or nearly always retrieved as wholes under the model. This difference between fragment grammars and the exemplar-based models is a consequence of the differing psychological assumptions underlying the two frameworks. Fragment grammars are intended to model the mental processes by which words are generated. Therefore, they assume that every word instance has a single, unique derivation, which represents the specific mental computations that were involved in producing or comprehending that word token. Furthermore, fragment grammars also make the assumption that there is a prior bias to store fewer items when possible. Thus, the model prefers to generate multiple instances of the same word using a single set of stored fragments. If a word is stored whole and represented by a single fragment, then all instances of that word will tend to be generated using that same fragment. If a word is composed using some particular set of fragments, then all instances of that word will tend to be composed using the same set of fragments.

By contrast, the exemplar-based models are not defined in terms of an explicit model of morphological computation that commits to a single derivation of each word token and that favors smaller lexica. Instead, they assume that every fragment consistent (in any way) with each input word is stored in memory, and that all possible derivations of a word participate simultaneously during computation. As a consequence, they predict that word forms will tend to be ambiguous with regards to decomposition and retrieval.

The tendency of fragment grammars to treat all tokens of a given word uniformly has a secondary consequence. Since all tokens of a word tend to be composed or retrieved, affixes that occur in those word tokens also tend to be uniformly composed, or uniformly stored together with their base. This, in turn, means that fragment grammars predict a sharper differentiation between productive and unproductive affixes than the exemplar-based models. Fragment grammars assign the highest probability to systems where affixes are either highly productive or highly unproductive. This bias in favor of disjoint sets of very productive and very unproductive morphemes seems to be crucial for explaining the sharp differences in productivity between derivational suffixes like *-ness* and *-th*, or the differences between regular and irregular inflectional processes in the English past tense.

7.3.3 Morpheme Combination and Ordering

In Section 6.3.7, we discussed the strong morpheme combination and ordering preferences exhibited by English derivational morphology and the fact that these preferences lead to a long-standing theoretical question know as the affix-ordering problem. In this section, we examine the ability of our five models to account for morpheme combination and affix-ordering phenomena.

The affix-ordering problem arises from the observation that morpheme sequences that occur in English words constitute a tiny subset of logically possible morpheme sequences. Our first analysis considers the ability of the models to predict the corpus frequencies of attested suffix pairs in the corpus of English suffix sequences compiled by Plag and Baayen (2009). In this analysis, the `MAG` (full-listing) model provides the best fit to the data, with the `FG` (inference-based) model coming a close second. We argue that this result is a consequence of the fact that the `MAG` (full-listing) model fully memorizes all words in the training set. By contrast, the `FG` (inference-based) model accounts for attested suffix pairs by a combination of both storage and composition. In our subsequent analyses, we therefore focus on composable structure in attested suffix sequences.

First, we demonstrate that the corpus of Plag and Baayen (2009) exhibits evidence of productive composition in the set of suffix pairs. We do this by showing that the `FG` (inference-based) model still predicts a large amount of the variance in the frequency of attested pairs, even when predictions are constrained to only consider suffix sequences in novel words. This contrasts with the `MAG` (full-listing) model, which fares much worse in this evaluation.

As we discussed in Section 6.3.7, one way in which models that mix productivity and reuse can generate suffix sequences is by composing productive suffixes with stored, morphologically complex bases. As highlighted by Hay and colleagues, this possibility leads to the productivity-and-ordering generalization. Since unproductive suffixes tend to occur as part of stored words, and productive suffixes can be composed with those stored words, more productive suffixes tend to occur after less productive suffixes in morphologically complex words. In our next analysis, we show that the `FG` (inference-based) model is strongly consistent with this generalization, predicting a correlation between the productivity of individual suffixes and the position in which they are expected to occur in morphologically complex words.

As we noted in Section 6.3.7, there are a large number of exceptions to the productivity-and-ordering generalization. One class of exceptions that has proven difficult to account for in earlier theories of affix ordering are paradoxical suffix combinations. These are cases where an unproductive suffix, such as *-ity*, can generalize productively when it follows certain other suffixes, such as *-able*. Generalizable combinations of suffixes, such as *-ability*, are predicted to be unlikely or impossible under several earlier theories of ordering such as the level-ordering generalization or Hay's complexity-based ordering (see, Section 6.3.7).

In our last set of analyses, we consider these generalizable combinations of suffixes. Because the `FG` (inference-based) model allows arbitrary subcomputations to be stored, it can learn that such combinations behave like single units, giving rise to derivations of the form [*stem* [*-sffx1 -sffx2*]]. We show that only the `FG` (inference-based) model is able to correctly predict the pattern of results for one pair of cases that has been studied experimentally (*-ivity*/*-iveness* versus *-ability*/*-ableness*). We give examples of other combinations of suffixes which are predicted to be generalizable by the `FG` (inference-based) model, and argue that these examples include the majority of difficult cases discussed in the literature. We conclude with a summary of how the `FG` (inference-based) model relates to earlier theories of morpheme combination and ordering.

7.3.3.1 Predicting the Frequency of Suffix Sequences

We first examine the ability of the five models to predict the pattern of attested English suffix sequences. A natural way to examine this question is to consider the degree to which the models are able to predict the frequency of different suffix combinations. Plag and Baayen (2009) studied the attested ordering of a set of thirty-one English suffixes in a corpus of English words. Their paper provides frequency information for each of the 961 possible pairs.

To predict these frequencies, we computed an approximation to the marginal probability of each suffix pair by summing the probabilities of all possible words consisting of a simple stem followed by the target pair of suffixes. For example, if the target suffixes were *-able* and *-ness*, we summed over all possible words of the form `stem` *-able -ness*. Note that for the models that allow a mixture of productivity and reuse, this quantity includes contributions from both stored word forms and from composed forms.

Model	`FG`	`DMPCFG`	`MAG`	`DOP1`	`ENDOP`
Correlation	0.682	0.367	**0.698**	0.456	0.422

Table 7.4

Correlation between Total Probability and Frequency of Suffix Pair: The Pearson correlation between the total (log) probability of each pair of suffixes and the (log) frequency of that pair as reported in Plag and Baayen (2009).

Table 7.4 shows the correlations between model predictions for each pair of suffixes and the (log) frequencies reported in Plag and Baayen (2009). Because most of the 961 possible suffix pairs are unattested in the Plag and Baayen (2009) sample—most for the principled reason that they are impossible—we have excluded frequency zero forms from this analysis.[9] The table demonstrates that the MAG (full-listing) model best predicts the frequencies of suffix pairs from the corpus. Nevertheless, the FG (inference-based) model also produces excellent predictions of corpus frequencies. The two exemplar-based models (i.e., DOP1 and ENDOP) do substantially worse, and the DMPCFG (full-parsing) model does extremely poorly.

As we demonstrated in Section 7.3.2.1, a majority of derived word forms in English are stored in memory. Furthermore, highly frequent, morphologically complex words account for a large part of the frequency of common suffix pairs. Therefore, the correlations just discussed are likely driven to a great extent by existing word forms. Because the MAG (full-listing) model obligatorily memorizes all words in the training set, it is able to provide very accurate predictions about pair frequency for these forms. The FG (inference-based) model also stores many morphologically complex words. However, it is also likely that under the FG (inference-based) model some of the variance in suffix-pair frequency was attributable to suffix pairs that were composed rather than fully stored. In our next analysis, we demonstrate that there is compositional structure in the set of suffix sequences.

7.3.3.2 Composition in the Set of Suffix Pairs

In order to examine whether composition accounts for the attestation of some suffix pairs in the data of Plag and Baayen (2009), we compute an alternative predictor of pair frequency. While our last analysis considered the total probability of each pair, here we compute the generalization probability of each pair. To do this, we sum over all ways in which a word can be formed from a novel, wug base and the pair of suffixes. For example, if our target suffixes were *-able* and *-ness*, we sum over words of the form *wug -able -ness*. Unlike the total probability, the generalization probability of a suffix pair does not include any contribution from stored word forms.

9. We are correlating log probabilities and frequencies. Since $\log(x)$ is undefined when $x = 0$ (i.e., $\lim_{x \to 0} \log(x) = -\infty$), it is customary to add 1 to frequency values in such analyses. However, since only 103 of the 931 possible pairs have non-zero frequencies in the corpus of Plag and Baayen (2009), doing this would highly distort the results. Of the 828 pairs with a frequency of 0, the majority are impossible. Adding 1 to the frequency of all 828 pairs would equate low-probability but possible pairs with impossible pairs. Furthermore, since the distribution of productively composed pairs is a large number of rare events distribution (Baayen, 2001; Khmaladze, 1987), even with large increases in sample size, only a small number of these pairs would be attested—and even then with low frequency.

Model	FG	DMPCFG	MAG	DOP1	ENDOP
Correlation	**0.527**	0.367	0.426	0.434	0.407

Table 7.5

Correlation of Generalization Probability and Frequency: The Pearson correlation between the generalization (log) probability of each pair of suffixes and the log frequency of that pair as reported in Plag and Baayen (2009).

Table 7.5 shows the correlations between model predictions and the logarithm of the pair frequencies from Plag and Baayen (2009), again excluding zero-frequency forms. The first thing to note about these correlations is that they are all somewhat lower than the correlations with total pair probability reported in Table 7.4. This is expected. Because the generalization probability does not include any contribution from stored forms, the variance in frequency between pairs due to morphologically complex, stored words cannot be predicted. Nevertheless, the FG (inference-based) model is still able to account for a large portion of the variance in suffix-pair frequencies. By contrast, the MAG (full-listing) model now does substantially worse. This result shows that some significant portion of the variance in pair frequency in the data set is the result of the composition of productive suffixes (and combinations of suffixes).

7.3.3.3 The Productivity and Ordering Generalization

In the last section, we showed that some of the suffix pairs attested in the corpus of Plag and Baayen (2009) are likely to be the result of productive composition. As we discussed above, the productive composition of individual suffixes with stored, morphologically complex bases leads to the productivity-and-ordering generalization: more productive suffixes tend to appear after less productive suffixes. In this section, we examine how consistent each of the five models is with the productivity-and-ordering generalization, showing that the FG (inference-based) model produces the most robust relationship between (single-suffix) productivity and ordering.

To examine the productivity-and-ordering generalization, we compared two quantities. First, we computed the log odds that each suffix appeared in the second suffix position in words with two suffixes. To compute this quantity, we considered the set of all possible word forms with exactly two suffixes (i.e., words of the form *stem -sffx1 -sffx2*). Based on this set, we computed the total (marginal) probability of each suffix appearing first or second in the suffix sequence. Using these values, we further calculated the log odds of each suffix appearing in the second suffix position. This value can be understood as the (log) number of times more likely it is that a particular suffix will appear in the second position rather than the first position (in forms with exactly two suffixes). We compare the log-odds that each suffix appears second with the estimates of $\Pi_{S,r}$ (absolute productivity) from Section 7.3.1.1. Table 7.6 shows the correlation between these two quantities for each of the five models.

Model	FG	DMPCFG	MAG	DOP1	ENDOP
Correlation	**0.323**	-0.023	0.173	-0.273	-0.064

Table 7.6

Correlation of Productivity of Suffixes and Ordering: The Pearson correlation between the (log) absolute productivity of a suffix ($\Pi_{S,r}$) and the log odds that it will appear second in a sequence of two suffixes.

Examining Table 7.6, we see that the FG (inference-based) model produces the most robust correlations between suffix order and productivity, and, thus, is most consistent with the productivity-and-ordering generalization. The MAG (full-listing) model also produces a positive, but much weaker correlation, while the other models produce correlations which contradict the empirical generalization that more productive suffixes tend to appear after less productive suffixes. It is especially notable that both of the exemplar-based models (DOP1 and ENDOP) predict a negative relationship between productivity and ordering, despite the fact that they allow both the storage of morphologically complex words and productive affixation—two crucial factors underlying the productivity-and-ordering generalization.

There are several different reasons that a suffix pair might be attested in an English word. First, some words containing two suffixes are simply stored as wholes in memory. This fact explains part of the difference in the ability of the FG (inference-based) model to predict the frequencies of suffix pairs using total probability—which includes contributions from stored forms—and generalization probability—which excludes contributions from stored forms—in the analyses in Sections 7.3.3.1 and 7.3.3.2. Second, some suffix pairs are the result of attaching a productive suffix to a stored base that ends in another (typically less productive) suffix. We have argued that this phenomenon is what gives rise to the fact that more productive suffixes tend to occur after less productive suffixes (i.e., the productivity-and-ordering generalization). In this section, by showing that the FG (inference-based) model is consistent with the productivity-and-ordering generalization, we have confirmed that it derives at least some suffix pairs in this way. However, the FG (inference-based) model also predicts a third way in which a suffix pair can be attested in English words: if the pair forms a single, generalizable unit. We consider this case in the next section.

7.3.3.4 Generalizable Combinations of Suffixes

As we discussed in Sections 6.3.7 and 6.4.3.3, a particularly difficult empirical phenomenon for some theories of ordering is paradoxical suffix combination. These are cases where an unproductive suffix, such as *-ity*, can be generalized productively when it occurs in the context of another (often more productive) suffix, such as *-able*. For theories that predict the order of affixes based on properties like morpheme productivity, composability (Hay's complexity-bases ordering) or phonology (stratal theories), combinations such as *-ability* represent a paradox. The outer suffix, *-ity*, which is generally less decomposable and more phonologically idiosyncratic, is predicted by these theories to occur before the inner suffix, *-able*, which is generally more decomposable and phonologically regular.

One solution to this problem is to reject the idea that order can be predicted (purely) on the basis of such properties, and, instead, assume that attested affix orders are simply the side-effect of a large number of affix-specific selectional restrictions. Under such selectional-restriction-based theories, the generalizability of suffix combinations like *-ability* can be modeled using morpheme-specific selectional restrictions that specify, for example, that *-ity* can attach either to one of a specific set of stems (e.g., *scarce*, *severe*, *senior*, etc.) or to any base ending in one of a specific set of suffixes (e.g., *-able*, *-ic*, *-al*; Hay and Plag, 2004; Plag, 1996, 2002, 2003).

We note two important points about such morpheme-specific selectional restrictions. First, while theories that allow these restrictions can model the generalizability of suffix combinations like *-ability*, they raise a significant learning problem. In principle, any morpheme might be restricted to co-occur with any other, and therefore the language learner must determine which of the many attested combinations of morphemes are so-restricted and which are not. Second, the use of morpheme-specific selectional restrictions is logically equivalent to the storage of fragments that contain an affix together with the morphemes it selects.[10] This second point suggests that the FG (inference-based) model may be a solution to the problem of learning morpheme-specific selectional restrictions.

In this section, we demonstrate that the FG (inference-based) model provides a viable theory of the learning of morpheme-specific selectional restrictions. We present two analyses in support of this conclusion. In the first analysis, we examine combinations of suffixes ending in *-ity*. There are several reasons why these suffix combinations are of special interest. First, *-ity* is common in English words, and, in particular, frequently appears after a number of other suffixes (e.g., *-ability*, *-osity*, *-anity*, *-icity*, *-ality*, *-ivity*). Second, these suffix combinations vary in their productivity; several can readily be used to form new words (e.g, *-ability*, *-icity*, *-ality*), while others are dispreferred in new words (e.g., *-ivity*, *-osity*). Third, *-ity* is closely matched in meaning with the highly productive suffix *-ness*. Since *-ness* can be used to form new words from any adjectival base, it is always available in cases where *-ity* generalizes (e.g., *googleability/googleableness*)—making the acceptability of the *-ity* forms more surprising. The similarity of *-ity* and *-ness* also means that it is possible to perform carefully controlled experiments comparing the generalizability of the two suffixes in different contexts. In a few cases, a preference for *-ity* or *-ness* has been demonstrated experimentally (*-ivity/-iveness* versus *-ability/-ableness* Anshen and Aronoff, 1981; Aronoff and Schvaneveldt, 1978; Cutler, 1980; Embick and Marantz, 2008). In the analysis in the next section, we demonstrate that only the FG (inference-based) model is able to capture the empirical pattern of generalization of *-ity* and *-ness* in combination with other suffixes. In our second analysis, we provide examples of other suffix combinations that are predicted to be generalizable by the FG (inference-based) model, including many difficult cases that have been discussed in the literature.

7.3.3.4.1 Generalizable Combinations of Suffixes Ending in –ity

In this section, we examine the differential generalizability of suffix combinations ending in *-ity*. Suffix combinations such as *-ability*, *-osity*, *-anity*, *-icity*, *-ality*, and *-ivity* are extremely common in existing English words. However,

10. More accurately, there is typically a translation from the fragment-based representation to the selectional-restriction-based representation, and vice versa. Of course, whether any two specific variants of the rule-/fragment-based representations are equivalent will depend on the exact details of each theory.

they vary in the readiness with which they give rise to new words. In general, given a novel adjective ending in a suffix such as *-ive* (e.g., *remortive*), English speakers prefer to form the corresponding abstract noun using *-ness* (*remortiveness* rather than *remortivity*; Aronoff and Schvaneveldt, 1978). This fact is unsurprising since, in general, *-ness* is far more productive than *-ity*. However, for novel adjectives ending in a small number of suffixes—including *-able*, *-ic*, and *-al*—speakers prefer to form the corresponding abstract noun using *-ity* (Anshen and Aronoff, 1981; Aronoff and Schvaneveldt, 1978; Cutler, 1980; Embick and Marantz, 2008; Hay, 2003; Hay and Plag, 2004; Plag, 1996, 2002, 2003).

In order to examine the ability of the five models to capture this pattern, we computed approximations to the probability that *-ity*/*-ness* suffix combinations generalize to form new words. This was done by computing the probability that each model assigned to words of the form [*wug -X -ity*] and [*wug -X -ness*], where *-X* was any suffix that appeared before *-ity* or *-ness* in the training corpus, and *wug* was a novel stem. Using these probabilities, we computed the log odds of the *-ity*-combination versus the *-ness*-combination for each preceding suffix. These values can be understood as measuring the preference of a model for the nominalization ending in *-ity* compared to the nominalization ending in *-ness* when a given novel base ends in another suffix (e.g., the preference for *remortivity* versus *remortiveness* when given *remortive*).

Figure 7.6 plots the preferences for *-ity* versus *-ness* on novel bases ending with the nine suffixes that most commonly preceded these morphemes in the training corpus. For comparison, we also include a bar showing the preference for *-ity* versus *-ness* on a (novel) unaffixed stem, labeled "none." Bars that extend to the left of the center line signify a preference for *-ity*. Bars that extend to the right of the center line signify a preference for *-ness*.

For most novel bases the FG (inference-based) model prefers *-ness* to *-ity*—consistent with the fact that *-ness* is generally a more productive suffix. In three cases—novel bases ending in *-ic*, *-able*, and *-al*—the FG (inference-based) model strongly prefers *-ity*, indicating that it has correctly identified that the suffix combinations *-icity*, *-ability*, and *-ality* productively generalize as single units. As we discussed above, these all represent instances of paradoxical suffix combinations, which have been discussed in the literature (Anshen and Aronoff, 1981; Aronoff and Schvaneveldt, 1978; Cutler, 1980; Embick and Marantz, 2008; Hay, 2003; Hay and Plag, 2004; Plag, 1996, 2002, 2003). Finally, for the case of novel bases ending in *-ous*, the FG (inference-based) model exhibits a slight preference for *-ity*. Aronoff and Schvaneveldt (1978)

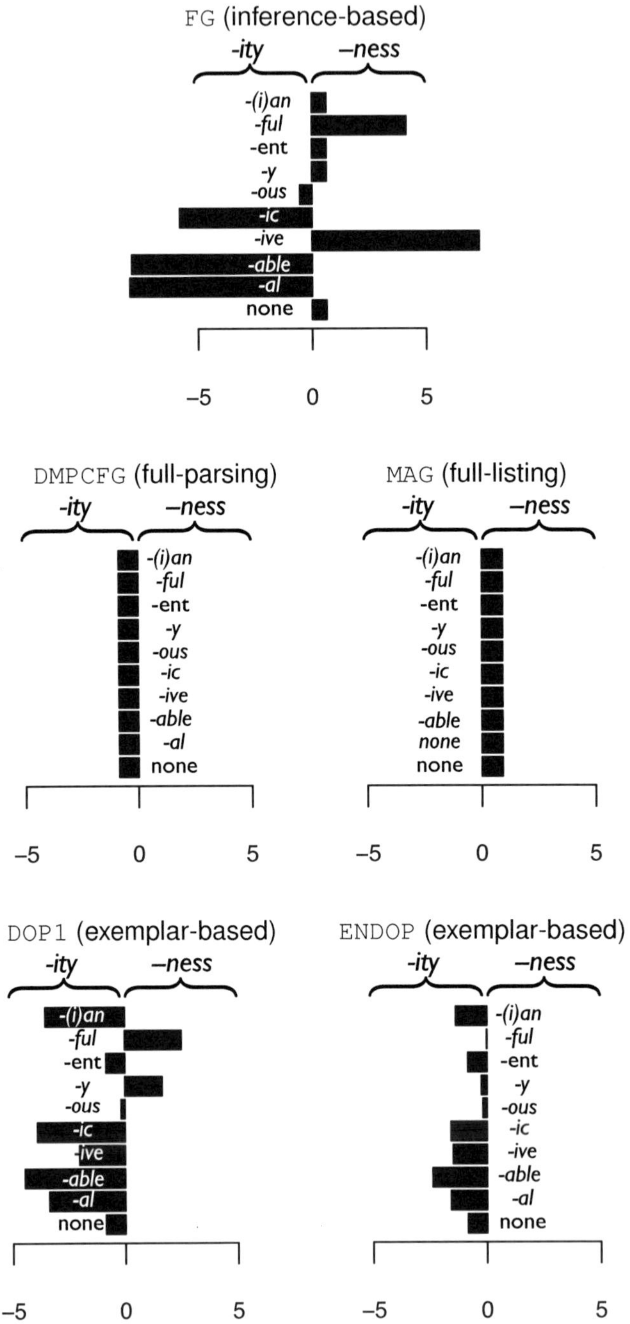

Figure 7.6

Preference for Suffix Combinations with *–ity* versus *–ness*: This plots shows the preferences of each model for *–ity* versus *–ness* on novel bases ending in the suffixes labeling each bar.

claim that *-ousness* is preferred to *-osity*, so this prediction may indicate a failure of the model. However, to our knowledge there is currently no experimental evidence testing or measuring the strength of *-osity/-ousness* preferences, and, therefore, we set this issue aside until the direction and strength of such preferences are measured experimentally.

Turning to the other models, neither the DMPCFG (full-parsing) model nor the MAG (full-listing) model exhibit any variation in preferences for *-ity* versus *-ness* in combination with other suffixes. The DMPCFG (full-parsing) model only stores minimal, abstract units, and estimates the probability of these units using their token frequencies. Since *-ity* is more frequent than *-ness* in words in the training corpus, the DMPCFG (full-parsing) model always exhibits a small preference for *-ity*. Similarly, the MAG (full-listing) model only stores complete derivation (sub)trees. When generating a novel form, it must use rules, such as N⟶Adj *-ness* or N⟶Adj *-ity*, from its underlying CFG. Because the probability of these rules is estimated using their type frequency across stored trees, and because *-ness* occurs in more words than *-ity* in the training corpus, the model exhibits a small overall preference for *-ness*.

The failure of the MAG (full-listing) model to capture the correct pattern of *-ity/ ness* generalization is of special importance. In several places in this book, we have seen that the MAG (full-listing) model provides reasonably accurate predictions about generalization for individual affixes. For example, it performed nearly as well as the FG (inference-based) model when generalizing the regular past-tense rule (see Section 5.3.2.1) and made reasonable predictions about individual suffix productivity in the analyses in Section 7.3.1. However, the ability of the MAG (full-listing) model to generalize correctly in these cases depended on the fact that (i) these cases concerned individual suffixes that could be represented using a single rule in the model's underlying CFG, and (ii) type frequency was a good approximation to the number of low-frequency forms in these datasets (see Section 6.3.4). However, because it cannot represent more complex generalizable structures, the MAG (full-listing) model is unable to learn that combinations of suffixes such as *-ability* are generalizable.

Finally, although the two exemplar-based models (DOP1 and ENDOP) exhibit some variation in preferences for *-ity* and *-ness*, they do not correctly distinguish paradoxical suffix combinations from other suffix combinations ending in *-ity*.

7.3.3.4.2 *Examples of Other Generalizable Combinations of Suffixes*

As we noted in Section 6.3.7, a number of other difficult, generalizable combinations of suffixes have been discussed in the literature, including *-ation*, *-istic*, and *-mental*. In this section, we present examples of the most generalizable suffix combinations of different lengths predicted by the `FG` (inference-based) model.[11]

To generate these examples, we performed a systematic search over all sequences of suffixes up to length four, computing the probability of the sequence appearing with a novel stem. Table 7.7 displays the suffix sequences of length two, three, and four which are predicted to be most generalizable by the `FG` (inference-based) model.

Table 7.7 includes a number of interesting examples. As we have mentioned, several of the combinations contained in the table—including *-ation*, *-istic*, and *-mental*—are paradoxical suffix combinations that have been mentioned in the literature. In fact, the table contains most of the combinations that have received extensive attention in the work reviewed in the last chapter. The suffix combination *-ation* is of particular interest. In Section 7.2.2, we discussed the decision on the part of the designers of the CELEX database to use a single suffix, *-ation*, to represent the combination *-ate⁀ion* in the analyses of words where the corresponding *-ate* form does not exist (e.g., *form/*formate/formation*). In order to provide the opportunity for our models to learn that this suffix combination should be stored as a single unit, we segmented *-ation* into *-ate* and *-ion* in our training corpus. Reassuringly, the `FG` (inference-based) model not only learned that this combination of suffixes should be stored as a single unit, but in fact predicts that *-ation* is the most productive combination of two suffixes in the English language.

Several other combinations in the table are interesting because they are pronounced in ways that are not strictly predictable from their parts. These include *-ify -ic* (i.e., *-ific*), *-ance -al*, (i.e., *-antial*), and *-ify -ate -ion* (*-ification*). The idiosyncratic pronunciation of these suffix combinations provides additional evidence that they may be stored as single units by native speakers of English. We emphasize, however, that the `FG` (inference-based) model was able to learn to store these combinations purely from their patterns of co-occurrence in

11. Because the `MAG` (full-listing) and `DMPCFG` (full-parsing) models cannot represent generalizable combinations of suffixes, they are not relevant in this analysis. In the case of the the two exemplar-based models (`DOP1` and `ENDOP`) performing the search described below is infeasible due to the massive size of the grammars of these models.

Length 2		Length 3		Length 4	
Sequence	Category	Sequence	Category	Sequence	Category
-ate -ion	N	*-ology -ic -al*	Adj	*-al -ize -ate -ion*	N
-ic -al	A	*-ate -ion -al*	Adj	*-ology -ic -al -ly*	Adv
-ate -ive	A	*-ment -ate -ion*	N	*-ist -ic -al -ly*	Adv
-al -ity	N	*-ify -ate -ion*	N	*-ist -ic -al -ly*	Adj
-al -ize	V	*-ize -ate -ion*	N	*-ology -ic -al -ness*	N
-ology -ist	N	*-ate -ion -al*	N	*-en -th -al -er*	N
-ment -al	A	*-ic -al -ly*	Adj	*-ate -ion -al -ly*	Adv
-able -ity	N	*-ate -ion -ary*	Adj	*-ity -ary -an -ism*	N
-ist -ic	A	*-al -ist -ic*	Adj	*-ology -ic -al -ity*	N
-ous -ity	N	*-ate -ion -ist*	N	*-ology -ic -al -ism*	N

Figure 7.7

Generalizable Sequences of Suffixes: This table shows the suffix sequences of length two, three, and four that the FG (inference-based) model predicts are most likely to occur in new word.

English words, without the aid of any phonological information, demonstrating again that these two sources of information often provide redundant predictions.

Many other intuitive combinations also appear on these lists. For example, the combinations *-ology*, *-ologist*, and *-ological* are devoted to terms having to do with various sciences. Predictions of the FG (inference-based) model like those in Table 7.7 should provide a rich source of hypotheses for future experimental work.

7.3.3.5 Summary and Discussion

In this section, we showed that, of the five models considered in this book, the FG (inference-based) model provides the best account of suffix combination and ordering in English derivational morphology. The model explains the pattern of possible English suffix sequences by a combination of three mechanisms: (i) the whole-form storage of morphologically complex words, which sometimes contain more than one suffix; (ii) the productive composition of individual suffixes with stored bases, which sometimes include other (typically less productive) suffixes—this mechanism explains the important empirical generalization that more productive suffixes tend to appear after less productive suffixes (the productivity-and-ordering generalization); (iii) the storage of generalizable combinations of suffixes such as *-ability*—explaining the existence of paradoxical suffix combinations, which have been problematic for earlier theories.

In Section 6.4.3, we discussed two approaches to affix-ordering introduced to repair empirical difficulties with classical, stratal theories: selectional-restriction based theories and Hay's complexity-based ordering. In the remainder of this section, we examine the relationship between the FG (inference-based) model and these theories in more detail.

7.3.3.5.1 Selectional-Restriction-Based Theories of Ordering

Selectional-restriction-based theories of affix ordering attempt to reduce order generalizations to a large number of affix-specific selectional restrictions on rules. Selectional restrictions come in several flavors. Some select for classes of bases that share some broad phonological or semantic property (e.g., monosyllabic or animate bases). As we have seen, however, restrictions also sometimes select for the presence of specific morphemes in bases (i.e., morpheme-specific selectional restrictions). From the point of view of learning, these two kinds of selectional restrictions are quite different. While general phonological or semantic properties can be identified from the phonological form and meaning of words, morpheme-specific selectional restrictions must be identified largely from the pattern of co-occurrence between morphemes in a language.

As we pointed out above, the stem and affix combinations represented by stored fragments under the fragment grammars model can be understood as a notational variant of the morpheme-specific selectional restrictions that annotate morphological rules in traditional linguistic theories. However, while these two systems are representationally equivalent, fragment grammars also provide a precise, mathematical theory of which combinations should be stored, how they should generalize, and how the learner might acquire this knowledge from the distribution of input forms. To our knowledge, such a theory of the learning of morpheme-specific selectional restrictions has never previously been offered in the literature. Fragment grammars address this theoretical gap.

7.3.3.5.2 Relationship to Complexity-Based Ordering

In Section 6.3.7, we discussed the important empirical generalization, observed by Hay and colleagues (Hay, 2002, 2003; Hay and Plag, 2004; Plag and Baayen, 2009), that more productive affixes tend to appear outside of less productive affixes (the productivity-and-ordering generalization). We have argued that the productivity-and-ordering generalization follows as a necessary consequence of the fact that words are often the result of the composition of productive affixes with stored bases containing less productive affixes. Hay (2002) offers a somewhat different account of this empirical fact—complexity-based ordering—which attempts to ground the productivity-and-ordering generalization in considerations of efficient lexical processing.

One important kind of evidence that has been adduced in favor of complexity-based ordering is the fact that many exceptions to the productivity-and-ordering generalization can be explained by the storage of words with high base-to-whole-form frequency ratios. For example, according to the productivity-and-ordering generalization, the suffix *-al* should tend to appear after the suffix *-ion* because the former is generally more productive than the latter. While *-al* appears after *-ion* in most words containing both suffixes (e.g., *confess**ional***), there are are a number of exceptional forms, such as *coloni**al**izat**ion***. Hay and Plag (2004) argued that such cases can be explained by Hay's relative frequency hypothesis. Since the base *colony* is less frequent than *colonial*, the latter is likely to be stored in memory under the relative frequency hypothesis. Words like *colonialization* are thus explained by the combination of this stored form with the productive suffix *-ize* and the productive suffix combination *-ation*.

However, this class of exceptions to the productivity-and-ordering generalization does not adjudicate between Hay's processing-based theory and our inference-based theory. As we discussed in Section 7.3.2.2, the FG (inference-based) model, like the relative frequency hypothesis, predicts storage for forms with a low base to whole-form frequency ratio. In fact, it goes beyond the relative frequency hypothesis, predicting a variety of frequency interactions between whole-words, part-words, and morphemes not even present in a word. Fragment grammars explain the exceptional order of *-al* and *-ion* in examples like *colonialization* in much the same way as complexity-based ordering, but without any appeal to lexical processing. In fact, a complete account of the order of *-al* and *-ion* in the word *colonialization* must appeal to the productivity of the suffix combination *-ation*. This form is a prototypical example of a paradoxical suffix combination that cannot be handled by complexity-based ordering, but whose generalizability is predicted by fragment grammars.

In short, the inference-based theory of productivity and reuse presented in this book—which makes no reference to processing efficiency—can provide an account for nearly all of the evidence that has been argued to favor the complexity-based ordering,[12] while explaining cases, such as paradoxical suffix combinations, which are difficult for that theory. Nevertheless, we are not aware of any direct evidence against a role for processing efficiency in theories of computation and storage. In fact, as we discussed in Section 1.2.3.1, language is replete with learning synergies—instances where several different kinds of evidence all conspire to make similar, redundant predictions about productivity and storage. Perhaps the mixture of productivity and reuse that best explains the distribution of forms in the input data is also the mixture that makes those forms easiest to process.

12. The one exception we are aware of is evidence for the differential storage of prefixes and suffixes (see, Hay, 2003). This appears to require the asymmetry inherent in the left-to-right processing of word forms.

7.4 Summary and Conclusion

In the preceding sections, we reported a series of analyses that tested the ability of the five computational models considered in this book to explain a number of phenomena in English derivational morphology. We first showed that only the fragment grammars model produces plausible predictions for the productivity of individual English suffixes. We furthermore showed that the single-suffix productivity predictions of fragment grammars converged with the predictions made by Baayen's hapax-based estimators $\mathcal{P}$ and $\mathcal{P}^*$. This finding shed light on the behavior of the fragment grammars model, showing that, in some cases, it behaves approximately like it is using the proportion of hapax forms to determine productivity. Furthermore, since fragment grammars is based on more psychologically plausible assumptions than Baayen's estimators, this result may also provide a theoretical justification for the use of $\mathcal{P}$ and $\mathcal{P}^*$ as approximate measures of productivity.

Next, we examined the predictions of the five models with regards to the decomposition and retrieval of existing words. We showed that fragment grammars make predictions consistent with Hay's relative frequency hypothesis—whether a word is decomposed or retrieved as a whole depends on the ratio of its part-word to whole-word frequencies. Furthermore, we argued that, in fact, such a relationship is necessary in any probabilistic model that (i) allows the storage of arbitrary fragments of structure in memory, (ii) defines the probability of a form as the product of the probabilities of the fragments used to generate it, and (iii) estimates the probability of fragments based on the frequency with which they are used. However, while fragment grammars are consistent with the relative frequency hypothesis, they go beyond it, making a much larger range of precise predictions about the interactions between words, word parts, and even morphemes that are not part of a word.

In our last set of analyses, we showed that fragment grammars provided the best explanation of the pattern of morpheme combination and ordering in English derivational morphology. The model produced predictions that were globally consistent with the empirical generalization that more productive affixes tend to appear outside of less productive affixes (productivity-and-ordering generalization), but was also able to account for exceptions to this principle. In particular, fragment grammars were the only model able to account for paradoxical suffix combinations, like *-ability*, that can generalize as single units. We argued that fragment grammars can therefore be understood as providing a theory of how the morpheme-specific selectional restrictions used in many linguistic theories can be learned from input data.

8 Conclusion

Linguistic creativity is made possible by a division of labor between an inventory of stored, reusable units and a set of computational processes that compose those units into an unbounded number of new expressions, allowing us to produce and comprehend previously unthought concepts and ideas with an ease that we usually take for granted. This book has addressed the deceptively simple-sounding questions, what are the stored units and which computational processes can give rise to new expressions?

Despite many years of intense research, the answers to these questions remain controversial. To a first approximation, it is clear that the inventory of stored items comprises (some representation of) morphemes, words, idioms, and, perhaps, other phrasal constructions. Larger structures are built on the fly using these primitives. However, which structures should be stored and which should be computed is not prima facie obvious from the linguistic evidence. Even in the two domains of word structure examined in this book—the English past tense, and English derivational morphology—it is far from trivial to draw a sharp boundary between storage and computation. English verbs can be inflected for pastness using a variety of processes that vary in whether they can rise to new forms, or can only be (re)used with certain sets of existing stems. These vary from the regular and productive +/d/ rule (e.g., *Google/Googled*), to semi-regular and semi-productive transformations such as /ɪ/ → /æ/ (e.g., *sing/sang*), to one-off, suppletive pairs of forms such as *go/went*. Similarly, English adjectives are turned into abstract nouns using one of several different suffixes: the productive and regular affix *-ness* (e.g., *coolness*), the contextually productive suffix *-ity* (e.g., *verticality*), and the unproductive *-th* (e.g., *warmth*).

Patterns such as this raise a fundamental problem. Whatever innate knowledge the child is born with, this knowledge does not include the inventory of primitive, stored units particular to their language. The child must determine which parts of input expressions correspond to reusable units that can be productively composed to generate new expressions, and which parts do not. This observation raises the fundamental questions that motivated the research in this book: what kind of computational system can account for the mixture of productivity and reuse found in natural language, and how can the child learn which units are productive and which are only reused as part of other forms?

In this book, we have proposed a precise, mathematical theory of productivity and reuse, and studied a specific instantiation of that theory: the fragment grammars model. The model is based on the idea that the correct pattern of storage and computation in a language can be determined (at least partially) by the result of a probabilistic inference based on the distribution of forms in the input data. We have shown that this idea, combined with standard assumptions from linguistics and psychology, leads to a model under which the productivity of a word-formation process is evidenced by the productive use of that process. Put another way, the model, like the linguist—and, by hypothesis, the child learner—concludes that a word-formation process is productive by observing it used to generate new forms. By contrast, the model infers the storage of morphologically complex forms by the repeated use of those forms. Again, this property of the model amounts to nothing more than the idea—standard in cognitive science—that more frequent forms are more likely to be memorized.

In Chapters 1–3, we discussed the psychological assumptions and mathematical underpinnings behind this model. We showed how the theory could be formalized using the tools of probabilistic generative modeling and probabilistic programming (Goodman et al., 2008; Johnson et al., 2007a). In particular, fragment grammars, a generalization of the adaptor grammars model (Johnson et al., 2007a), can be understood in terms of stochastic variants of two techniques from computer science: stochastic memoization and stochastically lazy evaluation. Furthermore, we showed that these tools can be generalized to allow the reuse of subcomputations in arbitrary probabilistic generative models. We also formalized three alternative approaches to productivity and reuse whose behavior we have compared and contrasted with fragment grammars throughout the book: the full-parsing approach (implemented by Dirichlet-Multinomial Context-Free Grammars, `DMPCFG`; Johnson et al., 2007b; Kurihara and Sato, 2006), the full-listing approach (implemented by Maximum a posteriori Adaptor Grammars, `MAG`; Johnson et al., 2007a), and the exemplar-based approach (implemented by two variants of data-oriented parsing, `DOP1` and `ENDOP`; Bod et al., 2003). Throughout our discussion, we have seen that the inference-based approach formalized by the fragment grammars model provides superior empirical coverage and a better explanation of the linguistic pattern of productivity and reuse than these other models.

In Chapters 4–7, we presented two detailed, empirical case studies of productivity and reuse in the domains of English verbal inflection and English derivational morphology. In those chapters, we reviewed a large number of empirical phenomena, theoretical models, and simulations results, and argued that the inference-based approach not only provided the best account of the phenomena we discussed, but also organized, explained, and unified a number of earlier proposals. These results and discussions can be categorized by considering seven correlated empirical properties of morphological processes: productivity, frequency, processing, combination and ordering, regularity, transparency, and selectivity.

Productivity, the ability to give rise to new word forms, varies widely between morphological processes in both English verbal inflection and English derivational morphology. We showed how fragment grammars can provide an explanation for both the sharp dichotomy between regular and irregular inflectional processes in English verbal morphology, and the difference between the absolute productivity of suffixes like *-ness* and the context-dependent productivity of suffixes like *-ity* (e.g., after *-able*) in derivational morphology. We argued that, in light of these results, fragment grammars can be understood as a theory for learning the morpheme-specific selectional restrictions commonly used in linguistic models.

Furthermore, we have emphasized throughout our discussion that the existence of cases where several word-formation processes can all apply to produce outputs with the same meaning, such as *splinged/splang* or *wugability/wugableness*, calls for a quantitative theory of competition which explains the degree of preference for different forms. We argued that both English verbal inflection and English derivational morphology resolve competition using the elsewhere condition, which states that a rule with more specific input conditions takes precedence over a rule with more general conditions, and showed how a generalized, probabilistic version of the elsewhere condition followed as a logically-necessary consequence of the assumptions of our model—without the need for further, ad hoc stipulations.

Despite these findings, we wish to emphasize that productivity in not an explicitly quantitative or scalar property in our model (in contrast, see, Bauer, 2001; Hay and Baayen, 2005). The fragment grammars model assumes that representations are either productive or they are not. Quantitative variation in productivity arises only as a result of the uncertainty facing the learner who is attempting to discover the reusable units in their language. For example, differences in preferences between *Googleability* and *Googleableness* are the result of ambiguity in the input concerning whether units like *-ity*, *-ness*, *-ability*, and *-ableness* can combine with arbitrary bases (whether existing or novel). In this way, our view is similar to traditional linguistic theories, where productivity is seen as an epiphenomenon of the discrete, recursive nature of the computational system underlying language.

Our theory also explains the (partial) correlations between productivity and other empirical properties of word-formation processes. Turning to frequency, we discussed the fact that productive morphemes tend to be associated with word-frequency distributions with a large number of low-frequency forms, while unproductive morphemes tend to be associated with word-frequency distributions with greater variance in frequency, often with a bias towards higher-frequency words. This property is a natural consequence of the fact that productive morphemes frequently give rise to new words. We have demonstrated that it is precisely this aspect of word-frequency distributions that the fragment grammars model uses to infer the productivity of individual affixes. Furthermore, we have argued that this suggests that the leading quantitative theory of productivity in the literature, Baayen's hapax-based approach (Baayen, 1992, 1993, 1994, 2001, 2003, 2006; Baayen and Lieber, 1991; Baayen and Renouf, 1996; Du and Zhang, 2010; Hay and Baayen, 2002; van Marle, 1992), can be understood as approximating the inference of the fragment grammars model.

Our model also explains the relationship between productivity and some well-known differences in the online processing of different words. Variance in processing latencies for high-frequency, morphologically complex words tends to be explained better by the whole-form frequency of the words, while variance between lower-frequency, morphologically complex forms tends to be explained better by the frequencies of word parts. In particular, the relationship between frequency and decomposition is codified by Hay's relative frequency hypothesis which states that whether a form is decomposed or retrieved whole from memory depends on the ratio a word's part-to-whole-form frequencies. We have shown that the relative frequency hypothesis must be true for any probabilistic model that makes assumptions similar to fragment grammars. However, while consistent with the relative frequency hypothesis, our approach differs from it in a number of respects. Most importantly, following Baayen (1993), we have argued that it is not surface frequencies that matter, but, rather, the frequencies with which structures have been used in previous instances of decomposition or retrieval—an inherently psychological quantity. Our approach also goes beyond the relative frequency hypothesis in predicting a large number of quantitative interactions between frequency and processing that are left unspecified by the relative frequency hypothesis.

Another empirical property associated with productivity is the distribution of morphemes with respect to combination and ordering. Essentially by definition, productive morphemes tend to appear attached to a greater variety of stems and bases than unproductive morphemes. In this book, we have shown that the fragment grammars model is able to learn the particular sets of stems and bases that can serve as the input for unproductive morphological processes such as irregular past tense generalizations (e.g., *sing/sang*) and unproductive derivational affixes (e.g., *warm/warmth*). Furthermore, differential preferences for various combinations of morphemes also lead to the important empirical observation, made by Hay and colleagues (Hay, 2002, 2003; Hay and Plag, 2004), that, in words containing more than one affix, more productive affixes tend to appear outside of less productive affixes—the productivity-and-ordering generalization. The fragment grammars model predicts the productivity-and-ordering generalization for a simple reason: all else being equal, a productive affix can attach to a stored, morphologically complex word containing a (generally) less productive affix as easily as to a simple stem. By contrast, a less productive affix is much less likely to attach to a stored base that contains a (typically) more productive affix. However, unlike earlier theories, fragment grammars are also able to provide a learning theory that accounts for an important exception to the productivity-and-ordering generalization: paradoxical suffix combinations. These are cases, such as *-ability*, where a typically unproductive suffix can be used productively when combined with another suffix (i.e., *-able*).

The remaining three empirical properties of word-formation processes—regularity, transparency, and selectivity—all concern phonological and/or semantic aspects of linguistic structure. In general, more productive affixes tend to be more phonologically regular, more semantically transparent, and to be associated with fewer categorical phonological and semantic selectional restrictions than less productive affixes.[1] However, because we did not explicitly model phonological or semantic structure in any of the simulations in this book, we were unable to make explicit quantitive predictions about how these properties influence the child's acquisition of the set of reusable units in their language.

However, we have argued that in most instances, the inclusion of phonological and semantic information will provide evidence which strengthens the inferences drawn by our model. For example, consider the suffix *-th* which occurs in words such as *warmth*, *width*, *sloth*, and *dearth*. The suffix cannot be used to form new words (**coolth*; productivity), occurs in a

1. Note that we are referring to selectional restrictions that make reference to broad phonological and semantics categories such as "monosyllabic" or "agentive," rather than morpheme-specific selectional restrictions, which are part of combination and ordering.

small number of forms with variable frequency (*truth* versus *dearth*; frequency), and that involve a limited set of stems (combination and ordering). Furthermore, in many instances, such as the word *sloth*, native speakers do not seem to recognize the presence of the suffix at all (processing).[2] The suffix also triggers a number of unpredictable phonological changes on bases (e.g., *young/youth*, *wide/width*; regularity), is often semantically opaque (e.g., ?*weal/wealth*, ?*dear/dearth*; transparency), and appears to only attach to monosyllables (selectivity).

In fact, several earlier theories of morphological learning have suggested that phonological and semantic information of this kind is a crucial determinant of productivity. For example, Marcus et al. (1992) and Marcus et al. (1995) proposed that a child might learn that the regular past tense ending +/d/ is productive by observing that it occurs with a phonologically heterogeneous set of stems (regularity/selectivity). This all suggests that the integration of phonological and semantic representations with the theory of productivity and reuse presented in this book should lead to an improvement in the ability of the model to correctly acquire morphological structure.

Finally, we wish to make some remarks about the generality of the theory presented in this book. There is nothing morphology-specific about the assumptions underlying our model of productivity and reuse. As we discussed in Section 2.4.4.1, there are several models from the natural language processing literature, developed simultaneously with the current framework, that can be understood as explicitly adopting the idea that productivity and reuse should be treated as a probabilistic inference (Cohn et al., 2010, 2009; Post and Gildea, 2009; Zuidema, 2007). These models have been evaluated primarily in the domain of syntax, on natural language processing parsing tasks. The performance of these models on these syntactic problems highlights the differences between the domains of morphology and syntax and the advantages of an inference-based approach in both. While inference-based models predict a high rate of storage of complex structures in morphology (see Section 7.3.2.1), they predict a higher rate of computation when applied to syntax (see, e.g., Cohn et al., 2010). This accords with the intuition that syntax is a domain where some storage of larger fragments is desirable—to capture collocational structure, idioms, and the like—but is inherently much more compositional than

2. Of course, this evidence of whole-form storage of the word *sloth* is anecdotal. But our experience with informal presentations of this material indicates that the morphologically complex nature of this word is surprising to most native speakers.

morphology. This highlights the advantage of the inference-based approach. In morphology, pervasive storage is appropriate, while in syntax more computation is appropriate. Only an inference-based approach to productivity and reuse can apply in both domains.

However, more exciting is the possibility that the approach to productivity and reuse advocated in this book could provide a domain-general theory of the integration of domain-specific knowledge and memory. In recent years, there has been a large amount of work showing how probabilistic generative models can explain the remarkable ways in which people can quickly and robustly generalize in the face of fragmentary and uncertain evidence, across many different areas of cognition. This work suggests that many domains of human knowledge may be built on the foundation of recursive, hierarchical computational systems—like that of language—that can capture an unbounded number of generalizations with finite means. However, it is also clear that the domain-specific knowledge possessed by humans is highly specialized to specific experiences. Most work on the psychology of memory has focused on the memorization of simple lists of items, rather than on how generative systems can be specialized by experience-specific input. However, our generalization of the framework in this book to arbitrary generative models (see Section 2.3.7), as well as recent work on memory using similar mathematical tools (Gershman, 2013), suggests that the ideas in this book may provide the first steps towards a domain-general theory of storage and computation.

A Past-Tense Inflectional Classes

In this appendix, we display the regular and irregular rules used for forms in the simulations presented in Chapter 5.

Tense	Stem Change	Suffix	Example
VBD	/u/ → /ɔ/	+/t/	*lost*
VBD	/ɪ/ → /æ/	∅	*rang*
VBD	/æ/ → /ʌ/	∅	*hung*
VBD	/u/ → /ɪ/	+/d/	*redid*
VBD	/i/ → /ɛ/	∅	*led*
VBD	/ej/ → /ʊ/	∅	*took*
VBD	/aj/ → /ɔ/	∅	*fought*
VBD	/ɪ/ → /ej/	∅	*gave*
VBD	/i/ → /ej/	∅	*ate*
VBD	go/went	go/went	*went*
VBD	/ʌ/ → /æ/	∅	*ran*
VBD	/itʃ/ → /ɔ/	+/t/	*taught*
VBD	/ɪŋ/ → /ɔ/	+/t/	*brought*
VBD	/oʊ/ → /u/	∅	*grew*
VBD	/aj/ → /ej/	∅	*lay*
VBD	/aj/ → /ɪ/	∅	*lit*
VBD	/ej/ → /ɛ/	+/d/	*said*
VBD	∅	∅	*beat*
VBD	/i/ → /ɔ/	∅	*foresaw*
VBD	/oʊ/ → /ɛ/	∅	*withheld*
VBD	/ɛ/ → /a/	∅	*got*
VBD	delete-final-consonant	+/d/	*made*
VBD	/ɪ/ → /ʌ/	∅	*won*

Table A.1

Inflectional Classes from SWBD Simulations

Tense	Stem Change	Suffix	Example
VBD	/aj/ → /u/	∅	*flew*
VBD	∅	+/d/	*experimented*
VBD	/æn/ → /ʊ/	∅	*understood*
VBD	/u/ → /a/	∅	*shot*
VBD	/a/ → /ɛ/	∅	*fell*
VBD	/ɛ/ → /oʊ/	+/d/	*sold*
VBD	/aj/ → /ɔ/	+/t/	*bought*
VBD	/aj/ → /aʊ/	∅	*wound*
VBD	/ʌ/ → /ej/	∅	*came*
VBD	/aj/ → /ʌ/	∅	*struck*
VBD	/ætʃ/ → /ɔ/	+/t/	*caught*
VBD	/i/ → /ɛ/	+/t/	*left*
VBD	/i/ → /ɛ/	+/d/	*heard*
VBD	/i/ → /oʊ/	∅	*stole*
VBD	/aj/ → /oʊ/	∅	*wrote*
VBD	/ɔ/ → /u/	∅	*drew*
VBD	/u/ → /oʊ/	∅	*chose*
VBD	/d/ → /t/	∅	*sent*
VBD	/ej/ → /oʊ/	∅	*woke*
VBN	/aj/ → /aʊ/	∅	*wound*
VBN	/ej/ → /oʊ/	+/n/	*worn*
VBN	/ɛ/ → /oʊ/	+/d/	*sold*
VBN	∅	+/d/	*incurred*
VBN	/ɛ/ → /a/	+/n/	*gotten*
VBN	/æn/ → /ʊ/	∅	*understood*
VBN	/ɪ/ → /ʌ/	∅	*won*
VBN	/oʊ/ → /ɛ/	∅	*withheld*
VBN	/itʃ/ → /ɔ/	+/t/	*taught*
VBN	/aj/ → /ɔ/	∅	*fought*
VBN	/i/ → /ɛ/	+/t/	*left*
VBN	/u/ → /a/	∅	*shot*

Table A.2
Inflectional Classes from SWBD Simulations

Tense	Stem Change	Suffix	Example
VBN	/æ/ → /ʌ/	∅	*hung*
VBN	/ætʃ/ → /ɔ/	+/t/	*caught*
VBN	/ej/ → /ɛ/	+/d/	*said*
VBN	/u/ → /oʊ/	+/n/	*chosen*
VBN	delete-final consonant	+/d/	*made*
VBN	/oʊ/ → /ɔ/	+/n/	*gone*
VBN	/aj/ → /ɪ/	+/n/	*ridden*
VBN	/ɪ/ → /æ/	∅	*sat*
VBN	/d/ → /t/	∅	*rebuilt*
VBN	/i/ → /ɛ/	∅	*led*
VBN	/i/ → /ɛ/	+/d/	*heard*
VBN	/aj/ → /oʊ/	+/n/	*flown*
VBN	/ɪŋ/ → /ɔ/	+/t/	*brought*
VBN	/i/ → /oʊ/	+/n/	*spoken*
VBN	/u/ → /ʌ/	+/n/	*overdone*
VBN	∅	+/n/	*drawn*
VBN	∅	∅	*run*
VBN	/aj/ → /ʌ/	∅	*struck*
VBN	/aj/ → /ɔ/	+/t/	*bought*
VBN	∅	+/t/	*burnt*
VBN	/u/ → /ɔ/	+/t/	*lost*
VB	∅	∅	*drive*
VBG	∅	+/ɪŋ/	*continuing*
VBP	∅	∅	*follow*
VBZ	/ej/ → /ɛ/	+/z/	*says*
VBZ	∅	+/z/	*favors*

Table A.3
Inflectional Classes from SWBD Simulations

B Derivational Suffixes

This Appendix contains a list of English derivational suffixes which appeared in at least 10 words types in the simulations reported in Chapter 7.

String Form	Input Category	Output Category
-able	BND	A
-able	N	A
-able	V	A
-age	N	N
-age	V	N
-al	A	A
-al	A	N
-al	BND	A
-al	BND	N
-al	N	A
-al	N	N
-al	V	A
-al	V	N
-an	A	A
-an	A	N
-an	BND	A
-an	BND	N
-an	N	A
-an	N	N
-ance	A	N
-ance	BND	N
-ance	V	N
-ant	BND	A

Table B.1
Derivational Suffixes

String Form	Input Category	Output Category
-ant	BND	N
-ant	N	A
-ant	V	A
-ant	V	N
-ary	BND	A
-ary	N	A
-ary	N	N
-ary	V	A
-ary	V	N
-ate	A	V
-ate	BND	V
-ate	N	A
-ate	N	N
-ate	N	V
-ate	V	A
-ate	V	V
-ation	N	N
-ation	V	N
-dom	A	N
-dom	N	N
-ed	N	A
-ee	N	N
-ee	V	N
-eer	N	N
-en	A	V
-en	BND	V
-en	N	A
-en	N	V
-en	V	A
-ence	A	N
-ence	BND	N
-ence	V	N
-ent	BND	A

Table B.2
Derivational Suffixes

String Form	Input Category	Output Category
-ent	N	A
-ent	V	A
-ent	V	N
-er	A	N
-er	BND	N
-er	N	N
-er	V	N
-ery	N	N
-ery	V	N
-ese	N	A
-ese	N	N
-ess	N	N
-ette	N	N
-ety	A	N
-fold	Q	A
ful	N	A
-ful	N	N
-ful	V	A
-hood	N	N
-ic	BND	A
-ic	BND	N
-ic	N	A
-ic	N	N
-ic	V	A
-ier	N	N
-ier	V	N
-ify	A	V
-ify	BND	V
-ify	N	V
-ion	A	N
-ion	BND	N

Table B.3
Derivational Suffixes

String Form	Input Category	Output Category
-ion	N	N
-ion	V	N
-ish	A	A
-ish	N	A
-ism	A	N
-ism	BND	N
-ism	N	N
-ist	A	N
-ist	BND	N
-ist	N	N
-ist	V	N
-itis	N	N
-ity	A	N
-ity	BND	N
-ity	N	N
-ity	V	N
-ive	A	A
-ive	BND	A
-ive	N	A
-ive	V	A
-ize	A	V
-ize	BND	V
-ize	N	V
-less	N	A
-less	V	A
-let	A	N
-let	N	N
-like	N	A

Table B.4
Derivational Suffixes

String Form	Input Category	Output Category
-ling	N	N
-ly	A	A
-ly	A	B
-ly	N	A
-ly	N	B
-ly	V	A
-ly	V	B
-ment	BND	N
-ment	N	N
-ment	V	N
-most	A	A
-ness	A	N
-ness	N	N
-oid	N	A
-oid	N	N
-ology	BND	N
-ology	N	N
-or	A	N
-or	BND	N
-or	N	N
-or	V	N
-ory	N	A
-ory	V	A
-ous	BND	A
-ous	N	A
-ous	V	A
-proof	N	A
-ry	N	N
-ry	V	N
-ship	N	N
-some	A	A
-some	N	A

Table B.5
Derivational Suffixes

String Form	Input Category	Output Category
-some	V	A
-ster	A	N
-ster	N	N
-th	A	N
-th	Q	Q
-th	V	N
-ty	A	N
-ty	Q	Q
-ward	B	A
-ward	N	A
-ward	N	B
-wise	N	B
-y	A	A
-y	A	N
-y	BND	N
-y	N	A
-y	N	N
-y	V	A
-y	V	N

Table B.6
Derivational Suffixes

Bibliography

Abelson, H. and Sussman, G. J. (1996). *The Structure and Interpretation of Computer Programs*. MIT Press, Cambridge, MA.

Albright, A. and Hayes, B. (2002). Modeling English past tense intuitions with minimal generalization. In *Proceedings of the 6th Meeting of the ACL Special Interest Group in Computational Phonology*.

Albright, A. and Hayes, B. (2003). Rules vs. analogy in English past tenses: A computational/-experimental study. *Cognition*, 90(2):119–161.

Alegre, M. and Gordon, P. (1999). Frequency effects and the representational status of regular inflections. *Journal of Memory and Language*, 40:41–61.

Allen, S. (1998). *Dumbth: The Lost Art of Thinking With 101 Ways to Reason Better & Improve Your Mind*. Prometheus Books, Amherst, NY.

Ambridge, B. (2010). Children's judgments of regular and irregular novel past-tense forms: New data on the English past-tense debate. *Developmental Psychology*, 46(6):1497–1504.

Anderson, J. R. (1990). *The Adaptive Character of Thought*. Lawrence Erlbaum Associates, Hillsdale, NJ.

Anderson, J. R. (1991). The place of cognitive architectures in a rational analysis. In Van Len, K., editor, *Architectures for Intelligence*. Lawrence Erlbaum Associates, Hillsdale, NJ.

Anderson, J. R. and Lebiere, C. (1998). *The Atomic Components of Thought*. Lawrence Erlbaum Associates, Mahwah, NJ.

Anderson, S. R. (1969). *West Scandinavian Vowel Systems*. PhD thesis, Massachusetts Institute of Technology.

Andrews, A. D. (1982). The representation of case in modern Icelandic. In Bresnam, J., editor, *The Mental Representation of Grammatical Relations*, pages 427–503. MIT Press, Cambridge, MA.

Andrews, A. D. (1990). Unification and morphological blocking. *Natural Language and Linguistic Theory*, 8(4):507–557.

Andrews, S. (1986). Morphological influences on lexical access: Lexical or nonlexical effects? *Journal of Memory and Language*, 25:726–740.

Angluin, D. (1980). Inductive inference of formal languages from positive data. *Information and Control*, 45:117–135.

Anshen, F. and Aronoff, M. (1981). Morphological productivity and phonological transparency. *Canadian Journal of Linguistics*, 26:63–72.

Anshen, F. and Aronoff, M. (1988). Producing morphologically complex words. *Linguistics*, 26:641–655.

Arnon, I. and Snider, N. (2010). More than words: speakers are sensitive to the frequency of multi-word sequences. *Journal of Memory and Language*, 62:67–82.

Aronoff, M. (1976). *Word Formation in Generative Grammar*. Linguistic Inquiry Monographs. MIT Press, Cambridge, MA.

Aronoff, M. (1983). Potential words, actual words, productivity and frequency. In *Proceedings of the XIIIth International Congress of Linguists*, pages 163–171, Tokyo.

Aronoff, M. and Fudeman, K. (2005). *What Is Morphology?* Blackwell Publishing, Malden, MA.

Aronoff, M. and Schvaneveldt, R. W. (1978). Testing morphological productivity. *Annals of the New York Academy of Sciences*, 318:106–114.

Aronoff, M. and Sridhar, S. N. (1983). Morphological levels in English and Kannada or atarizing Reagan. In Richardson, J. F., Marks, M., and Chukerman, A., editors, *Papers from the Parasession on the Interplay of Phonology, Morphology and Syntax*, pages 3–16, Chicago, IL.

Autebert, J.-M., Berstel, J., and Boasson, L. (1997). Context-free languages and pushdown automata. In *Handbook of Formal Languages*, volume 1, pages 111–172. Springer-Verlag, Berlin.

Baayen, R. H. (1992). Quantitative aspects of morphological productivity. In Booij, G. and van Marle, J., editors, *Yearbook of Morphology 1991*, pages 109–149. Kluwer, Dordrecht.

Baayen, R. H. (1993). Discussion: On frequency, transparency, and productivity. In *Yearbook of Morphology 1992*. Kluwer, Dordrecht.

Baayen, R. H. (1994). Productivity in language production. *Language and Cognitive Processes*, 9(3):447–469.

Baayen, R. H. (2001). *Word Frequency Distributions*. Springer, Berlin, Germany.

Baayen, R. H. (2003). Probabilistic approaches to morphology. In Bod, R., Hay, J., and Jannedy, S., editors, *Probabilistic Linguistics*. MIT Press, Cambridge, MA.

Baayen, R. H. (2006). Corpus linguistics in morphology: Morphological productivity. In Lüdeling, A. and Kytö, M., editors, *Corpus Linguistics: An International Handbook*. Mouton de Gruyter, The Hague.

Baayen, R. H., Burani, C., and Schreuder, R. (1997a). Effects of semantic markedness in the processing of regular nominal singulars and plurals in Italian. In Booij, G. and van Marle, J., editors, *Yearbook of Morphology 1996*. Kluwer, Dordrecht.

Baayen, R. H., Dijkstra, T., and Schreuder, R. (1997b). Singulars and plurals in Dutch: Evidence for a parallel dual-route model. *Journal of Memory and Language*, 37(1):94–117.

Baayen, R. H., Feldman, L. B., and Schreuder, R. (2006). Morphological influences on the recognition of monosyllabic monomorphemic words. *Journal of Memory and Language*, 55:290–313.

Baayen, R. H. and Lieber, R. (1991). Productivity and English derivation: A corpus-based study. *Linguistics*, 29:801–843.

Baayen, R. H. and Lieber, R. (1996). Word frequency distributions and lexical semantics. *Computers and the Humanities*, 30(4):281–291.

Baayen, R. H., McQueen, J. M., Dijkstra, T., and Schreuder, R. (2003). Frequency effects in regular inflectional morphology: Revisiting Dutch plurals. In Baayen, R. H. and Schreuder, R., editors, *Morphological Structure in Language Processing*. Mouton de Gruyter, The Hague.

Baayen, R. H., Piepenbrock, R., and van Rijn, H. (1993). The CELEX lexical database. Linguistic Data Consortium, University of Pennsylvania.

Baayen, R. H. and Renouf, A. (1996). Chronicling the Times: Productive lexical innovations in an English newspaper. *Language*, 72(1):69–96.

Baayen, R. H. and Schreuder, R. (1999). War and peace: Morphemes and full forms in a non-interactive activation parallel dual–route model. *Brain and Language*, 68:27–32.

Baayen, R. H. and Schreuder, R. (2000). Towards a psycholinguistic computational model for morphological parsing. *Philosophical Transactions of the Royal Society: Mathematical, Physical and Engineering Sciences*, 358(1769):1281–1293.

Baayen, R. H., Schreuder, R., and Sproat, R. (2000). Morphology in the mental lexicon: A computational model for visual word recognition. In *Lexicon Development for Speech and Language Processing*, pages 267–293. Kluwer, Dordrecht.

Baayen, R. H., Wurm, L. H., and Aycock, J. (2007). Lexical dynamics of low-frequency complex words: A regression study across tasks and modalitites. *The Mental Lexicon*, 2(3):419–463.

Balota, D. A., Law, M. B., and Zevin, J. D. (2000). The attentional control of lexical processing pathways: Reversing the word frequency effect. *Memory and Cognition*, 28(7):1081–1089.

Bandi-Rao, S. and Murphy, G. L. (2007). The role of meaning in past-tense inflection: Evidence from polysemy and denominal derivation. *Cognition*, 104:150–162.

Bannard, C. and Matthews, D. (2008). Stored word sequences in language learning: The effect of familiarity on children's repetition of four-word combinations. *Psychological Science*, 19(3):241–248.

Bar-Hillel, Y. (1953a). On recursive definitions in empirical science. In *Proceedings of the XIth International Congress of Philosophy*, volume 5–9, pages 160–165.

Bar-Hillel, Y. (1953b). A quasi-arithmetical notation for syntactic description. *Language*, 29(1):47–58.

Barnwell, B. (2010). Syntactic productivity in the transition to multi–word speech. Master's thesis, University of California Santa Barbara.

Barton, G. E., Berwick, R. C., and Ristad, E. S. (1987). *Computational Complexity and Natural Language*. MIT Press, Cambridge, MA.

Bauer, L. (1992). Scalar productivity and *-lily* adverbs. In *Yearbook of Morphology 1991*, pages 185–191. Kluwer, Dordrecht.

Bauer, L. (2001). *Morphological Productivity*. Cambridge University Press, Cambridge.

Bauer, L. (2005). Productivity: Theories. In *Handbook of Word-Formation*. Springer, Berlin.

Bauer, L., Lieber, R., and Plag, I. (2013). *The Oxford reference guide to English morphology*. Oxford University Press, Oxford.

Beck, M.-L. (1997). Regular verbs, past tense and frequency: tracking down a potential source of NS/NNS competence differences. *Second Language Research*, 13(2):93–115.

Beesley, K. R. and Karttunen, L. (2003). *Finite State Morphology*. CSLI Publications, Stanford, CA.

Berant, J., Caldwell-Harris, C., and Edelman, S. (2008). Tracks in the mind: Differential entrenchment of common and rare liturgical and everyday multiword phrases in religious and secular Hebrew speakers. In *Proceedings of the 30th Annual Meeting of the Cognitive Science Society*, Washington, DC.

Berko, J. (1958). The child's learning of English morphology. *Word*, 14:150–177.

Bernardo, J. M. and Smith, A. F. M. (1994). *Bayesian Theory*. John Wiley and Sons, Chichester.

Bertram, R., Baayen, R. H., and Schreuder, R. (2000a). Effects of family size for complex words. *Journal of Memory and Language*, 42:390–405.

Bertram, R., Laine, M., Baayen, R. H., Schreuder, R., and Hyönä, J. (2000b). Affixal homonymy triggers full-form storage even with inflcted words, even in a morphologically rich language. *Cognition*, 74:B13–B25.

Bertram, R., Laine, M., and Karvinen, K. (1999). The interplay of word formation type, affixal homonymy, and productivity in lexical processing: Evidence from a morphologically rich language. *Journal of Psycholinguistic Research*, 28(3):213–225.

Bertram, R., Schreuder, R., and Baayen, R. H. (2000c). The balance of storage and computation in morphological processing: The role of word formation type, affixal homonymy, and productivity. *Journal of Experimental Psychology: Learning, Memory and Cognition*, 26(2):489–511.

Berwick, R. C. (1985). *The Acquisition of Syntactic Knowledge*. The MIT Press, Cambridge, Massachusetts and London, England.

Biederman, I. (1987). Recognition-by-components: A theory of human image understanding. *Psychological Review*, 94(2):115–147.

Bird, S., Klein, E., and Loper, E. (2009). *Natural Language Processing with Python*. O'Reilly Media Inc., Sebastopol, CA.

Bishop, C. M. (1995). *Neural Networks for Pattern Recognition*. Oxford University Press, Oxford.

Bloom, L. (1970). *Language Development: Form and Function in Emerging Grammars*. MIT Press, Cambridge, MA.

Bloom, L. (1973). *One Word at a Time: The Use of Single-Word Utterances before Syntax*. Mouton, The Hague.

Bloomfield, L. (1933). *Language*. Henry Holt, New York.

Bobalijk, J. D. (2008). Paradigms (optimal and otherwise): A case for skepticism. In Bachrach, A. and Nevins, A., editors, *Inflectional Identity*. Oxford University Press, Oxford.

Bod, R. (1995). *Enriching Linguistics with Statistics: Performance Models of Natural Language*. PhD thesis, Universiteit van Amsterdam.

Bod, R. (1998). *Beyond Grammar: An Experience-Based Theory of Language*. CSLI Publications, Stanford, California.

Bod, R. (2001). Sentence memory: Storage vs. computation of frequent sentences. Talk presented at CUNY.

Bod, R. (2003). An efficient implementation of a new DOP model. In *Proceedings of the 10th Conference of the European Chapter of the Association for Computational Linguistics*, volume 1, pages 19–26, Budapest, Hungary.

Bod, R. (2006). An all-subtrees approach to unsupervised parsing. In *Proceedings of the 21st International Conference on Computational Linguistics and the 44th annual meeting of the Association for Computational Linguistics*, pages 865–872, Sydney, Australia.

Bod, R., Scha, R., and Sima'an, K., editors (2003). *Data-Oriented Parsing*. CSLI, Stanford, CA.

Bradley, D. C. (1978). *Computational Distinctions of Vocabulary Type*. PhD thesis, Massachusetts Institute of Technology, Cambridge, MA.

Bradley, D. C. (1980). Lexical representation of derivational relation. In *Juncture: A Collection of Original Papers*, pages 37–55. Anma Libri, Satatoga, California.

Braine, M. D. S. (1976). *Children's First Word Combinations*, volume 42 of *Monographs of the Society for Research in Child Development*. Blackwell Publishing, Malden, MA.

Bresnan, J. (1982). The passive in lexical theory. In *The Mental Representation of Grammatical Relations*, pages 3–86. The MIT Press, Cambridge, MA.

Bresnan, J. (2001). Explaining morphosyntactic competition. In *Handbook of Contemporary Syntactic Theory*. Blackwell, Oxford.

Brown, R. W. (1973). *A First Language: The Early Stages*. Harvard University Press, Cambridge, MA.

Burani, C. and Caramazza, A. (1987). Representation and processing of derived words. *Language and Cognitive Processes*, 2(3/4):217–227.

Burani, C., Salmaso, D., and Caramazza, A. (1984). Morphological structure and lexical access. *Visible Language*, 18(4):342–352.

Burzio, L. (2000). Missing players: Phonology and the past-tense debate. *Lingua*, 112(3):157–199.

Bybee, J. L. (1985). *Morphology: A Study of the Relation between Meaning and Form*. John Benjamins, Amsterdam.

Bybee, J. L. (1988). Morphology as lexical organization. In Hammond, M. and Noonan, M., editors, *Theoretical Morphology: Approaches in Modern Linguistics*, chapter 7, pages 119–141. Academic Press, San Diego.

Bybee, J. L. (1995). Regular morphology and the lexicon. *Language and Cognitive Processes*, 10(5):425–455.

Bybee, J. L. and Hopper, P. J. (2001). *Frequency and the Emergence of Linguistic Structure*. John Benjamins, Amsterdam.

Bybee, J. L. and Moder, C. L. (1983). Morphological classes as natural categories. *Language*, 59(2):251–270.

Bybee, J. L. and Slobin, D. I. (1982). Rules and schemas in the development and use of the English past tense. *Language*, 58(2):265–289.

Caramazza, A. (1997). How many levels of processing are there in lexical access? *Cognitive Neuropsychology*, 14(1):177–208.

Caramazza, A., Laudana, A., and Romani, C. (1988). Lexical access and inflectional morphology. *Cognition*, 28:297–332.

Carstairs-McCarthy, A. (2001). Paradigmatic structure: Inflectional paradigms and morphological classes. In *Handbook of Morphology*. Blackwell, Oxford.

Cazden, C. B. (1968). The acquisition of noun and verb inflections. *Child Development*, 39(2):433–448.

Chamberlain, A. F. (1906). Preterite-forms, Etc., in the language of English-speaking children. *Modern Language Notes*, 21(2):42–44.

Chen, S. and Goodman, J. (1998). An empirical study of smoothing techniques for language modeling. Technical Report TR-10-98, Harvard University.

Chialant, D. and Caramazza, A. (1995). Where is morphology and how is it processed? The case of written word recognition. In *Morphological Aspects of Language Processing*, pages 55–78. Lawrence Erlbaum Associates, Mahwah, NJ.

Chomsky, N. (1956). Three models for the description of language. *IEEE Transactions on Information Theory*, 2(3):113–124.

Chomsky, N. (1965). *Aspects of the Theory of Syntax*. The MIT Press, Cambridge, MA.

Chomsky, N. (1966). *Cartesian Linguistics*. Harper & Row.

Chomsky, N. (1970). Remarks on nominalization. In Jacobs, R. and Rosenbaum, P., editors, *Readings in English Transformational Grammar*. Ginn and Company, Waltham, MA.

Chomsky, N. (1975, 1955). *The Logical Structure of Linguistic Theory*. Plenum Press, New York, NY.

Chomsky, N. (1981). *Lectures on Government and Binding*. Foris, Dordrecht.

Chomsky, N. and Halle, M. (1968). *The Sound Pattern of English*. Harper & Row, New York, NY.

Church, A. (1932). A set of postulates for the foundation of logic. *The Annals of Mathematics*, 33(2):346–366.

Church, A. (1933). A set of postulates for the foundation of logic. *The Annals of Mathematics*, 34(4):839–864.

Clahsen, H., Hadler, M., and Weyerts, H. (2004). Speeded production of inflected words in children and adults. *Journal of Child Language*, 31:683–712.

Clark, E. V. (1978). Awareness of language: Some evidence from what children say and do. In Sinclair, R. J. A. and Levelt, W. J. M., editors, *The Child's Conception of Language*. Springer-Verlag, Berlin.

Cohn, T., Blunsom, P., and Goldwater, S. (2010). Inducing tree-substitution grammars. *Journal of Machine Learning Research*, 11:3053–3096.

Cohn, T., Goldwater, S., and Blunsom, P. (2009). Inducing compact but accurate tree-substitution grammars. In *Proceedings of Human Language Technologies: The 2009 Annual*

Conference of the North American Chapter of the Association for Computational Linguistics, Boulder, CO.

Colé, P., Beauvillain, C., and Segui, J. (1989). On the representation and processing of prefixed and suffixed derived words: A differential frequency effect. *Journal of Memory and Language*, 28:1–13.

Colé, P., Segui, J., and Taft, M. (1997). Words and morphemes as units for lexical access. *Journal of Memory and Language*, 37:312–330.

Collins, M. and Duffy, N. (2002). Convolution kernels for natural language. In *Advances in Neural Information Processing Systems*.

Conklin, K. and Schmitt, N. (2008). Formulaic sequences: Are they processed more quickly than nonformulaic language by native and nonnative speakers? *Applied Linguistics*, 29(1):72–89.

Coppock, E. (2007). Toward a true theory of the periphery: Why Culicover's "odd prepositions" aren't that odd. In *Proceedings of the 33rd Meeting of the Berkeley Linguistics Society*, Berkeley, CA.

Cox, R. T. (1946). Probability, frequency, and reasonable expectation. *American Journal of Physics*, 14(1):1–13.

Croft, W. (2001). *Radical Construction Grammar: Syntactic Theory in Typological Perspective*. Oxford University Press, Oxford, England.

Culicover, P. (1999). *Syntactic Nuts: Hard Cases in Syntax*. Foundations of Syntax. Oxford University Press, Oxford.

Culicover, P. and Jackendoff, R. (2005). *Simpler Syntax*. Oxford University Press, Oxford.

Cutler, A. (1980). Productivity in word formation. In *Papers from the Sixteenth Regional Meeting, Chicago Linguistic Society*, Chicago, IL.

Cutler, A., Hawkins, J. A., and Gilligan, G. (1985). The suffixing preference: A processing explanation. *Linguistics*, 23:723–758.

Daelemans, W. and van den Bosch, A. (2005). *Memory-Based Language Processing*. Cambridge University Press, Cambridge.

Daugherty, K., MacDonald, M. C., Petersen, A. S., and Seidenberg, M. S. (1993). Why no mere mortal has ever flown out to center field but people often say they do. In *Proceedings of the 15th Annual Conference of the Cognitive Science Society*, Bloomington, IN.

Daugherty, K. G. and Hare, M. L. (1993). What's in a rule? The past tense by some other name might be called a connectionist net. In *Proceedings of the 1993 Connectionist Models Summer School*, Boulder, CO.

Daugherty, K. G. and Seidenberg, M. S. (1992). Rules or connections? The past tense revisited. In *Proceedings of the 14th Annual Conference of the Cognitive Science Society*, pages 259–264.

de Finetti, B. (1974). *Theory of Probability*. Wiley, London.

De Marcken, C. (1996a). Linguistic structure as composition and perturbation. In *Proceedings of the 34th annual meeting on Association for Computational Linguistics*, pages 335–341. Association for Computational Linguistics.

De Marcken, C. (1996b). *Unsupervised Language Acquisition*. Dissertation, Massachusetts Institute of Technology.

de Vaan, L., Schreuder, R., and Baayen, R. H. (2007). Regular morphologically complex neologisms leave detectable traces in the mental lexicon. *The Mental Lexicon*, 2:1–23.

Demetras, M. (1989a). Changes in parents' conversational responses: A function of grammatical development. In *American Speech-Language-Hearing Association*, St. Louis, MO.

Demetras, M. (1989b). Working parents conversational responses to their two-year-old sons. Technical report, University of Arizona. Working Paper.

Demuth, K., Culbertson, J., and Alter, J. (2006). Word–minimality, epenthesis, and coda licensing in the acquisition of English. *Language and Speech*, 49:137–174.

Derwing, B. L. and Baker, W. J. (1979). Recent research on the acquisition of English morphology. In Fletcher, P. and Garman, M., editors, *Language Acquisition: Studies in First Language Development*, pages 209–223. Cambridge University Press, Cambridge.

Derwing, B. L. and Skousen, R. (1994). Productivity and the English past tense: Testing Skousen's analogy model. In *The Reality of Linguistic Rules*, pages 193–218. John Benjamins, Amsterdam.

Di Sciullo, A. M. and Williams, E. (1987). *On the Definition of Word*. MIT Press, Cambridge, MA.

Diez, F. C. (1838). *Grammatik der romanischen Sprachen*, volume 2. E. Weber, Bonn.

Domínguez, A., Cueto, F., and Segui, J. (2000). Morphological processing in word recognition: A review with particular reference to Spanish data. *Psicológica*, 21:375–401.

Du, L. and Zhang, X. (2010). A survey of the measurements of morphological productivity. *English Language Teaching*, 3(1):60–63.

Eddington, D. (2000). Analogy and the dual–route model of morphology. *Lingua*, 110:281–298.

Eliot, G. (1995, 1876). *Daniel Deronda*. Penguin Books, New York, NY.

Embick, D. and Marantz, A. (2008). Architecture and blocking. *Linguistic Inquiry*, 39(1):1–53.

Emonds, J. (1970). *Root and Structure Preserving Transformations*. PhD thesis, Massachusetts Institute of Technology, Cambridge, MA.

Ervin, S. M. (1964). Imitation and structural change in children's language. In *New Directions in the Study of Language*. MIT Press, Cambridge, MA.

Ervin, S. M. and Miller, W. R. (1963). Language development. In *Child Psychology: The Sixty Second Yearbook of the National Society for the Study of Education*. University of Chicago Press, Chicago.

Erwig, M. and Kollmansberger, S. (2006). Probabilistic functional programming in Haskell. *Journal of Functional Programming*, 16(1):21–34.

Evert, S. and Lüdeling, A. (2001). Measuring morphological productivity: Is automatic preprocessing sufficient? In *Proceedings of the Corpus Linguistics 2001 Conference*, Atlanta, GA.

Fabb, N. A. J. (1988). English suffixation is constrained only by selectional restrictions. *Natural Language and Linguistic Theory*, 6(4):527–539.

Feldman, L. B. and Soltano, E. G. (1999). Morphological priming: The role of prime duration, semantic transparency, and affix position. *Brain and Language*, 68:33–39.

Fellbaum, C. (1998). *Wordnet: An Electronic Lexical Database*. MIT Press, Cambridge, MA.

Fernández-Domínguez, J., Díaz-Negrillo, A., and Štekauer, P. (2007). How is low morphological productivity measured? *Atlantis*, 29(1):29–54.

Fillmore, C. J., Kay, P., and O'Connor, M. C. (1988). Regularity and idiomaticity in grammatical constructions: The case of let alone. *Language*, 64(3):501–538.

Flanagan, D. and Matsumoto, Y. (2008). *The Ruby Programming Language*. O'Reilly Media Inc.

Forrester, N. and Plunkett, K. (1994). The inflectional morphology of the Arabic broken plural: A connectionist account. In *Proceedings of The 16th Annual Conference of the Cognitive Science Society*.

Forster, K. I. (1976). Accessing the mental lexicon. In *New Approaches to Language Mechanisms: A Collection of Psycholinguistic Studies*. North-Holland Publishing Company, Amsterdam.

Forster, K. I. (1979). Levels of processing and the structure of the language processor. In Cooper, W. E. and Walker, E. C. T., editors, *Sentence Processing: Psycholinguistic Studies Presented to Merrill Garrett*, pages 27–81. Lawrence Erlbaum Associates, Hillsdale, NJ.

Forster, K. I. (1981). Frequency blocking and lexical access: One mental lexicon or two? *Journal of Verbal Learning and Verbal Behavior*, 20:190–203.

Forster, K. I. and Azuma, T. (2000). Masked priming for prefixed words with bound stems: Does *submit* prime *permit*? *Language and Cognitive Processes*, 15(4/5):539–561.

Forster, K. I. and Bednall, E. S. (1976). Terminating and exhaustive search in lexical access. *Memory and Cognition*, 4(1):53–61.

Fowler, C. A., Napps, S. E., and Feldman, L. B. (1985). Relations among regular and irregular morphologically related words in the lexicon as revealed by repetition priming. *Memory and Cognition*, 13(3):241–255.

Frauenfelder, U. H. and Schreuder, R. (1991). Constraining psycholinguistic models of morphological processing and representation: The role of productivity. In *Yearbook of Morphology 1991*, pages 165–183. Springer, Dordrecht, The Netherlands.

Freer, C. E. and Roy, D. M. (2009). Computable exchangeable sequences have computable de Finetti measures. In Ambos-Spies, K., Löwe, B., and Merkle, W., editors, *Mathematical Theory and Computational Practice: Fifth Conference on Computability in Europe, CiE 2009*, Heidelberg, Germany.

Freer, C. E., Roy, D. M., and Tenenbaum, J. B. (2012). Towards common-sense reasoning via conditional simulation: Legacies of Turing in Artificial Intelligence. In *Legacies of Alan Turing*. Cambridge University Press, Cambridge.

Gaeta, L. and Ricca, D. (2006). Productivity in Italian word formation: A variable-corpus approach. *Linguistics*, 44(1):57–89.

Gale, W. A. and Sampson, G. (1995). Good-Turing frequency estimation without tears. *Journal of Quantitative Linguistics*, 2(3):217–237.

Gelman, A., Carlin, J. B., Stern, H. S., and Rubin, D. B. (2003). *Bayesian Data Analysis, Second Edition*. Chapman & Hall/CRC, Boca Raton, FL.

Gershman, S. J. (2013). *Memory Modification in the Brain: Computational and Experimental Investigations*. PhD thesis, Princeton University.

Giegerich, H. J. (1999). *Lexical Strata in English: Morphological Causes, Phonological Effects*. Cambridge University Press, Cambridge.

Gillespie, J. H. (2004). *Population Genetics: A Concise Guide*. Johns Hopkins University Press, Baltimore.

Gleason, J. B. (1980). Parent-child interaction and the acquisition of lexical information during play. *Developmental Psychology*, 16:404–409.

Godfrey, J., Holliman, E., and McDaniel, J. (1992). Switchboard: Telephone speech corpus for research and development. In *IEEE International Conference on Acoustics, Speech, and Signal Processing (ICASSP)*, pages 517–520.

Goldberg, A. E. (2006). *Constructions at Work*. Oxford University Press, Oxford.

Goldwater, S., Griffiths, T. L., and Johnson, M. (2006). Interpolating between types and tokens by estimating power-law generators. In *Advances in Neural Information Processing Systems 18*.

Goldwater, S., Griffiths, T. L., and Johnson, M. (2011). Producing power-law distributions and damping word frequencies with two-stage language models. *Journal of Machine Learning Research*, 12:2335–2382.

Goodman, J. (1998). *Parsing Inside-Out*. PhD thesis, Harvard University.

Goodman, J. (2003). Efficient parsing of DOP with PCFG-reductions. In *Data-Oriented Parsing*. CSLI Publications, Stanford, CA.

Goodman, N. D., Mansinghka, V. K., Roy, D., Bonawitz, K., and Tenenbaum, J. B. (2008). Church: A language for generative models. In *Uncertainty in Artificial Intelligence*, Helsinki, Finland.

Gordon, P. and Alegre, M. (1999). Is there a dual system for regular inflections? *Brain and Language*, 68:212–217.

Gordon, P. and Miozzo, M. (2008). Can word formation be *understood* or *understanded* by semantics alone? *Cognitive Psychology*, 56:30–72.

Griffiths, T. L., Lieder, F., and Goodman, N. D. (2015). Rational use of cognitive resources: Levels of analysis between the computational and the algorithmic. *Topics in Cognitive Science*, 7(2):217–299.

Gropen, J., Pinker, S., Hollander, M., Goldberg, R., and Wilson, R. (1989). The learnability and acquisition of the dative alternation in English. *Language*, 65(2):203–257.

Grünwald, P. D. (2007). *The Minimum Description Length Principle*. The MIT Press, Cambridge, MA.

Gurevich, O. and Goldberg, A. E. (2010). Incidental verbatim memory for language. *Language and Cognition*, 2(1):45–78.

Halle, M. (1973). Prolegomena to a theory of word formation. *Linguistic Inquiry*, 4(1):3–16.

Halle, M. (1997). Distributed morphology: Impoverishment and fission. *MIT Working Papers in Linguistics 30: Papers at the Interface*, 30:425–449.

Halle, M. and Marantz, A. (1993). Distributed morphology and the pieces of inflection. In Hale, K. L. and Keyser, S. J., editors, *The View from Building 20*, pages 111–176. MIT Press, Cambridge, MA.

Halle, M. and Mohanan, K. P. (1985). Segmental phonology of modern English. *Linguistic Inquiry*, 16:57–116.

Halpern, J. Y. (1999). Cox's theorem revisited. *Journal of Artificial Intelligence Research*, 11:429–435.

Hare, M. L., Elman, J. L., and Daugherty, K. G. (1995). Default generalization in connectionist networks. *Language and Cognitive Processes*, 10(6):601–630.

Hay, J. (2001). Lexical frequency in morphology: Is everything relative? *Linguistics*, 39(6):1041–1070.

Hay, J. (2002). From speech perception to morphology: Affix ordering revisited. *Language*, 78(3):527–555.

Hay, J. (2003). *Causes and Consequences of Word Structure*. Routledge, New York, NY.

Hay, J. and Baayen, R. H. (2002). Parsing and productivity. In *Yearbook of Morphology 2001*, volume 35, pages 203–236. Springer, Dordrecht, The Netherlands.

Hay, J. and Baayen, R. H. (2005). Shifting paradigms: Gradient structure in morphology. *Trends in Cognitive Sciences*, 9(7):342–348.

Hay, J. and Plag, I. (2004). What constrains possible suffix combinations? On the interaction of grammatical and processing restrictions in derivational morphology. *Natural Language and Linguistic Theory*, 22:565–596.

Higginson, R. P. (1985). *Fixing-Assimilation in Language Acquisition*. PhD thesis, Washington State University.

Hindley, J. R. and Seldin, J. P. (2008). *Lambda-Calculus and Combinators, an Introduction*. Cambridge University Press, Cambridge.

Hockett, C. F. (1954). Two models of grammatical description. *Word*, 10:210–231.

Hockett, C. F. (1955). *A Manual of Phonology (International Journal of American Linguistics, Memoir 11)*. Waverly Press, Baltimore.

Hoeffner, J. (1992). Are rules a thing of the past? The acquisition of verbal morphology by an attractor network. In *Proceedings of the 14th Annual Conference of the Cognitive Science Society*, pages 861–866.

Hoeffner, J. (1996). *Are Rules a Thing of the Past? A Single Mechanism Account of English Past Tense Acquisition and Processing*. PhD thesis, Carnegie Mellon University.

Horn, L. R. (1984). Toward a new taxonomy for pragmatic inference: Q–based and R–based implicature. In *Meaning Form and Use in Context: Linguistic Applications (Georgetown University Round Table on Languages and Linguistics)*, pages 11–42, New York, NY.

Huang, Y. T. and Pinker, S. (2010). Lexical semantics and irregular inflection. *Language and Cognitive Processes*, 25(10):1411–1461.

Huybregts, R. (1985). The weak inadequacy of context-free phrase structure grammars. In de Haan, G. J., Trommelen, M., and Zonneveld, W., editors, *Van Periferie naar Kern*, pages 81–99. Foris Publications, Dordrecht.

Jackendoff, R. (1975). Morphological and semantic regularities in the lexicon. *Language*, 51(3):639–671.

Jackendoff, R. (2002a). *Foundations of Language*. Oxford University Press, New York.

Jackendoff, R. (2002b). What's in the lexicon? In Nooteboom, S., Weerman, F., and Wijnen, F., editors, *Storage and Computation in the Language Faculty*. Kluwer Academic Press, Dordrecht.

Jaynes, E. T. (2003). *Probability Theory: The Logic of Science*. Cambridge University Press, Cambridge.

Johnson, M. (1995). Memoization in top-down parsing. *Computational Linguistics*, 21(3):405–417.

Johnson, M. (2002). The DOP estimation method is biased and inconsistent. *Computational Linguistics*, 28(1):71–76.

Johnson, M. (2008a). Unsupervised word segmentation for Sesotho using Adaptor Grammars. In *The Tenth Workshop of the Association for Computational Linguistics' Special Interest Group in Morphology and Phonology*, Columbus, OH.

Johnson, M. (2008b). Using Adaptor Grammars to identify synergies in the unsupervised acquisition of linguistic structure. In *Proceedings of the 46th Annual Meeting of the Association for Computational Linguistics*, Columbus, OH.

Johnson, M. and Goldwater, S. (2009). Improving nonparameteric Bayesian inference: Experiments on unsupervised word segmentation with Adaptor Grammars. In *Human Language Technologies: The 2009 Annual Conference of the North American Chapter of the ACL*, pages 317–325, Boulder, CO.

Johnson, M., Griffiths, T. L., and Goldwater, S. (2007a). Adaptor Grammars: A framework for specifying compositional nonparametric Bayesian models. In *Advances in Neural Information Processing Systems 19*, Cambridge, MA. MIT Press.

Johnson, M., Griffiths, T. L., and Goldwater, S. (2007b). Bayesian inference for PCFGs via Markov chain Monte Carlo. In *Proceedings of the North American Conference on Computational Linguistics*, Rochester, New York.

Joshi, A. K. and Schabes, Y. (1997). Tree-adjoining grammars. In *Handbook of Formal Languages*, volume 3, pages 69–124. Springer, Berlin.

Joshi, A. K., Vijay-Shanker, K., and Weir, D. (1991). The convergence of mildly context-sensitive grammar formalisms. In Sells, P., Shieber, S. M., and Wasow, T., editors, *Foundational Issues in Natural Langauge Processing*. The MIT Press, Cambridge, Ma.

Jurafsky, D. and Martin, J. H. (2000). *Speech and Language Processing: An Introduction to Natural Language Processing, Computational Linguistics, and Speech Recognition*. MIT Press, Cambridge, MA.

Katz, L., Rexer, K., and Lukatela, G. (1991). The processing of inflected words. *Psychological Research*, 53:25–32.

Kay, P. (2002). An informal sketch of a formal architecture for construction grammar. *Grammars*, 5(1):1–19.

Kay, P. and Fillmore, C. J. (1999). Grammatical constructions and linguistic generalizations: The what's X doing Y? construction. *Language*, 75(1):1–33.

Kempley, S. T. and Morton, J. (1982). The effects of priming with regularly and irregularly related words in auditory word recognition. *British Journal of Psychology*, 73:441–454.

Khmaladze, E. V. (1987). The statistical analysis of large numbers of rare events. Technical Report MS-R8804, Department of Mathematical Statistics, CWI, Amsterdam, The Netherlands.

Kielar, A., Joanisse, M. F., and Hare, M. L. (2008). Priming English past tense verbs: Rules or statistics? *Journal of Memory and Language*, 58:327–346.

Kim, J. J., Marcus, G. F., Pinker, S., Hollander, M., and Coppola, M. (1994). Sensitivity of children's inflection to grammatical structure. *Journal of Child Language*, 21:173–209.

Kim, J. J., Pinker, S., Prince, A., and Prasada, S. (1991). Why no mere mortal has ever flown out to center field. *Cognitive Science*, 15:173–218.

Kiparsky, P. (1973). "Elsewhere" in phonology. In Anderson, S. R. and Halle, M., editors, *A Festschrift for Morris Halle*. Holt, Rinehart & Winston of Canada Ltd.

Kiparsky, P. (1982a). *Explanation in Phonology*. Foris, Dorderecht.

Kiparsky, P. (1982b). From cyclic phonology to lexical phonology. In van der Hulst, H. and Smith, N., editors, *The Structure of Phonological Representations*, volume 1, pages 131–175. Foris, Dorderecht.

Kiparsky, P. (1982c). Word-formation and the lexicon. In *Proceedings of the 1982 Mid-America Linguistic Conference*.

Kiparsky, P. (1985). Some consequences of lexical morphology. *Phonology*, 2(1):85–138.

Kiparsky, P. (2005). Blocking and periphrasis in inflectional paradigms. In *Yearbook of Morphology 2004*, pages 113–135. Springer, Berlin.

Kiselyov, O. and Shan, C.-c. (2009). Embedded probabilistic programming. *Lecture Notes In Computer Science*, 5658:360–384.

Kuczaj, S. A. (1977). The acquisition of regular and irregular past tense forms. *Journal of Verbal Learning and Verbal Behavior*, 16:589–600.

Kuperman, V. (2008). *Lexical Processing of Morphologically Complex Words: An Information-Theoretical Perspective*. PhD thesis, Universiteit Nijmegen.

Kuperman, V., Bertram, R., and Baayen, R. H. (2010). Processing trade-offs in the reading of Dutch derived words. *Journal of Memory and Language*, 62:83–97.

Kurihara, K. and Sato, T. (2006). Variational bayesian grammar induction for natural language. In *The 8th International Colloquium on Grammatical Inference (ICGI-2006)*.

Lachter, J. and Bever, T. G. (1988). The relation between linguistic structure and associative theories of language learning—A constructive crtitique of some connectionist learning models. *Cognition*, 28:195–247.

Lambek, J. (1958). The mathematics of sentence structure. *The American Mathematical Monthly*, 65(3):154–170.

Laplace, Simon-Pierre, M. d. (1951, 1814). *A Philosophical Essay on Probabilities*. Dover Publications, New York, New York.

Laudana, A. and Burani, C. (1985). Address mechanisms to decoposed lexical entries. *Linguistics*, 23:775–792.

Levy, R. (2008). Expectation-based syntactic comprehension. *Cognition*, 106:1126–1177.

Lewis, R. L., Howes, A., and Singh, S. (2014). Computational rationality: Linking mechanism and behavior through bounded utility maximization. *Topics in Cognitive Science*, 6(2):279–311.

Liang, P., Jordan, M. I., and Klein, D. (2010). Type-based MCMC. In *Human Language Technologies: The Annual Conference of the North American Chapter of the Association for Computational Linguistics*, Upsala, Sweden.

Lieber, R. (1980). *On the Organization of the Lexicon*. PhD thesis, Massachusetts Institute of Technology, Cambridge, MA.

Lieberman, E., Michel, J.-B., Jackson, J., Tang, T., and Nowak, M. A. (2007). Quantifying the evolutionary dynamics of language. *Nature*, 449(7163):713–716.

Ling, C. X. and Marinov, M. (1993). Answering the connectionist challenge: A symbolic model of learning the past tenses of English verbs. *Cognition*, 49:235–290.

MacKay, D. G. (1978). Derivational rules and the internal lexicon. *Journal of Verbal Learning and Verbal Behavior*, 17:61–71.

MacKay, D. G. (1979). Lexical insertion, inflection, and derivation: Creative processes in word production. *Journal of Psycholinguistic Research*, 8(5):477–498.

MacKay, D. J. C. (2003). *Information Theory, Inference, and Learning Algorithms*. Cambridge University Press.

MacWhinney, B. (1978). *The Acquisition of Morphophonology*, volume 43 of *Monographs of the Society for Research in Child Development*. Blackwell Publishing.

MacWhinney, B. (2000). *The CHILDES Project: Tools for Analyzing Talk*. Lawrence Erlbaum Associates, Mahwah, NJ.

MacWhinney, B. (2008). Enriching CHILDES for morphosyntactic analysis. In *Corpora in Language Acquisition Research: History, Methods, Perspectives*, volume 6, pages 165–98.

MacWhinney, B. and Leinbach, J. (1991). Implementations are conceptualizations: Revising the verb learning model. *Cognition*, 40:121–157.

Mahowald, K., Tenenbaum, J. B., and O'Donnell, T. J. (under review). Quantifying availability through linguistic productivity.

Manelis, L. and Tharp, D. A. (1977). The processing of affixed words. *Memory and Cognition*, 5(6):690–695.

Manning, C. D. and Schütze, H. (1999). *Foundations of Statistical Natural Language Processing*. MIT Press, Cambridge, MA.

Manova, S. (2010). Suffix combinations in Bulgarian: Parsability and hierarchy-based ordering. *Morphology*, 20:267–296.

Manova, S. and Aronoff, M. (2010). Modeling affix order. *Morphology*, 20:109–131.

Mansinghka, V. K. (2009). *Natively Probabilistic Computation*. PhD thesis, MIT.

Manzini, R. and Wexler, K. (1987). Parameters, binding theory, and learnability. *Linguistic Inquiry*, 18(3):413–444.

Marantz, A. (1997). No escape from syntax: Don't try morphological analysis in the privacy of your own lexicon. *University of Pennsylvania Working Papers in Linguistics*, 4(2):201–225.

Maratsos, M. (2000). More overregularizations after all: New data and discussion on Marcus, Pinker, Ullman, Hollander, Rosen, & Xu. *Journal of Child Language*, 27:183–212.

Marcus, G. F. (1995). The acquisition of the English past tense in children and multilayered connectionist networks. *Cognition*, 56:271–279.

Marcus, G. F. (2001). *The Algebraic Mind: Integrating Connectionism and Cognitive Science*. MIT Press, Cambridge, MA.

Marcus, G. F., Brinkmann, U., Clahsen, H., Wiese, R., and Pinker, S. (1995). German inflection: The exception that proves the rule. *Cognitive Psychology*, 29:189–256.

Marcus, G. F., Pinker, S., Ullman, M. T., Hollander, M., Rosen, T. J., and Xu, F. (1992). *Overregularization in Language Acquisition*. Monographs of the society for research in child development. University of Chicago Press, Chicago, IL.

Marcus, M. P., Santorini, B., Marcinkiewicz, M. A., and Taylor, A. (1999). Treebank-3. Technical report, Linguistic Data Consortium, Philadelphia.

Marr, D. (1982). *Vision*. Henry Holt and Co., New York, NY.

Marslen-Wilson, W. D. and Tyler, L. K. (1997). Dissociating types of mental computation. *Nature*, 387:592–594.

Marslen-Wilson, W. D., Tyler, L. K., Waksler, R., and Older, L. (1994). Morphology and meaning in the English mental lexicon. *Psychological Review*, 101(1):3–33.

Marslen-Wilson, W. D., Zhou, X., and Ford, M. (1997). Morphology, modality and lexical architecture. In Booij, G. and van Marle, J., editors, *Yearbook of Morphology 1996*, pages 117–134. Kluwer Academic Publishers, Dordrecht.

McCarthy, J. (1960). Recursive functions of symbolic expressions and their computation by machine, part 1. *Communications of the Association for Computing Machinery*, 3(4):184–195.

McCawley, J. (1978). Conversational implicature and the lexicon. In Cole, P., editor, *Syntax and Semantics 9: Pragmatics*, pages 245–259. Academic Press, New York.

McCloskey, J. and Hale, K. L. (1983). On the syntax of person-number inflection in modern irish. *Natural Language and Linguistic Theory*, 1(4):487–533.

Meijer, E., Fokkinga, M., and Paterson, R. (1991). Functional programming with bananas, lenses, envelopes and barbed wire. *Lecture Notes In Computer Science*, 523/1991:124–144.

Meunier, F. and Segui, J. (1999). Morphological priming effect: The role of surface frequency. *Brain and Language*, 68:54–60.

Michie, D. (1968). "memo" functions and machine learning. *Nature*, 218:19–22.

Mohanan, K. P. (1986). *The Theory of Lexical Phonology*. D. Reidel Publishing Company, Dordrecht.

Mohri, M. (1997). Finite-state transducers in language and speech processing. *Computational Linguistics*, 23(2):269–311.

Molnar, R. A. (2001). "Generalize and sift" as a model of inflection acquisition. Master's thesis, Massachusetts Institute of Technology, Cambridge, MA.

Mooney, R. J. and Califf, M. E. (1995). Induction of first-order decision lists: Results on learning the past tense of English verbs. *Journal of Artificial Intelligence Research*, 3:1–24.

Morton, J. (1969). Interaction of information in word recognition. *Psychological Review*, 76(2):165–178.

Morton, J. (1970). A functional model for memory. In *Models of Human Memory*, pages 203–254. Academic Press.

Moscoso del Prado Martín, F., Deutsch, A., Frost, R., Schreuder, R., de Jong, N. H., and Baayen, R. H. (2005). Changing places: A cross-language perspective on frequency and family size in Dutch and Hebrew. *Journal of Memory and Language*, 53:496–512.

Moscoso del Prado Martín, F., Kostić, A., and Baayen, R. H. (2004). Putting the bits together: An information theoretical perspective on morphological processing. *Cognition*, 94:1–18.

Münte, T. F., Say, T., Clahsen, H., Schiltz, K., and Kutas, M. (1999). Decomposition of morphologically complex words in English: Evidence from event-related brain potentials. *Cognitive Brain Research*, 7:241–253.

Murphy, G. L. (2004). *The Big Book of Concepts*. MIT Press, Cambridge, MA.

Murray, W. S. and Forster, K. I. (2004). Serial mechanisms in lexical access: The rank hypothesis. *Psychological Review*, 111(3):721–756.

Nakisa, R. C. and Hahn, U. (1996). Where defaults don't help: the case of the german plural system. In *Proceedings of the 18th Annual Conference of the Cognitive Science Society*, pages 177–183.

Napps, S. E. (1989). Morphemic relationships in the lexicon: Are they distinct from semantic and formal relationships? *Memory and Cognition*, 17:729–739.

Nelson, K. (1989). *Narratives from the Crib*. Harvard University Press, Cambridge, MA.

Nooteboom, S., Weerman, F., and Wijnen, F., editors (2002). *Storage and Computation in the Language Faculty*. Springer, Berlin, Germany.

Norvig, P. (1991). Techniques for automatic memoization with applications to context–free parsing. *Computational Linguistics*, 17(1):91–98.

Nosofsky, R. M. (1990). Relations between exemplar-similarity and likelihood models of classification. *Journal of Mathematical Psychology*, 34:394–418.

Noyer, R. and Harley, H. (1999). Distributed morphology. *Glot International*, 4(4):3–9.

Nunberg, G., Sag, I. A., and Wasow, T. (1994). Idioms. *Language*, 70(3):491–538.

O'Donnell, T. J. (2011). *Productivity and Reuse in Language*. PhD thesis, Harvard University, Cambridge, MA.

O'Donnell, T. J., Goodman, N. D., and Tenenbaum, J. B. (2009). Fragment grammars: Exploring computation and reuse in language. Technical Report MIT-CSAIL-TR-2009-013, MIT Computer Science and Artificial Intelligence Laboratory Technical Report Series, Cambridge, MA.

O'Donnell, T. J., Snedeker, J., Tenenbaum, J. B., and Goodman, N. D. (2011). Productivity and reuse in language. In *Proceedings of the 33rd Annual Conference of the Cognitive Science Society*, Boston, MA.

O'Reilly, R. C. (1996). *The LEABRA Model of Neural Interactions and Learning in the Neocortex*. PhD thesis, Carnegie Mellon University, Pittsburgh, PA.

Orlitsky, A., Santhanam, N. P., and Zhang, J. (2003). Always Good Turing: Asymptotically optimal probability estimation. *Science*, 302(5644):427.

Peterson, R. R., Burgess, C., Dell, G. S., and Eberhard, K. M. (2001). Dissociation between syntactic and semantic processing during idiom comprehension. *Journal of Experimental Psychology: Learning, Memory, and Cognition*, 27(5):1223–1237.

Pfeffer, A. (2001). IBAL: A probabilistic rational programming language. In *Proceedings of the International Joint Conferences on Artificial Intelligence*.

Pierrehumbert, J. B. (2001). Exemplar dynamics: Word frequency, lenition, and contrast. In *Frequency Effects and the Emergence of Lexical Structure*, pages 137–157. John Benjamins, Amsterdam.

Pinker, S. (1984). *Language Learnability and Language Development*. Harvard University Press, Cambridge, MA.

Pinker, S. (1989). *Learnability and Cognition: The Acquisition of Argument Structure*. MIT Press, Cambridge, MA.

Pinker, S. (1991). Rules of language. *Science*, 253(5019):530–535.

Pinker, S. (1999). *Words and Rules*. HarperCollins, New York, NY.

Pinker, S. and Prince, A. (1988). On language and connectionism: Analysis of a parallel distributed processing model of language acquisition. *Cognition*, 28:73–193.

Pinker, S. and Prince, A. (1996). The nature of human concepts: Evidence from an unusual source. *Communication and Cognition*, 29(3/4):307–362.

Pitman, J. (2002). Combinatorial stochastic processes. Technical report, Department of Statistics University of California Berkeley.

Pitman, J. and Yor, M. (1995). The two-parameter Poisson-Dirichlet distribution derived from a stable subordinator. Technical report, Department of Statistics University of California, Berkeley.

Plag, I. (1996). Selectional restrictions in English suffixation revisited: A reply to Fabb (1988). *Linguistics*, 34:769–798.

Plag, I. (2002). The role of selectional restrictions, phonotactics and parsing in constraining suffix ordering in English. In *Yearbook of Morphology 2001*. Foris, Dordrecht.

Plag, I. (2003). *Word-Formation in English*. Cambridge University Press, Cambridge.

Plag, I. (2004). Productivity. In *Encyclopedia of Language and Linguistics*. Elsevier.

Plag, I. and Baayen, R. H. (2009). Suffix ordering and morphological processing. *Language*, 85(1):109–152.

Plunkett, K. and Juola, P. (1999). A connectionist model of the English past tense and plural morphology. *Cognitive Science*, 23(4):463–490.

Plunkett, K. and Marchman, V. A. (1991). U-shaped learning and frequency effects in a multilayered perceptron: Implications for child language acquisition. *Cognition*, 38:43–102.

Plunkett, K. and Marchman, V. A. (1993). From rote learning to system building: Acquiring verb morphology in children and connectionist nets. *Cognition*, 48:21–69.

Plunkett, K. and Marchman, V. A. (1996). Learning from a connectionist model of the acquisition of the English past tense. *Cognition*, 61(3):299–308.

Plunkett, K. and Nakisa, R. C. (1997). A connectionist model of the Arabic plural system. *Language and Cognitive Processes*, 12(5):807–836.

Poser, W. J. (1992). Blocking of phrasal constructions by lexical items. In Sag, I. A. and Szabolcsi, A., editors, *Lexical Matters*, pages 111–130. CSLI Publications, Stanford, CA.

Post, E. L. (1943). Formal reductions of the general combinatorial decision problem. *American Journal of Mathematics*, 65(2):197–215.

Post, E. L. (1944). Recursively enumerable sets of positive integers and their decision problems. *Bulletin of the American Mathematical Society*, 50:284–316.

Post, M. and Gildea, D. (2009). Bayesian learning of a tree substitution grammar. In *Proceedings of the Joint Conference of the 47th Annual Meeting of the ACL and the 4th International Joint Conference on Natural Language Processing of the AFNLP*.

Prasada, S. and Pinker, S. (1993). Generalisation of regular and irregular morphological patterns. *Language and Cognitive Processes*, 8(1):1–56.

Prasada, S., Pinker, S., and Snyder, W. (1990). Some evidence that irregular forms are retrieved from memory but that regular forms are rule-generated. In *Proceedings of the 31st Annual Meeting of the Psychonomics Society*.

Pullum, G. K. and Gazdar, G. (1982). Natural languages and context-free languages. *Linguistics and Philosophy*, 4:471–504.

Rainer, F. (1988). Towards a theory of blocking: The case of Italian and German quality nouns. In *Yearbook of Morphology*. Springer, Berlin, Germany.

Ramscar, M. (2002). The role of meaning in inflection: Why the past tense does not require a rule. *Cognitive Psychology*, 45:45–94.

Rissanen, J. (1978). Modeling by shortest data description. *Automaticata*, 14(5):465–471.

Rissanen, J. (2007). *Information and Complexity in Statistical Modeling*. Springer, New York, New York.

Roark, B. and Sproat, R. (2007). *Computational Approaches to Morphology and Syntax*. Oxford University Press, Oxford.

Roy, D. M. (2011). *Computability, inference and modeling in probabilistic programming*. PhD thesis, Massachusetts Institute of Technology.

Rubenstein, H., Garfield, L., and Millikan, J. A. (1970). Homographic entries in the internal lexicon. *Journal of Verbal Learning and Verbal Behavior*, 9(5):487–494.

Rumelhart, D. E. and McClelland, J. L. (1986). On learning the past tenses of English verbs. In *Parallel Distributed Processing: Explorations in the Microstructure of Cognition.*, volume 2, pages 216–271. Bradford Books/MIT Press, Cambridge, MA.

Sachs, J. (1983). Taling about the there and then: The emergence of displaced reference in parent-child discourse. In Nelson, K. E., editor, *Children's language*, volume 4. Lawrence Erlbaum Associates.

Saffran, J. R. (2002). Constraints on statistical language learning. *Journal of Memory and Language*, 47:172–196.

Saffran, J. R. (2003). Statistical language learning: Mechanisms and constraints. *Current Directions in Psychological Science*, 12(4):110–114.

Sag, I. A., Wasow, T., and Bender, E. M. (2003). *Syntactic Theory: A Formal Introduction*. CSLI, Stanford, CA, second edition.

Sandra, D. (1990). On the representation and processing of compound words: Automatic access to constituent morphemes does not occur. *The Quarterly Journal of Experimental Psychology*, 42A(3):529–567.

Sandra, D. (1994). The morphology of the mental lexicon: Internal word structure viewed from a psycholinguistic perspective. *Language and Cognitive Processes*, 9(3):227–269.

Sangati, F. and Zuidema, W. (2011). Accurate parsing with compact tree-substitution grammars: Double-dop. In *Proceedings of the 2011 Conference on Empirical Methods in Natural Language Processing*, pages 84–95.

Scha, R. (1990). Taaltheorie en taaltechnologie; competence en performance. In de Kort, R. and Leerdam, G., editors, *Computertoepassingen in de Neerlandistiek*, pages 7–22. Landelijke Vereniging van Neerland.

Scha, R. (1992). Virtuele grammatica's en creatieve algoritmes. *Gramma/TTT*, 1(1):57–77.

Schreuder, R. and Baayen, R. H. (1995). Modeling morphological processing. In *Morphological aspects of language processing*, chapter 6, pages 131–157. Lawrence Erlbaum Associates, Mahwah, NJ.

Schreuder, R. and Baayen, R. H. (1997). How complex simplex words can be. *Journal of Memory and Language*, 37:118–139.

Schultink, H. (1992). Herkomst en groei van het begrip 'produktiviteit' als morfologisch fenomeen. In Klein, M., editor, *Nieuwe eskapades in de neerlandistiek*, pages 187–201. Wolters-Noordhoff, Groningen, Netherlands.

Seidenberg, M. S. and Bruck, M. (1990). Consistency effects in the generation of past tense morphology. In *Proceedings of the 31st Annual Meeting of the Psychonomics Society*.

Seidenberg, M. S. and Hoeffner, J. (1998). Evaluating behavioral and neuroimaging data on past tense processing. *Language*, 74(1):104–122.

Seki, H. (1991). On multiple context-free grammars. *Theoretical Computer Science*, 88:191–229.

Selkirk, E. (1982). *The Syntax of Words*. MIT Press, Cambridge, MA.

Sereno, J. A. and Jongman, A. (1997). Processing of English inflectional morphology. *Memory and Cognition*, 25(4):415–437.

Sethuraman, J. (1994). A constructive definition of Dirichlet priors. *Statistica Sinica*, 4(2):639–650.

Shenkman, K. D. (1994). *Structure Sensitivity and Language Processing in Adult Learners of English*. PhD thesis, University of Rochester.

Shieber, S. M. (1985). Evidence against the context-freeness of natural language. *Linguistics and Philosophy*, 8:333–343.

Siegel, D. (1974). *Topics in English Morphology*. PhD thesis, Massachusetts Institute of Technology.

Sivia, D. (1996). *Data Analysis: A Bayesian Tutorial*. Clarendon Press, Oxford.

Skousen, R. (1992). *Analogy and Structure*. Springer, Berlin, Germany.

Skousen, R., Lonsdale, D., and Parkinson, D. B. (2002). *Analogical Modeling*. John Benjamins Publishing Company, Amsterdam, Netherlands.

Spencer, A. (1991). *Morphological Theory*. Blackwell, Malden, MA.

Sprenger, S. A., Levelt, W. J., and Kempen, G. (2006). Lexical access during the production of idiomatic phrases. *Journal of Memory and Language*, 54:161—184.

Sproat, R. (1992). *Morphology and Computation*. MIT Press, Cambridge, MA.

Stabler, E. P. (1997). Derivational minimalism. In *Logical Aspects of Computational Linguistics*. Springer, Berlin, Germany.

Stanners, R. F., Neiser, J. J., Hernon, W. P., and Hall, R. (1979a). Memory representation for morphologically related words. *Journal of Verbal Learning and Verbal Behavior*, 18:399–412.

Stanners, R. F., Neiser, J. J., and Painton, S. (1979b). Memory representation for prefixed words. *Journal of Verbal Learning and Verbal Behavior*, 18:733–743.

Stemberger, J. P. (1983). Inflectional malaproprisms: Form-based errors in English morphology. *Linguistics*, 21:573–602.

Stemberger, J. P. and MacWhinney, B. (1986). Frequency and the lexical storage of regularly inflected forms. *Memory and Cognition*, 14(1):17–26.

Strauss, S. L. (1982). On "relatedness paradoxes" and related paradoxes. *Linguistic Inquiry*, 13(4):694–700.

Stump, G. T. (2001a). Inflection. In *Handbook of Morphology*. Blackwell, Oxford, England.

Stump, G. T. (2001b). *Inflectional Morphology: A Theory of Paradigm Structure*. Number 93 in Cambridge Studies in Linguistics. Cambridge University Press, Cambridge.

Suppes, P. (1974). The semantics of children's language. *American Psychologist*, 29:103–114.

Sweet, H. (1913). *Collected Papers of Henry Sweet*. Oxford University Press, Oxford.

Taatgen, N. A. (2001). Extending the past-tense debate: A model of the German plural. In *Proceedings of the 23rd Annual Conference of the Cognitive Science Society*.

Taatgen, N. A. and Anderson, J. R. (2002). Why do children learn to say "Broke"? A model of learning the past tense without feedback. *Cognition*, 86(2):123–155.

Taatgen, N. A. and Dijkstra, M. (2003). Constraints on generalization: Why are past-tense irregularization errors so rare? In *Proceedings of the 25th Annual Conference of the Cognitive Science Society*.

Taatgen, N. A. and Lee, F. J. (2003). Production compilation: A simple mechanism to model complex skill acquisition. *Human Factors: The Journal of the Human Factors and Ergonomics Society*, 45(1):61–76.

Taft, M. (1979). Recognition of affixed words and the word frequency effect. *Memory and Cognition*, 7(4):263–272.

Taft, M. (1994). Interactive-activiation as a framework for understanding morphological processing. *Language and Cognitive Processes*, 9(3):271–294.

Taft, M. and Forster, K. I. (1975). Lexical storage and retrieval of prefixed words. *Journal of Verbal Learning and Verbal Behavior*, 14:638–647.

Taft, M. and Forster, K. I. (1976). Lexical storage and retrieval of polysyllabic words. *Journal of Verbal Learning and Verbal Behavior*, 15:607–620.

Taft, M., Hambly, G., and Kinoshita, S. (1986). Visual and auditory recognition of prefixed words. *The Quarterly Journal of Experimental Psychology*, 38A(3):351–365.

Teh, Y. W. (2006). A hierarchical Bayesian language model based on Pitman-Yor processes. In *Proceedings of the 21st International Conference on Computational Linguistics and 44th Annual Meeting of the Association for Computational Linguistics*, pages 985–992, Sydney, Australia. Association for Computational Linguistics.

Townsend, J. T. (1974). Issues and models concerning the processing of a finite number of inputs. In *Human Information Processing: Tutorials in Performance and Cognition*. Lawrence Erlbaum Associates, Mahwah, NJ.

Tremblay, A. and Baayen, R. H. (2010). Holistic processing of regular four-word sequences: A behavioral and ERP study of the effects of structure, frequency, and probability on immediate free recall. In *Perspectives on Formulaic Language: Acquisition and Communication*, pages 151–173. Continuum.

Turing, A. M. (1937). On computable numbers, with an application to the Entscheidungsproblem. *Proceedings of the London Mathematical Society*, 42:230–265.

Ullman, M. T. (1993). *The Computation of Inflectional Morphology*. PhD thesis, Massachusetts Institute of Technology.

Ullman, M. T. (1999). Acceptability ratings of regular and irregular past-tense forms: Evidence for a dual-system model of language from word frequency and phonological neighborhood effects. *Language and Cognitive Processes*, 14(1):47–67.

Ullman, M. T. (2001). The declarative/procedural model of lexicon and grammar. *Journal of Psycholinguistic Research*, 30(1):37–69.

Van Gael, J., Saatci, Y., Teh, Y. W., and Ghahramani, Z. (2008). Beam sampling for the Infinite Hidden Markov Model. In *Proceedings of the 25th International Conferene on Machine Learning*.

van Marle, J. (1986). The domain hypothesis: The study of rival morphological processes. *Linguistics*, 281–283(3):601–628.

van Marle, J. (1992). The relationship between morphological productivity and frequency: A comment on Baayen's performance–oriented conception of morphological productivity. In Booij, G. and van Marle, J., editors, *Yearbook of Morphology 1991*, volume 208, pages 151–163. Kluwer Academic Publishers, Dordrecht.

Vannest, J. and Boland, J. E. (1999). Lexical morphology and lexical access. *Brain and Language*, 68:324–332.

Vitányi, P. M. B. and Li, M. (2008). *An Introduction to Kolmogorov Complexity and Its Applications*. Springer, Berlin, Germany, 3 edition.

von Humboldt, W. (1999, 1836). *On Language: On the Diversity of Human Language Construction and Its Influence on the Mental Development of the Human Species (Über die Verschiedenheit des menschlichen Sprachbaus und ihren Einfluss auf die geistige Entwicklung des Menschengeschlechts)*. Cambridge University Press, Cambridge.

Weist, R. M., Pawlak, A., and Hoffman, K. (2009). Finiteness systems and lexical aspect in child Polish and English. *Linguistics*, 47(6):1321–1350.

Westermann, G. (1998). Emergent modularity and U-shaped learning in a constructivist neural network learning the English past tense. In *Proceedings of The 20th Annual Conference of the Cognitive Science Society*.

Westermann, G. and Goebel, R. (1995). Connectionist rules of language. In *Proceedings of the 17th Annual Conference of the Cognitive Science Society*, pages 236–241.

Wexler, K. and Manzini, M. R. (1987). Parameters and learnability in binding theory. In Roeper, T. and Williams, E., editors, *Parameter Setting*, pages 41–46. Reidel, Dordrecht, Netherlands.

Wheeler, C. J. and Schumsky, D. A. (1980). The morpheme boundaries of some English derivational affixes. *Glossa*, 14:3–34.

Williams, E. (1981). On the notions "lexically related" and "head of a word". *Linguistic Inquiry*, 12(2):245–274.

Wood, D. (2010). *Perspectives on Formulaic Language: Acquisition and Communication*. Continuum.

Xu, F. and Pinker, S. (1995). Weird past tense forms. *Journal of Child Language*, 22:531–556.

Yang, C. (2005). On productivity. In *Linguistic Variation Yearbook 5*. John Benjamins, Amsterdam.

Yang, C. (2009). Who's afraid of George Kingsley Zipf? Unpublished manuscript.

Yang, C. (2010). Three factors in language variation. *Lingua*, 120:1160–1177.

Yang, C. D. (2002). *Knowledge and Learning in Natural Language*. Oxford linguistics. Oxford University Press, New York.

Yip, K. and Sussman, G. J. (1996). A computational model for the acquisition and use of phonological knowledge. Technical Report AIM-1575, Massachusetts Institute of Technology – Artificial Intelligence Laboratory, Cambridge, MA.

Yip, K. and Sussman, G. J. (1997). Sparse representations for fast, one-shot learning. Technical Report AIM-1633, Massachusetts Institute of Technology – Artificial Intelligence Laboratory.

Zipf, G. K. (1935). *The Psycho-Biology of Language: An Introduction to Dynamic Philology*. Houghton Mifflin Company, Boston.

Zipf, G. K. (1949). *Human Behavior and the Principle of Least Effort*. Addison-Wesley Publishing Company, Cambridge.

Zirkel, L. (2010). Prefix combinations in English: Structural and processing factors. *Morphology*, 20:239–266.

Zuidema, W. (2007). Parsimonious data-oriented parsing. In *Proceedings of the 2007 Joint Conference on Empirical Methods in Natural Language Processing and Computational Natural Language Learning (EMNLP–CoNLL 2007)*.

Index